Energy Capitals

History of the Urban Environment

Martin V. Melosi and Joel A. Tarr, Editors

ENERGY CAPITALS

Local Impact, Global Influence

Edited by
JOSEPH A. PRATT,
MARTIN V. MELOSI,
and
KATHLEEN A. BROSNAN

University of Pittsburgh Press

Published by the University of Pittsburgh Press, Pittsburgh, Pa., 15260
Copyright © 2014, University of Pittsburgh Press
All rights reserved
Manufactured in the United States of America
Printed on acid-free paper
10 9 8 7 6 5 4 3 2 1

ISBN 10: 0-8229-6266-7
ISBN 13: 978-0-8229-6266-3

Cataloging-in-Publication data is available at the Library of Congress.

For Felix and Theo, Gianna, and Jake

CONTENTS

ACKNOWLEDGMENTS

The Center for Public History (CPH) at the University of Houston (UH) sponsored the workshop "Energy Capitals: Local Impact, Global Influence" on May 21–22, 2010. That workshop led directly to this edited volume through the timely financial support of the National Science Foundation and supplemental support from the CPH and the UH Division of Research.

Aside from the authors in *Energy Capitals,* we need to acknowledge other participants in the workshop who contributed papers, chaired sessions, and provided excellent commentary. They include John Byrne, Tim Ezzell, Victor Flatt, Jim Granato, Tracy Hester, Karl Ittmann, Kairn Klieman, Robert Lifset, Tyler Priest, and Nancy Young. Carolyn Melosi and Adria Melosi McDonald provided a fine catered dinner at the end of the workshop, and Kristin Deville and Julie Cohn from the CPH were tireless in making all the arrangements for the workshop with what appeared to be effortless grace.

A final thanks to the staff of the University of Pittsburgh Press and the reviewers of the manuscript for helping us produce an edited volume we can proudly share with all of you.

INTRODUCTION

The cities in this volume represent important energy capitals in the fossil-fuel era. Indeed, in their own ways they have played, and some still do play, important roles in the production, processing, and transfer of hydrocarbon energy sources. In turn, fossil fuels have had significant impacts on them. The relationship between the coal and petroleum industries and these cities/regions not only represents an economic connection with a global reach, but major political, social, and environmental links as well.

Energy Capitals: Local Influence, Global Impact speaks to the intersection of fossil-fuel production and use and urbanization in specific locations around the world. The immediate results of this intersection, largely in the form of generating huge supplies of energy and large amounts of capital, often masks long-term local effects including the transformation of regional economies, fundamental changes in labor markets and educational institutions, high social costs in the areas of environmental quality and health, the shaping of regional infrastructure and the transformation of urban space, and changes in cultural attitudes. Studying the evolution of energy capitals reveals similarities and differences useful in understanding historical patterns of energy-led development where it takes place, with special emphasis on political, economic, technical, and social variables that influenced those patterns and shaped the environment in which energy industries emerged, grew, and in some cases, declined. A study of energy capitals can contribute a long-term perspective to current debates about the best ways to capture the benefits while managing the costs of such energy development.

This volume is the result of "Energy Capitals: Local Impact, Global Influence," a workshop held at the University of Houston (UH) on May 21–22, 2010, and sponsored by the National Science Foundation and UH's Center for Public History. With the exception of the essay on Pittsburgh, all of the remaining essays are versions of papers delivered at the workshop. The event was meant to bring together historians, social scientists, and other experts whose work on energy history and its intersection with urban and environmental history did not have a well-defined home in the current maze of academic associations. The subfield of energy history has remained rather dormant for several years since the end of the "energy crisis" in the 1970s, but has been reinvigorated recently because of the growing public awareness of a wide array of energy issues. The speakers and commentators at the workshop utilized their presentations as a departure point to begin a conversation on the study of energy history and its place within a variety of societal contexts. A subsequent workshop titled "Energy Resources: Europe and Its Former Colonies," was held in cooperation with the Rachel Carson Center for Environment and Society at the Ludwig-Maximilians-Universität in Munich, Germany, in October 2012. In all cases, the emphasis on energy history through the workshops has meant from the start to be global in scale.

Energy in the Fossil-Fuel Era

On the most fundamental level, energy is "the available capacity of a body to do work." Humans have unleashed energy by using the power of their own muscles or that of other animals. Humans also have exploited various forms of energy that nature stores in several forms such as gravitational potential (hydropower, tidal power), heat (geothermal), nuclear (fission, fusion), kinetic (windmills), waves, radiation (solar), and chemical (fossil fuels, wood, fuel cells).[1] All preindustrial societies relied on muscle power or energy sources that were relatively short-term transformations of solar radiation, such as flowing water or wind.[2] The Industrial Revolution, first in England and then elsewhere, swiftly began replacing muscle power with other forms of energy, sources that essentially were nonrenewable unlike those immediately derived from solar radiation. Since 1900, biomass [organic matter], coal, and oil supplied large quantities of energy, which at the time seemed endless and infinitely better at producing huge amounts of power.[3]

From the 1890s onward, fossil fuels overshadowed biomass as an industrial fuel even though the majority of the global population did not use them directly. In essence, fossil fuels effectively dominated worldwide energy production and use in the late nineteenth and twentieth centuries, resulting in a massive upsurge of total energy use.[4] Coal was the earliest modern fuel. Burning coal took place in ancient and medieval societies, but was never more than a minor source of heat or power in those years. In the seventeenth century, the Netherlands became the first country to shift to fossil fuels by burning peat, but England was the first to extensively utilize coal. Beginning as early as the 1540s (some two hundred years before the Industrial Revolution), the English began to mine all their major coalfields, shipping substantial quantities to London. By 1900 coal accounted for about 95 percent of the global primary energy supply, but slipped to about 23 percent in 2000. In the twentieth century, it was oil and not coal that captured broad scale energy use worldwide and it has been the dominant fossil fuel to this day.[5]

As historian John McNeill stated, "No other century—no millennium—in human history can compare with the twentieth for its growth in energy use. We have probably deployed more energy since 1900 than in all of human history before

1900."[6] Historian Alfred Crosby added, "We lurched into the fossil fuel era some two to three hundred years ago with the invention of the steam engine." This innovation allowed us, along with the internal combustion engine, "to tap the concentrated energies of ancient biomass which subterranean heat and pressure have transformed into coal, oil, and natural gas."[7] Electrification from major increases in the use of coal and then oil boosted demand for industrial and residential power needs, and these same fossil-fuel sources dramatically transformed motive power as well: locomotives, steamboats, automobiles, trucks, and airplanes. Probably the most important characteristic of society based on fossil fuels is "the exponential increase in per capita energy consumption."[8] Crosby concluded that "our technological civilization as it now exists would be impossible without the enormous consumption of these fossil fuels. Modern civilization is the product of an energy binge."[9]

Cities and Fossil Fuels

The most obvious and significant concentration of energy production and consumption occurred in and around cities. It makes good sense that coal, oil, and natural gas industries were attracted to cities as centers of production, processing, and transfer. Energy regimes in twentieth-century cities built upon arrangements whereby energy sources could be extracted near them, transported to them or close to them, stored, processed, and delivered to customers.[10] Beginning in the nineteenth century and extending into the twentieth, a combination of technological advances (such as extraction techniques and the steam engine) and sociopolitical processes (societal negotiations and choices made over energy systems) produced access to cheap and plentiful fossil fuels, which in turn stimulated economic growth and development. Urbanization was particularly influenced by the intense exploitation of fossil fuels, economically and environmentally.[11]

Before the Industrial Revolution, when biomass and muscle power fueled much of the world, the global urban population was around 3 percent, and the human, animal, and solar radiation sources that constituted much of the energy base limited the size of cities. Maybe somewhat exaggerated but with much truth, some experts stated that "[t]he high levels of contemporary urbanization owe their existence and continued growth entirely to fossil fuels," which takes into account power for farmlands that provide food, transportation, construction, industrial systems, and various household necessities and amenities.[12]

Cities and their hinterlands as foci of fossil-fuel production and consumption, therefore, are central historical phenomena of the twentieth century. Urban development per se exerts great pressure on local environments and the surrounding regions, particularly exaggerated by fossil-fuel production, processing, and transfer, and fossil-fuel dependency for power generation and transportation.[13] Cities also exhibit characteristics as flows of energy, whereby sources provide heat, light, power, and transport, and in so doing leave a large ecological footprint. Impacts include—within cities or related to urban needs—land for extraction, transportation, conversion of fuels, generation and transmission of electricity, and extensive use of water.[14]

Cities are major modifiers of the physical environment. "Their existence," geographer Ronald J. Johnston noted, "can influence the course of basic physical processes, such as the hydraulic cycle."[15] Urbanization removes much of the filtering capacity of soil and rapidly channels precipitation into available watercourses, thus encouraging flooding. Building cities affects the atmosphere by increasing airborne pollutants and also creating "heat islands" where temperatures are greater than the surrounding area. Various urban activities produce huge volumes of waste products that require complex disposal mechanisms. As geographers Thomas Detwyler and Melvin Marcus concluded, "Unfortunately, the urban ecosystem seldom treats air and water resources by riparian standards; that is, they are not returned to the ecosphere in the same condition in which they were received."[16] As such, this footprint measures the quantity of land, water, and air utilized to sustain the human population, resources consumed, and the waste absorbed.[17]

Cities for a variety of purposes also consume vast amounts of energy. According to one expert, "Collectively, buildings are either the largest or second largest consumers of energy (behind industrial conversions) in all rich societies." In the United States in 2000, buildings consumed 40 percent of all fuels and 75 percent of electricity.[18] In the first half of the twentieth century, consumption of fuel oil grew rapidly, not only because of industrialization but also because of home heating needs in cities. These two sources represented the greatest growth area for fuel oil use.[19] The production and consumption of energy by cities emphatically resulted in substantial environmental challenges—pollution, changes in land use, more infrastructure, and population concentration. Fossil-fuel cultures have left a very large environmental footprint in the form of land claim by extraction, transportation, and in the generation and transmission of electricity. Through combustion, fossil fuels oxidize carbon and heat their surroundings, produced most recently by automobiles and power plants in the case of air emissions or in extensive removal of land cover.[20] Cities require vast inflows of raw materials and structural components (such as concrete, metals, and wood products). Material inputs required to maintain high-energy societies are far greater than what preceded them. In general, fossil-fuel-supplied cities play a major role in key biochemical cycles by producing air and water pollution and significantly contributing to climate change on a regional and even global level.[21] Along with the exchange of energy by the oceans and atmosphere and solar energy, fossil-fuel emissions are a basic cause of climate change.[22] As environmental expert Vaclav Smil stated, "A century of fossil-fueled industrialization, urbanization, and subsidized farming changed both the extent and the rates of environmental intervention. . . . By the 1960s, when environmental concerns emerged as a major preoccupation of industrial civilization, there was no doubt that energy industries and energy use were the leading causes of environmental degradation and pollution."[23]

Energy Capitals

We define energy capitals as cities/regions with strong ties to energy industries and with strong roles in energy production, energy distribution, and/or energy technology. They also play a vital role in resource development and the provision of attendant services. When we think about energy capitals we normally think about them as centers for financial capital accumulation (profit centers) generating wealth for corporate entities or governments that draw that wealth from the production and sale of energy and then distribute it beyond the community where it was generated. This perspective is too narrow. Energy-led development (in this case during the fossil-fuel era) has shaped the evolution of many cities and regions, influencing metropolitan growth while changing patterns of energy consumption and concentrating the environmental impacts of energy production locally as well as in areas of consumption far removed from production facilities.

Cities such as Houston, Texas; Los Angeles, California; Baton Rouge/New Orleans, Louisiana; Tampico, Mexico; Calgary, Canada; Stavanger, Norway; Perth, Australia; and Port Gentil, Gabon (also Oklahoma City/Tulsa, Oklahoma, and Aberdeen, Scotland, neither of which are included in this volume), are representative of contemporary cities which deserve the moniker of energy capital. In the past, the production, transportation, and intensive use of energy also strongly influenced the development of such cities as Pittsburgh, Pennsylvania—a coal and oil center. At some point in their development, these cities or their surrounding regions became important parts of the complex of economic activities needed to produce and distribute energy to broader markets. The idea of energy capitals is not time specific or limited to regions that remain dominated by energy production. In the past, for example, oil strongly affected such cities as Los Angeles and Tampico, Mexico, in one phase of their development before waning in influence. Indeed, the most temporary form of an energy capital—the boom community, produced by the frenzied development of large newly discovered oil fields (such as Oil Creek, Pennsylvania, or Spindletop, Texas)—has attracted some attention by scholars, but not much beyond their role as catalysts for such booms.

The strong and complex connections at the intersection of energy-led development, urban growth, energy use, and environmental impacts in energy capitals are intuitively obvious. Yet they are largely missing from the existing historical literature. Perhaps the connections are simply too deeply embedded to be easily analyzed. Also, the study of energy history has not yet developed as fully as the vibrant fields of the history of technology, urban history, and environmental history. One way to begin to examine more fully these related issues, therefore, is to focus on extreme cases which show most dramatically the relationship between energy, environment, and urbanization. These extreme cases are the energy capitals.

Of course, it is not accurate to limit energy capitals to oil centers only. Other forms of energy have greatly influenced the development of cities over the centuries such as Pittsburgh or Manchester, England (coal); Oak Ridge, Tennessee (nuclear power); or hydropower in a variety of locations, including the Tennessee Valley and the Pacific Northwest in the United States, Austria, and throughout Scandinavia. For coherence, however, *Energy Capitals* focuses on fossil fuels in a very significant period of human history.

Energy capitals, in addition, have not emerged in a historical void. Their study has much to draw upon as well as contribute to the study of urban development in general and case studies of regions that have been shaped by different common economic influences, such as a variety of industrial products or even finance. Regional development driven by energy shares several important attributes with modern urban development in general. The most obvious is the movement of large numbers of migrants from rural areas to industrial jobs in and around cities. Equally important in the process of change is the connection of city and hinterland. As William Cronon, Kathleen Brosnan, and others have demonstrated, we understand the urbanization process to entail regional impacts which extend the influence of urban development beyond politically constructed borders, while creating interdependence between built and natural features.[24]

The production and transportation of large quantities of fossil fuels has greatly affected many parts of the world, a process often accompanied by the introduction of advanced technologies from large companies based outside the respective region. The long-term impacts have included the transformation of regional economies, population growth fueled by mass migration to the opportunities presented by growth, fundamental changes in labor markets and educational infrastructure, high social costs in the areas of environmental quality and health, the shaping of infrastructure and the transformation of urban space, and changes in cultural attitudes.[25]

Impacts of Energy Capitals

By their very nature, all cities are energy intensive. A high concentration of people in a limited space demands energy use for heating and cooling, for transportation, for work—for almost any activity one could imagine. However, energy capitals, in particular, are historically significant because of the roles they play and have played in both production *and* consumption of energy. Concentration of human and material resources for purposes of survival, construction of infrastructure, and the production and consumption of goods and services are essential characteristics of communal living. Energy-led development has shaped the evolution of many cities and regions, influencing metropolitan growth while changing patterns of energy consumption and concentrating the environmental impacts of fossil-fuel production locally as well as in areas of consumption far removed from production facilities.

Recent scholarship on global cities, especially in sociology, appears to provide some insight.[26] As sociologist Saskia Sassen noted, "Economic globalization, accompanied by the emergence of a global culture, has profoundly altered the social, economic, and political reality of nation-states, cross-national regions, and . . . cities."[27] While useful, the focus on transnational networks of cities and "transnational spaces for economic activity" in this study and elsewhere primarily focuses on the global economy, which is not the primary concern of *Energy Capitals*. Studies dealing with "the resource curse" or "the oil curse" also provide an additional basis for analysis by emphasizing the difficulties of emerging oil-producing nations as they seek to develop by exporting oil produced primarily by foreign companies.[28] But even here, the question of "the oil curse" is not relevant to all of the cities we study, nor does it give primary attention to the intersection between fossil-fuel production and use and urban growth. Nevertheless, more comparative work on energy-led development over time and place is welcome no matter what the specific emphases.

Energy Capitals attempts to help fill the gap in the historical literature especially with respect to the energy/city nexus. Cities and regions that reap the long-term economic benefits of energy production are often physically transformed (or at least modified) by the burgeoning energy industries and, at times, pay high social and physical costs. Demands for water, wastewater systems and solid-waste disposal systems, communication, transportation networks and facilities, and external sources of power (particularly electricity) put pressure on cities to expand their infrastructure, and in some cases, energy industries compete directly with municipal infrastructure needs. The historical impact of fossil fuels reaches beyond the conversion of resources to stationary and motive power, such as the illumination of streets and interiors; the movement of trains, cars, buses, and trucks; and the generation of heat and refrigerated air. Broadly understood, energy encompasses all processes of production and consumption that allow people to function in the physical world. One general impact is clear: changes in energy supply and demand in the past have greatly affected the economic and physical contexts within which cities and regions have grown.

New technologies using new sources of energy have shaped the transportation and communication revolutions that transformed the world economy in the last two centuries. To restate a main theme of the book: The impact of new sources of energy on the American economy has been particularly pronounced in the years since the mid-nineteenth century, when the widespread use of fossil fuel and urbanization accelerated. Fossil fuels helped transform the modern city worldwide (which is especially obvious in energy capitals), altering the physical environment in new and significant ways. The most obvious impact was a fundamental change in land-use patterns in and around cities, which reached out and absorbed the once rural land surrounding the sprawling urban centers. The concentrated use of fossil fuels in the production of goods brought a new scale of industrial pollution to cities; growing energy use for transporting people and goods added another layer of pollution to the mix, particular in cities that grew rapidly only after the advent of the automobile. In these and many other ways, as the lure of jobs and better opportunities from urban industrial growth attracted ever larger populations, the environmental impacts of increasing energy use also grew dramatically.

Thus, the most visible social and physical cost of energy industry development is the concentration of environmental and health risks from the production of fossil fuels. In this sense, energy capitals often have been forced to absorb substantial local costs for producing energy sold in national and international markets, becoming sacrifice zones of sorts. The response to such costs of energy-led industrialization often has been indifference or neglect. This is the case, in part at least, because technologies of energy production have been the historical focus of coal, oil, and natural gas industries, while investment has lagged substantially in technologies of pollution control. The political and legal processes for negotiating societal solutions to such problems also shaped regional responses to social and physical costs, and such processes have differed sharply across time and place.

The following questions provide a departure point for a comparative analysis of energy capitals. As a group, the chapters in this book address many of these issues in whole or in part. The authors, however, were free to explore the historical questions they believed were most pertinent or most relevant to their expertise.

- What economic benefits have accrued to the region over time because of fossil-fuel production?
- How have the needs of the energy industry shaped urban infrastructure, particularly industrial and municipal demand for water supply and wastewater systems, transportation and communication networks, disposal facilities, and scientific and technical educational systems?
- What have been the most obvious social costs? In particular, what have been the primary environmental impacts on the region of the specialized technology used in fossil-fuel production and processing? In the broadest sense, have energy capitals been

treated like "sacrifice zones" that carry the environmental burden of a product used far away from the point of production?[29]
- What about the broad implications of energy consumption, particularly questions related to availability of cheap energy close to the source and its impact on urban growth and development?
- What has been the impact of energy-related growth on political systems as they sought to make public policy about the costs and benefits of energy-led development?
- What has been the impact of the interaction between urbanization and energy development on culture, including labor, education, race and ethnicity, gender, and a variety of other social concerns?
- How has migration of workers to the jobs in energy-related manufacturing altered the demographics and spatial organization of the region, expanding the influence of the urban center out into its hinterland?

The current chapters do not give substantial attention to questions of specific technology and science related to energy development in the cities discussed, but future studies should consider these issues more fully. Most energy capitals share the common formative experience of rapid growth after the introduction of advanced technologies increased the regional production of supplies of fossil fuels destined for national and international markets. At times a cycle of boom and bust ensued, leaving a region vulnerable to changes in energy supply and demand beyond its control. In some cases, regions have been successful in absorbing new technologies and building economies capable of diversifying in the face of fundamental changes in the energy. Consider the following questions:

- What is the relationship between specific production technologies and environmental pollution?
- What are the specialized technologies that have grown out of these fossil-fuel industries? For example, such discussions can include refinery technologies of fuel production and petrochemicals, and various heating and cooling equipment.
- What has been the role of science in energy production and how does the application of various scientific processes affect demand for resources and potential impacts on urban life? This includes everything from chemical recycling to waste disposal.

Each of the queries stated above generates a large subset of related questions for analysis; together they offer potential opportunities for viewing energy capitals well beyond the passing attention they have received in the existing literature. Indeed, the same set of questions might be profitably used to compare urban development among a wide variety of major cities.

Energy Capitals and the Future

Many energy capitals share a life cycle of rapid growth, maturity, and decline. Differences may occur, of course, depending on when a resource is discovered and developed. If a region succeeds in absorbing the economic growth generated by fossil-fuel development, it will often enjoy an extended period of maturity in which the growth of demand for its products encourages economic expansion. Sooner or later, this era gives way to a time of decreased demands for its energy-related products, leading to a period of economic decline unless the region finds the will and the resources to diversify its economy. During this cycle of boom and bust, a region is vulnerable to changes in energy supply and demand beyond its control. In some cases, regions have been successful in absorbing new technologies and building economies capable of diversifying in the face of fundamental changes in the energy industries. We need to know more fully why some regions have been able to absorb the technologically advanced processes needed for energy production and some have not. We also need to understand the environmental implications of the production/consumption cycles.

The presumption of a post-petroleum world—or even more broadly a post-fossil-fuel world—confronting us sooner than later, can benefit from a deeper understanding of how energy capitals emerged and evolved throughout the world. Such studies can help us to understand how certain energy sources become essential to economic growth, but also how they shaped their physical surroundings in such a way as to develop mutual dependencies between industries and urban areas. Such interdependencies, in many ways, shape (and also constrain) transitions to new energy eras. Such matter-of-fact notions as "a post-petroleum era" are not something that will occur outside of their historical context. For us to understand the rise, growth, and fall of energy capitals is to better understand the role energy plays in society at large—not simply as a source of power, but as an engine of change.

Energy Capitals

PART I   Blessed by Fossil Fuels?

Pittsburgh, Houston, Louisiana,
and Los Angeles

Historically, cities have built up around exploitable resources. Urban entrepreneurs competed to control the harvesting, processing, and distribution of the earth's mineral wealth. Nearby salt mines, for example, allowed Salzburg (Austria) to dominate regional commerce for centuries, while "instant cities" such as San Francisco and Denver (United States) appeared in the mid-nineteenth century to manage the trade associated with the gold rushes in their respective hinterlands.[1] What distinguishes energy capitals from other resource capitals has been the transformative and persistent power of fossil fuels. In *Something New under the Sun,* John McNeill argues that by the twentieth century, humans became the overwhelmingly dominant driving force in environmental change around the globe, in large part, due to the availability of inexpensive energy. Cheap fossil fuels made it less costly to develop other resources, helped to create and extend materials-based infrastructure, and allowed industrial societies to metamorphose and grow rapidly.[2]

The emergence of the United States as the world's leading industrial nation in the late nineteenth century is directly attributable to the exploitation of fossil fuels. Coal initially spurred the creation and extension of a materials-based infrastructure in the United States as well as other nations. The combustion provided by coal allowed the growth of industries from textiles to pig iron. It fueled the trains and steamboats that redefined concepts of space and allowed energy capitals to spread their influence beyond their immediate hinterlands. Pittsburgh emerged as the nation's energy center in the nineteenth century through the extensive mining of the rich deposits of bituminous coal located in and around the city and by utilizing this cheap and easily accessible fossil fuel in the production of iron and steel.

Using Pittsburgh as a case study, Joel Tarr and Karen Clay remind us that the most persistent energy capitals are those that successfully make transitions from one energy regime to another while maintaining control of the production, processing, storage, or distribution of the key resources. While Pittsburgh is most closely associated with "coal as a source of both industrial progress and of environmental degradation," the city's energy history entailed diverse activities such as coal-oil processing and petroleum refining. Perhaps the most significant development in Pittsburgh's energy history, however, involved transitions from coal to natural gas. The first transition to natural gas in the late nineteenth century proved short-lived due in part to declining supplies, but local desires to avoid the smoke and other externalities that accompanied the use of coal and gave Pittsburgh the moniker of "Smoky City" prompted new interest in the development of alternative energy resources. A combination of private-sector initiatives and municipal initiatives led to the passage of a smoke control ordinance in 1941, although the city delayed its vigorous enforcement during the war. By 1947, new natural gas pipelines reached the city while local utilities expanded their underground storage pools for natural gas. The shift from coal to natural gas for home and commercial heating, along with the railroads' switch to diesel engines, improved air quality and facilitated the Pittsburgh Renaissance, an early attempt by an industrial city to renew itself.

Oil, as Daniel Yergin reminds us, "has meant mastery throughout the twentieth century."[3] Perhaps no city has more thoroughly mastered oil nor has been more thoroughly transformed by its role as an energy capital than the city of Houston. At the dawn of the twentieth century Houston operated as a small commercial and rail hub for its region's primarily agricultural economy. The discovery of oil at nearby Spindletop in 1901, however, placed the city on a new path. The case study presented by Joseph Pratt and Martin Melosi confirms that when a city enters the energy business is essential to its emergence as an energy capital—both in terms of the city's history and the industry's history. For Houston, proximity to the resource mattered, but the fact that the city at this time possessed a transportation network and aggressive business elite allowed it to dominate the Texas petroleum industry and play a major role in a relatively young global industry. Houston developed its refineries as the new automobile culture created a new and rapidly expanding demand for oil. Petrochemical companies and other manufacturing operations grew up around the Houston Ship Channel, a human-made conduit created through the investment of federal dollars and the dredging and expansion of Buffalo Bayou and Galveston Bay. Jobs drew a diverse group of workers to the growing metropolis. The oil and gas industry changed the city's social makeup as well as its physical landscape.

Craig Colten's study alternatively identifies a region rather than a city as an energy capital, although Louisiana's petrochemical corridor, like Houston, confirms the importance of some level of economic and political maturity at the onset of energy-led development. At the dawn of the twentieth century, the landscape between the political center in Baton Rouge and the economic and cultural center in New Orleans was filled with agricultural plantations and small-scale industry to refine raw sugar and press cotton prior to their shipment from the latter city's international port. Barred by antitrust laws from operating in Texas, Standard Oil built its first refinery in Louisiana in 1909 and launched the local industry there. Hugging the Mississippi River, the Louisiana corridor provided access to nearby oil fields and water transportation advantages. And as in Houston, government infrastructure investments fostered energy-led development. Federally financed levees initially constructed to protect farms made the Louisiana corridor an attractive location for the oil and gas industry. Yet development there has taken a different form. Standard Oil built the first refinery, and one of the largest, in Baton Rouge, but many energy companies subsequently dispersed their operations along the Mississippi River in smaller, rural communities. After World War II, the emergence and expansion of offshore drilling in the Gulf of Mexico spurred the growth of related activities in the coastal parishes that participated in the oil and gas economy. The industry transformed the social landscape and physical environment of Louisiana, but the changes were not universally celebrated. While jobs drew a mobile and better-paid workforce, the Louisiana corridor, much like the Houston region, has borne a disproportionate portion of the environmental and health consequences of the nation's reliance on fossil fuels.

In both the Louisiana corridor and Houston, a political culture of accommodation facilitated economic growth by giving energy companies great leeway in the development of resources and the transformation of local landscapes. Sarah Elkind reinforces this important theme in her persuasive and well-researched chapter on Los Angeles, the energy capital of the United States west of the Rocky Mountains. Oil speculation in the Los Angeles region dates back to the late nineteenth century, but it was the identification of new sources under downtown Los Angeles that sparked a new boom. Oil wells soon dotted the landscape, and over the next few decades the construction of pipelines and refineries made the city and its environs a center of oil production, refining, and transportation. The fact that the oil fields lay beneath existing residential communities presented unique challenges for the industry, forcing companies to negotiate leases with thousands of individual homeowners and invest more money in equipment than they would have with larger leases. Companies packed derricks into neighborhoods, drilling quickly and aggressively to extract oil as fast as possible. Residents faced a variety of environmental hazards, including the constant clatter of drills, spilled oil, fires, explosions, and runaway wells that sprayed oil hundreds of feet. The location of these fields in residential neighborhoods prompted the early emergence of grassroots opposition to energy industry expansion and the passage of local ordinances designed to limit drilling and protect parklands. However, government also facilitated development. Major oil companies used state and federal legislation to trump local regulations. California voters, for example, approved a 1936 ballot initiative, written by Standard Oil, which effectively legalized slant drilling because it earmarked oil royalties for public park acquisitions. And despite the resistance of some Angelenos, many local residents and officials agreed to the streamlining of new drilling permits and the relaxing of regulatory oversight over production. The federal construction of the Port of Los Angeles in the San Pedro neighborhood gave the city a deepwater harbor and enhanced the local industry's international influence. Finally, the port and the energy industry spurred the growth of other local businesses and factories by providing ample, cheap fuel. New migrants filled out the suburbs that surrounded the oil fields and were in large part the result of available oil, a growing car culture, and eventually the federal highway system. In the end, oil-led development changed the region's relatively open racial milieu, creating and entrenching patterns of racial segregation through suburbanization and other factors.

Los Angeles shared with Pittsburgh, Houston, and the Louisiana corridor certain attributes that allowed it to emerge as an energy capital. Along with good access to the essential fossil fuel, these energy capitals possessed the financial capital, business elites, and transportation networks that allowed them to capture an early lead in the energy industry. Moreover, they built on those initial advantages, frequently with the assistance of the public sector, to expand their energy-related services and their physical infrastructure, allowing them to participate in and sometimes dominate almost all phases of the industry.

1

Pittsburgh as an Energy Capital

Perspectives on Coal and Natural Gas
Transitions and the Environment

Joel A. Tarr and Karen Clay

Throughout most of its history Pittsburgh has been closely identified with the fossil fuel coal as a source of both industrial progress and of environmental degradation. Located on top of the high-quality Pittsburgh bituminous coal seam, the city's businesses, industries, residents, railroads, and steamboats benefited from the high-energy and easily available fuel. Coal has shaped the pattern of industrial development, settlement, population, and labor force composition. Its mining and consumption also drove the environmental contamination and physical alteration of land and water, as well as seriously polluting the air. Without coal and the advantages of its location, Pittsburgh would not have become one of the world's great industrial powers in the late nineteenth and early twentieth centuries.[1] The tentacles of the city's industrial and financial interests and its ecological footprint spread throughout the region, tying its hinterland ever closer to the city.

The full energy history of Pittsburgh, however, encompasses more than coal. The city, for instance, played a prominent role in the coal-oil industry in the late 1850s and early 1860s and in the petroleum industry soon after that, becoming the nation's oil refining center for approximately fifteen years. More recently, in the post–World War II decades, it played a prominent role in the development of nuclear energy.[2] But the most significant aspects of Pittsburgh's energy history are those involving transitions between coal and natural gas. The first of these transitions was relatively short-lived, taking place approximately between 1880 and 1890. The second has been more lasting, and began in the years after World War II and continues today.

These energy transitions had significant impacts upon the city's population, industries, and environment. They also shaped public policy and public health.[3] The transitions were primarily driven by costs and supply factors, but they were aided by the city's attempt to devise policies to escape from coal's environmental externalities and to find a substitute that would provide its energy needs. Today extensive debates about both coal and natural gas development in terms of their effects on both the environment and the public health have reemerged, reminding us of the importance of understanding our energy history.

This chapter is organized in the following manner. The first section, The Landscapes of Coal, will consider the development of coal mining, infrastructure necessary for its operation, and the evolution of transportation systems for moving coal to national and international markets. The second section, The Landscapes of Industry, will examine those industries most tightly linked to coal, especially iron and steel. The third section, The Environmental Effects of Coal will explore coal from mining through consumption. The fourth section will discuss The First Transition to Natural Gas. The fifth section will consider The Environmental Effects of Natural Gas. The sixth section will present The Return to Coal and Smoke, after natural gas supplies dropped, and the seventh section, Smoke Control, will examine the second shift from coal to natural gas that occurred from 1945 to 1960 and its impacts on the city and the region from different perspectives. The final section looks at the effects of this energy transition.

The Landscapes of Coal

The Pittsburgh coal seam before its exploitation was approximately 14,200 square miles in area, covering parts of twelve counties. The seam extended northward from the southwest corner of the state to Lake Erie and the southern New York State boundary and to the Allegheny Front on the east, and was composed of a number of different beds. The bituminous coal lay primarily in horizontal seams approximately three to four feet thick, located on the side of hills and at the bottom of valleys, with relatively little deep mining initially required to extract it.[4] The coal seam was first exposed on Pittsburgh hillsides, and further rich deposits were soon discovered extending into the city's hinterland along the Monongahela and Allegheny Rivers and their tributaries.[5] Bituminous production expanded greatly after the Civil War, as water and rail connections improved.[6] During the nineteenth and early twentieth centuries, Pennsylvania was the leading producer of bituminous coal, most of it coming from the Pittsburgh coal seam.

The center of the Western Pennsylvania coal district was the city of Pittsburgh. Mining and local consumption in Pittsburgh had begun as early as 1784.[7] The four key counties that produced bituminous coal in the nineteenth century were Allegheny, Fayette, Washington, and Westmoreland. The peak of bituminous coal production in the four-county region came in 1918, when 101,959,000 tons of coal was produced from over 1,000 mines. In the twentieth century the counties of Cambria, Clearfield, Indiana, and Somerset joined the four as major bituminous coal–producing areas.[8]

Over time the industry became increasingly mechanized and in 1924 about one-third of Pennsylvania's bituminous coal was produced by hand and about two-thirds by machine. In the years from 1920–24, the number of employees in Pennsylvania's bituminous coal mines averaged over 170,000.[9] In 1960, the number of miners remained about the same but because of mechanization they produced four times as much coal.[10]

Throughout the Pittsburgh region a landscape devoted to the mining of coal developed. Its most recognizable feature was the coal tipple and head frame, located near the mine entrance and protruding from riverbanks or near railroad tracks. Here the coal was sorted, weighed, and loaded for transport. Other buildings comprising the mining complex included a hoist house for deep mines to lower miners into coal shafts, a boiler house and electrical power house, a repair shop, a fan house for ventilation, mine office, lamp house, and storage sheds. A large "gob" pile or dump of slate and other refuse from coal cleaning usually overlooked the mine complex.

Coal companies built "patch towns" in order to house their workers near isolated mine sites and to provide a stable and controllable labor supply. They erected rows of identical semidetached wooden houses for workers and their families on unpaved roads leading to the mine or on a grid plan across from the railroad tracks serving the mine. The companies operated company stores and provided water through hydrants or outdoor pumps and open sewers for wastes and storm water. One writer described them in 1946 as being "pretty much alike, springing out of the fields, creeping up to slate dumps, climbing hilltops in rows of identical double houses with narrow frame porches divided in half, fenced back yards, and outdoor toilets."[11]

Bituminous coal from Western Pennsylvania found important markets in the region itself as well as nationally and internationally. Pittsburgh's network of rivers, including the major bodies of water, the Allegheny, Monongahela, and Ohio, and tributary streams such as the Beaver, Clarion, and Youghiogheny, were the primary avenues used for the transportation of coal before railroads entered the region in the 1850s. Many of the mines were located on their banks, and millions of tons of coal were shipped by barge to the city. In 1855, for instance, the city itself consumed 935,714 tons while exporting 570,649 tons. In 1917, 60,441,000 tons or 52.6 percent of coal mined in Western Pennsylvania was sold to markets in Pennsylvania (many in the Pittsburgh district).[12]

Most coal shipments were on the Monongahela, but until the private Monongahela Navigation Company completed a system of locks and dams to Brownsville, Pennsylvania (60 miles) in 1846, coal transport was limited by both low water conditions and by freezing rivers.[13] Navigation on the Ohio River improved in 1885 with the opening of the Davis Island Locks and Dam. Major improvements on Pittsburgh's third river, the Allegheny, came more slowly and had to wait until the city and county raised thirty-two low-lying bridges in the 1920s.

In 1852 the Pennsylvania Railroad linked Pittsburgh with Philadelphia, providing the first through rail route over the Alleghenies. Other mainline trunk roads followed, as did a number of small feeder roads that penetrated into the coalfields. Many of the region's railroad carriers such as the Pittsburgh Chartiers & Youghiogheny, the Montour, the Monongahela Southern, and the Unity Railroads owned mines and were dedicated solely to the transport of coal and coal products. Coal companies also invested directly in railroads as feeder lines. These lines grew rapidly, from five in 1860 to seventy-eight in 1898, about two-thirds of which were connected to mainline railroads.[14] All mainline railroads were located on flat ground along the major rivers. Feeder lines tended to hug the shores of tributary streams.

Coal had a special relationship with the nation's railroads but especially those in the Pittsburgh region. It was the largest single item of revenue tonnage for the roads, constituting about 20 percent of loaded freight cars. By 1880 bituminous made up over 90 percent of railroad fuel, peaking in 1918 at 34,210,000 tons for all carriers. It remained at that level through the 1920s. The onset of the Depression brought sharp operating cutbacks, and coal consumption declined until wartime demands generated a revival. In the postwar period, however, the adoption of the diesel-electric locomotive produced an even sharper decline and within a decade coal consumption by railroads was negligible.

Railroad transport of bituminous coal was typically greater than that by water. The best year for rail transport during the first half of the twentieth century was 1918, when railroads carried 128,518,000 tons of bituminous and barges carried only 8,985,000 tons. From the 1920s on, coal shipment by water encroached on rail, with an especially sharp drop in rail shipments after 1949 when coal consumption fell sharply. The cost of water transport, although slower, was usually about 18 percent of the rail rate. The margin over rail was most marked when the coal could be moved entirely by river from origin to destination such as mills and utilities sited on the rivers.[15]

The Landscapes of Industry

Coal provided the necessary energy source to power Pittsburgh's growing industrial sector in the nineteenth and twentieth centuries. As the writer Willard Glazier wrote in 1883, all of Pittsburgh's industry was "rendered possible by the coal which abounds in measureless quantities in the immediate neighborhood of the city."[16] Cheap fuel provided energy not only for the iron and steel industries, but also for glassworks, textiles, breweries, machine shops, and salt works. Iron and steel were the signature products manufactured by Pittsburgh industry. By the 1830s iron products had emerged as the young city's leading industry, with eighteen iron foundries and nine rolling mills; in twenty-five years the total number of plants had more than doubled. These iron manufacturing works initially depended on supplies of pig iron that came from charcoal blast furnaces based on rural iron plantations. By the late 1850s, however, they had shifted to iron furnaces using bituminous coal located in or near the city.[17]

The shift to coke as an iron-making fuel was made possible by the adoption in southwestern Pennsylvania of the beehive coke oven, a technology in which bituminous coal was baked in brick ovens for approximately forty-eight hours to drive out the oils, gases, and tars. The process produced a hard porous coke of silvery luster with few impurities. The best coking coal, known as Connellsville Coke, was found in a narrow forty-mile-long seam located about thirty-five miles south of the city. The ovens were usually located close to the coal mines and the resulting coke was shipped by water or rail to Pittsburgh blast furnaces. By the end of the 1870s coke had become the preferred fuel to produce pig iron and by 1900 ninety coke plants with nearly 21,000 ovens operated in the coke district; by 1920 this number had almost doubled.

An urban network arose to service the coal coke industry. Coke district towns like Greensburg, Connellsville, and Uniontown contained administrative offices, financial services, retail and wholesaling services, and associated manufacturing firms like foundries, machine shops, or mining equipment producers. The rivers and rails tied the district physically to Pittsburgh, while the steel corporations and allied financial interests connected it to the city through ownership of the major mines, coke plants, and coke ovens.[18]

The large integrated iron and steel mills that consumed the coke first emerged in the 1870s. They were initially formed by the entrepreneur Andrew Carnegie and his partners as well as by several independent managers with long experience in the iron industry.[19] In the early 1880s, Henry Clay Frick, the dominant figure in coke production, joined Carnegie Steel as a partner, providing a vital link between their complementary coke and steel interests. This alliance guaranteed Carnegie an uninterrupted flow of the high-quality coke necessary to the making of steel.[20] Because of the greater availability of large sites on the bends or meanders of the Monongahela River, the largest plants were first located there. As these sites filled up, the steel companies built mills along the Alleghany and its tributaries and down the Ohio River.

A brilliant manager, Carnegie believed in the importance of economies of scale for production and pricing as well as for the efficient movement of materials throughout the integrated plants. These values were reflected in the large integrated mills covering hundreds of acres of ground that sometimes occupied both banks of the rivers on which they were sited, with hot metal bridges connecting the blast furnaces on one bank and the Bessemer and open hearth ovens on the other. Carnegie and other iron and steelmakers such as Jones and Laughlin, and the National Tube Works built their plants in the flat land in the Monongahela River valley. As these sites filled up the steel companies built mills along the Allegheny and Ohio Rivers. The major integrated iron and steel mills were joined by dozens of more independent metals firms in the river valleys and an array of specialized enterprises, including sintering plants, bridge fabricators, machine shops, and barge builders, to form industrial corridors.[21]

Many of the steel operations presented a chaotic assemblage of huge brick and metal sheds, towering blast furnaces, hot ovens, Bessemer converters, open-hearth furnaces, rolling mills, giant ore loaders, and ore and coal yards, all framed by river and rail. Connecting the mills to neighboring towns and the city was a matrix of pipelines; electric, telegraph, and telephone wires; railroad lines; extensive gas pipe lines; and mile-long river barge tows that dominated the river valleys.[22] Railroads and rivers were key elements in this matrix. Rails, like industrial plants, were located largely on flat land, in floors of the river valleys, while mills were sited between railroad and river. This positioning facilitated the easy movement of goods both in the region and to and from the East Coast and the ever-expanding western markets. It also provided managers like Carnegie with the opportunity to play off the rivalry between the two modes of transport to secure cheaper rates. By World War I six major trunk lines and sixteen industrial and switching railroads served the city and its industries.[23]

In the first decades of the twentieth century by-product coke ovens, which were sited close to the mills rather than near the mines, added to the industrial assemblage. U.S. Steel opened the first full by-product coking plant in the region in 1916 at Clairton on the Monongahela River, twenty miles south of Pittsburgh. These works occupied a riverfront site 5,200 feet long and 1,800 feet in width; it was the world's largest plant at the time of its construction. The Jones and Laughlin Iron & Steel Corporation constructed a major by-product plant in 1918–20 in the Pittsburgh neighborhood of Hazelwood to serve its adjoining iron and steel works. Before World War I more than 20 percent of the bituminous coal mined in Pennsyl-

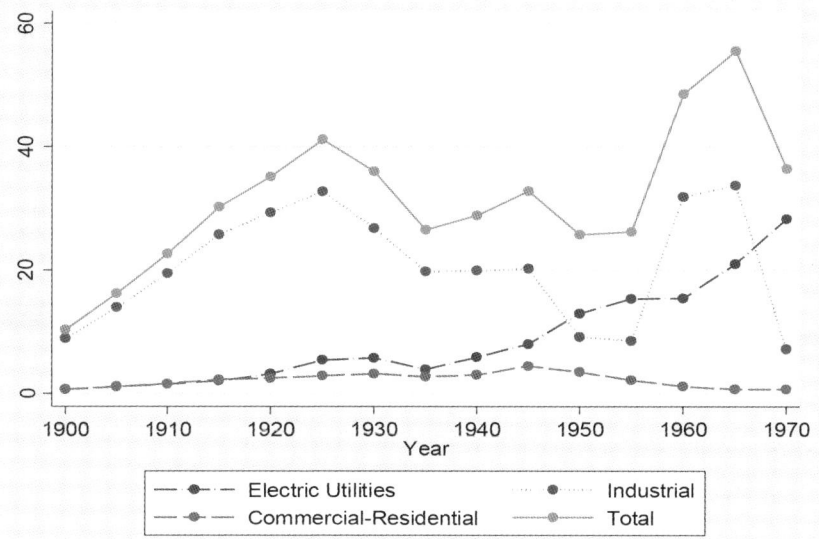

Figure 1.1. Coal consumption in Pennsylvania by type, 1900–1970.

Source: G. Gschwandtner, K. Gschwandtner, and K. Eldridge, "Historical Emissions of Sulfur and Nitrogen Oxides in the U.S. from 1900–1980," EPA report, October 1983 (Washington, DC: EPA).

vania was made into coke at the mine by beehive ovens, but the substitution of the by-product for the beehive oven caused the percentage to drop to 5 percent in 1929 and below 1 percent in 1958.[24]

While the iron and steel industry and allied coke works were the greatest users of coal in the Pittsburgh region, other manufacturing and transportation activities also consumed large amounts of the fuel as is shown in figure 1.1. Other consumers included the glass industry, metal products manufacturers, and cement and brick-makers. Electrical utilities became major coal consumers, especially after World War I. Quantities for retail sales of coal, primarily domestic, were available only on a national basis. At that time they constituted about 16 percent of national consumption but one can estimate that they probably exceeded that in the Pittsburgh region.[25]

The Environmental Effects of Coal

In spite of the economic benefits of the cheap fuel, coal had large environmental and human costs at every stage of its development. Thus, its cost and benefits should be evaluated from the point of its removal from the ground through different mining processes and to its end use.

Environmental damages in the coal-mining areas represent an extension of

Pittsburgh's footprint because of the linkages between the mines and the city's industrial and domestic consumption. Coal mining, by its very nature, whether surface or deep, alters the land, impacts geological strata, groundwater, and vegetation. Before 1977 coal mining had disturbed approximately 220,000 acres of Pennsylvania land. A smaller acreage had serious safety and environmental hazards.[26] Environmental problems included disrupted water supplies, streams filled with mining debris, and land subsidence. Mining left huge heaps of mining wastes or gob piles near the mines that loomed over the patch towns, occasionally collapsing with loss of life.[27]

The formation of acid mine drainage was a particular problem with long-lasting and difficult-to-remedy consequences. Mine acid results when iron sulfates interact with the sulfides found in coal and with groundwater entering the mine to form sulfuric acid. These acidic waters contaminated aquifers, often coming to the surface in rivers many miles from the original source. The high acidity of the water eradicated aquatic life and vegetation and often the iron oxide imparted a deep red or brown color to streams. Fish populations disappeared, ending stream use by commercial and recreational fishermen. Public health agencies worried that acid wastes in disinfected urban water supplies might force consumers to unknowingly drink from infected sources.[28] Water authorities were concerned with the corrosive nature of the contaminated water that damaged pumps and pipes and forced them to seek out new and protected sources. High acid concentrations also had large costs for industry, damaging boilers, pipes, and infrastructures. Industries that needed clean water for their processes constructed treatment facilities to ensure supplies and railroads built reservoirs in protected watersheds.[29]

Although mine acid drainage severely impacted a multitude of water users, no legal remedy existed for many decades. In the 1870s, in the landmark case of *Sanderson v. the Pennsylvania Coal Company* involving mine acid damages to private property, the Pennsylvania Supreme Court ruled that the inconvenience of an individual landowner must give way to the greater community's economic interests. The Court maintained that the production of acid drainage was a "natural and necessary result" of coal mining and that the plaintiff could not collect damages for harm. The Court further argued that burdening mine owners with the costs of limiting mine acid drainage would hamper economic development by raising the price of coal. The Sanderson precedent remained in force in Pennsylvania from the 1870s through the 1920s, when legal and societal attitudes concerning industrial pollution shifted and the courts began holding coal companies legally responsible for damages from mine acid.[30] Mine acid, however, especially from abandoned mines, was difficult to halt and still continues to be a major problem today.

The Connellsville Coke Region experienced particularly severe environmental damage as thousands of beehive ovens roasted coal and emitted volumes of smoke

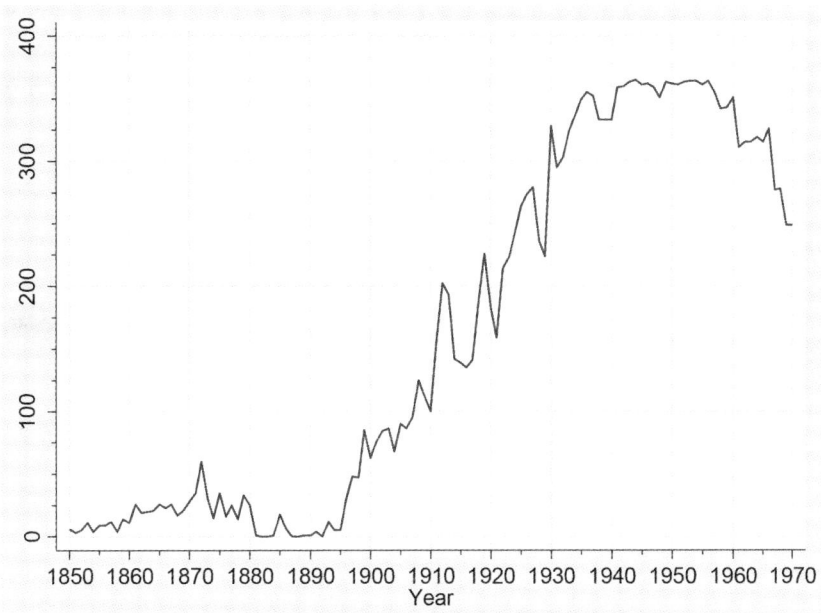

Figure 1.2. Number of days each year in which light, medium, or heavy smoke was recorded in Pittsburgh, 1850–1970.

Sources: Cliff I. Davidson, "Air Pollution in Pittsburgh: A Historical Perspective," APCA Journal 29 (October 1979): 1035–41; and Cliff I. Davidson and Debra Davis, "A Chronology of Airborne Particulate Matter in Pittsburgh," in *History and Reviews of Aerosol Science*, eds. G. J. Sem, D. Boulard, P. Brimblecombe, D. S. Ensor, J. W. Gentry, J. C. M. Marijnissen, and O. Preining (Portland: American Association for Aerosol Research, 2005), 347–70.

Note: Points prior to 1905 should be considered lower limits.

and fumes. "Cloud by day and fire by night" was a phrase used to characterize the region. In 1900 a Pennsylvania state botanist noted that the region's most conspicuous feature was "the general wretchedness of everything of the nature of shrub or tree," as sulfur fumes and hydrocarbons killed and stunted trees and crops and left a layer of ash on surrounding land surfaces. Operators at coke works dumped coal wastes into nearby creeks, damming their flow, damaging stream ecology, and causing flooding.[31]

Pittsburgh itself had been plagued with heavy smoke resulting from coal consumption since early in its history. Because of its easy availability and low costs, coal served as the city's primary fuel for domestic heating, commerce, and manufacturing. When railroads entered the city in the 1850s they added to its smoke burden. Meteorological conditions contributed to reducing air quality, since the city was subject to atmospheric inversions, a process in which warm air prevents the upward movement of air below and the dispersal of emissions. Conditions were especially severe in the river valleys where industries, railroads, and riverboats were concentrated.[32] In 1868 James Parton published an article in the *Atlantic Monthly* in which he named Pittsburgh the "Smoky City," where "every object is black. Smoke, smoke, smoke, everywhere smoke."[33] Nineteenth-century writers, however, often talked about the "salubrity" of the city's environment, arguing that the smoke possessed "anti-miasmatic" qualities. As late as the 1870s and 1880s the Pittsburgh Board of Health linked the city's low consumption death rate to the healthfulness of Pittsburgh's smoky air.[34]

Figure 1.2 shows the evolution of the number of days per year in which smoke was recorded. Compared to its very earliest days and the surrounding countryside, Pittsburgh was indeed getting smokier over the course of the 1860s and 1870s. Public officials took little action against the smoke. In 1861 the city council enacted an ordinance to control smoke from domestic sources and workshops by regulating chimney heights; in 1864, they banned construction of coke ovens within the city's boundaries; and in 1869 they forbid locomotives burning either bituminous coal or wood from entering the city. These were weak and ineffective laws that were poorly enforced and smoke density continued to increase.[35]

The First Transition to Natural Gas

A transition to natural gas, rather than public policy, produced the city's first clean air period. This fuel shift occurred from approximately 1880 to the early 1890s, when the substitution of natural gas for coal gave Pittsburgh a decade of clean air.[36] Driving the transition was the cheapness of natural gas, its ease of combustion, its cleanliness, and its high Btu value. Natural gas had been discovered in the northwestern corner of the state during the oil boom of the 1860s, and in the mid-1870s a major field was discovered at Murrysville, Pennsylvania, about seventeen miles outside of the city. In 1875 and 1876 gas was piped in from the Murrysville field to several Pittsburgh area ironworks, and in 1883 Andrew Carnegie began substituting gas for coal at his Union Mills, Edgar Thomson Bessemer Works, and Homestead Steel Works.[37] That year Pittsburgh Plate Glass as well as several other glass manufactures also shifted to gas.[38] During the 1880s and 1890s the city was able to draw gas from several gas fields relatively near, as well as from wells within the city.[39]

Development of gas resources in Pittsburgh for domestic use began in 1884, when George Westinghouse sunk a productive well on his homestead in the Point Breeze section of Pittsburgh. Westinghouse proceeded to organize the Philadelphia Company to supply gas to homes and industries in Pittsburgh and its outlying areas.[40] A number of other entrepreneurs formed companies to compete for the

Pittsburgh market and by 1886 sixteen natural gas companies operated in the city, supplying thousands of domestic and industrial consumers. Numerous derricks erected throughout Pittsburgh altered its landscape while natural gas standpipes throughout the city flared through the night.[41]

The widespread substitution of natural gas for coal, however, resulted in a sharp reduction in the number of days of heavy smoke as natural gas replaced dirty coal in industries and homes. Coal production went down by twenty million tons in Pittsburgh region mines, causing widespread unemployment in the minefields.[42] Writers in newspapers and magazines marveled at the city's cleaner air and changed appearance, noting "Fresh green shade trees . . . , newly painted buildings [and] cream colored granite walls," that were contrasted with "dingy-looking" buildings covered with "years of soot and smoke and . . . dyed to a uniform black."[43] In 1889 a county history applauded the benefits of natural gas and the disappearance of the "black pall-like cloud" that had previously hung over the city, while an 1892 article in *Harper's Weekly* observed that a "peaceful revolution" had taken place in Pittsburgh due to natural gas and it had lost its "Smoky City" title.[44] City boosters lauded the benefits of natural gas and boasted about its "almost incomprehensible quantities" and "inexhaustible" nature.[45]

The Environmental Effects of Natural Gas

The development of natural gas, like coal, had marked environmental impacts on its points of consumption and production and along its pipeline routes.[46] Most noted was the reduction in the smoke burden that plagued Pittsburgh because of the city's heavy coal use. This was a major accomplishment although, as will be seen, it was of relatively short duration. On the negative side, for several years after natural gas was introduced the city experienced a number of explosions resulting in deaths, fires, and the destruction of property. These were due to pipeline leaks and to a lack of consumer familiarity with the inflammability of the fuel. A *Pittsburgh Daily Post* editorial, "Death in the Streets," warned that "[s]ave in a state of war we don't believe any large city in the world was ever in a more perilous situation than Pittsburgh is today owing to the dangers of natural gas explosions."[47] Protest meetings were held throughout the city and the city councils appointed a Natural Gas Commission to investigate the causes of the explosions. Testimony from gas company officials before the commission revealed that gas pipes were inadequately tested for leaks. In response to its findings, in 1885 the councils passed a Natural Gas Act that made the transportation and supply of natural gas for public consumption a public service open to regulation. Later in the year councils passed an act setting specific standards for the laying and testing of pipe, for required pressures, for pressure gauges, and for a system of safety pipes. In 1886, however, Peoples Natural Gas challenged those ordinances giving the city engineer control over the manner in which pipes were laid, their location, and by whom, and the Pennsylvania Supreme Court declared these sections invalid.[48]

In addition to the wells drilled within the city, during the late nineteenth and into the twentieth century thousands of additional natural gas and oil wells were sunk throughout the western Pennsylvania region. Like coal, natural gas and oil wells could cause environmental damages at various stages of their development and these damages represented a further extension of Pittsburgh's environmental footprint.[49] Unfortunately, only a partial record exists of the effects of these wells during their operations. In addition, aside from wells with known locations, today there are many thousands of abandoned of so-called orphan wells whose locations are unknown. Many wells were abandoned when they were exhausted and an unknown number were not properly plugged. These orphan wells often leak gas, presenting an explosion hazard, and can be a conduit for salt brine that pollutes ground and surface waters.[50]

Those fragmentary records that do exist concerning the environmental effects of past well operations suggest the type of problems that could be created. For instance, in 1908 the Pennsylvania Department of Health (DOH) noted that operations at gas and oil wells often polluted water bodies with substances such as "sand pumpings," oil and grease wastes, and "burnt glycerine." It warned that contaminated wastewater ("brine") leaking from wells not properly cased or from abandoned wells left unplugged could contaminate surface streams and endanger urban water supplies, explaining that the problem often ensued when well owners removed the casing to use elsewhere when wells became dry. In addition, it warned that if wells were not properly plugged, surface and subsurface water could both fill up other gas and oil reservoirs and pollute water bodies.

The threat that these wells posed to water quality was reflected in a series of statutes passed by the Pennsylvania legislature first in 1878 and in successive years, requiring operators to plug their wells when they ceased to be productive. In 1885 the legislature approved a statute formalizing, "the incorporation and regulation of natural gas companies," which included specific regulations concerning the plugging of abandoned wells. The act stipulated a $200.00 fine in case of violations but no enforcement mechanism was specified. In 1891, the legislature passed new regulations requiring that abandoned wells be plugged in order to "prevent the pollution of springs, water wells and streams by water escaping from abandoned oil wells and gas wells." Violators of the law were guilty of a misdemeanor and required to pay fines of up to $1,000 or face up to six months imprisonment. In 1907, the legislature approved further statutes that required the clearing of brush and trees around natural gas and oil wells in order to prevent fires. A 1921 act provided more specific information about the manner in which wells were to be plugged and existing wells protected from water entering the gas strata from new well drilling.[51]

The passage of the laws indicates an awareness of the environmental problems these wells posed but their enforcement appears to have been inconsistent. Some companies, when abandoning wells, apparently plugged the wells following state law but many others probably did not. The Pennsylvania DOH was responsible for enforcing the law in cases of water pollution but was largely reactive to complaints rather than regularly inspecting well sites. The reports of the DOH and of the Sanitary Water Board (1923–71) note various complaints about gas and oil pollution of water supplies from runoff from drilling sites.[52] In 1906, for instance, the Clarion Water Company in Clarion County complained that the development of natural gas and oil wells in the watershed had polluted the sources of water from which they supplied the town, and asked for permission to extend their water-gathering area to a clean source. The DOH investigated, finding the company's supplies indeed to be polluted and "prejudicial to the public health." The department reported that the "waste material produced in the operation of drilling the wells, in shooting them, and in cleaning them out, is deposited on the surface of the ground round about and eventually gets into the main stream of the water supply." The DOH ordered the water company to either filter the water or find a new source. It also ordered the company not to allow any gas or oil drilling on its lands and to regularly inspect the wells in the vicinity of the borough in order to prevent saltwater from the wells to contaminate water supplies.[53]

In addition to state enforcement of the statutes relating to natural gas and oil statutes, the courts considered nuisance cases generated by gas and oil pollution of private drinking-water wells. Few of these cases got to the courts but when they did they were usually decided in the favor of the plaintiffs. In an 1890 case the Pennsylvania Supreme Court affirmed damages against a natural gas company for permitting saltwater to contaminate a private drinking-water well because of inadequate casing, noting that "when the salt water is allowed to mingle with the fresh, it will spoil the whole neighborhood." The court also held that the Sanderson opinion on mine acid drainage, discussed earlier, did not apply because the driller knew that such contamination could occur and did not take actions to prevent it.[54] Other pollution cases of private wells appear to have been settled out of court by the gas companies. The Pew Papers at the Hagley Museum and Library, for instance, contain several letters to Joseph N. Pew, president of the Peoples Natural Gas Company, from attorneys representing clients complaining of damage to their water supplies and injuries to livestock from leaks in gas wells and pipe lines; these complaints appear to be have been settled.[55] The Peoples Natural Gas Company was aware that problems ensuing from poor casing could cause water pollution. It required that contractors ensure that casing be inspected and if water was found "the well

. . . be thoroughly drained and sand pumped until all drillings and sediments are removed."[56] It is unknown, however, if other gas companies and especially, smaller and fly-by-night drillers, followed the regulations.

Return to Coal and Smoke

Natural gas resulted in cleaner air but it was not inexhaustible. By 1890, fluctuating and declining supplies negatively affected industrial users and they began shifting back to coal.[57] The depletion of natural gas supplies was a constant concern of utilities, and they unsuccessfully experimented with methods to produce manufactured gas from bituminous coal cheaply enough to compete with coal.[58] In 1892 a speaker at a meeting of the Western Pennsylvania Engineering Society woefully observed, "We are going back into the smoke. We had four or five years of wonderful cleanliness for Pittsburg [sic], and we have all had a taste of knowing what it is to be clean. We all felt better, we all looked better, we all were better. But we are back into the smoke. It is growing worse day by day."[59]

The return of the smoke caused the city's first major smoke control effort. The Ladies Health Protective Association of Allegheny County, an organization composed mostly of upper-class women, drove the campaign. Recruiting allies from among the engineering and business communities, the antismoke forces pushed for effective regulatory legislation in succeeding years. In 1892 and 1895 the city councils passed smoke control ordinances but they were only weakly enforced. In 1902 the courts declared the 1895 ordinance unconstitutional.[60] During the following years, as smoke pollution increased in severity, reformers drove the city council to approve new ordinances.

Several research efforts contributed to the pressure for such legislation. The Pittsburgh Survey, conducted by the Russell Sage Foundation in 1907–8, aroused Pittsburgh to various social and environmental issues that affected the quality of life in the industrial city, including smoke.[61] Following this, between 1911 and 1914, the Mellon Institute of Industrial Research conducted a Smoke Investigation, which published ten bulletins written by experts on the effects of smoke on architecture and building materials, weather, vegetation, human health, psychology, and Pittsburgh's economy. These studies showed smoke's negative impacts on many aspects of the city's life but were inconclusive in regard to its effect on health. Still, the reports contained enough information concerning the costs of smoke to drive the city council in 1914 to approve the toughest smoke control ordinance enacted to that time.[62]

Enforcement of the ordinance plus an increase in natural gas availability from West Virginia wells via pipelines improved Pittsburgh air quality during the immediate prewar years, but this source was also depleted in several years by waste and over use. Between 1914 and 1921, for instance, the amount of gas produced by

Figure 1.3. Smoke in downtown Pittsburgh, 1940.
Source: Archives Service Center, University of Pittsburgh.

the Equitable Gas Company, Pittsburgh's largest provider drawing from its West Virginia wells, declined from 43,189 to 13,359 million cubic feet.[63] The war years brought a further deterioration of air quality due to increases in industrial emissions. In addition, a collapse in rail transportation beginning in late 1916 and extending into the early 1920s generated a fuel crisis, putting pressure on coal and natural gas availability. The U.S. Fuel Administration, created in 1917, imposed natural gas rationing on some communities and industries and forbid natural gas delivery without a license.[64] A hearing held by the Pennsylvania Public Service Commission in Pittsburgh in 1919 emphasized the great waste in natural gas production, transport, and consumption.

The uncertainty of gas supplies caused major industries, especially in the metals sector, to shift back to coal in the 1920s, repeating the pattern of the 1890s.[65] Gas consumption by the iron and steel industry was sharply reduced due to the lower price of coal and concern about the consistency of gas supply. Domestic consumers were the largest natural gas users, constituting about 15 to 20 percent of total Pittsburgh households. The glass industry also continued to rely on natural gas due to its high caloric content and its freedom from ash and dust that provided special advantages for quality glass making.[66] Although thousands of new wells were drilled in Pennsylvania in the 1920s and 1930s, total gas production did not increase because older wells lost pressure.[67]

During the 1920s, engineers and smoke regulators involved in smoke control concluded that domestic smoke from coal consumption as well as that from industries and railroads had to be controlled. They argued that residential chimneys emitted smoke at low levels, where it could not be dispersed by the winds.[68] How to devise and implement such a policy, however, was politically difficult and involved both technological and fuel issues. But before the new strategy could take effect, the Great Depression hit and Pittsburgh experienced sharp industrial slowdowns and high unemployment. Households and industries such as steel and railroads cut back on fuel expenditures and the coal industry sharply curtailed production. The air became cleaner, but clear skies and unemployment was a combination that Pittsburghers found unacceptable.[69] Many wished for a return to smoking stacks and productive mills, as the old equation between smoke and jobs was reinforced. In a defiant reflection of this belief, in 1939 the city council voted to eliminate the Smoke Control Bureau.[70]

Smoke Control, the Second Transition to Natural Gas and Railroad Dieselization

While Pittsburgh was reducing its efforts to control smoke, however, another smoky city, St. Louis, was moving to improve its air quality. The mayor of St. Louis had appointed Raymond R. Tucker, a combustion engineer and former faculty member at Washington University of St. Louis, his secretary with the responsibility of devising a means to "clarify the air." Tucker hit on a strategy requiring either the use of clean fuel or smokeless mechanical equipment. He gradually achieved his goal by persuading the St. Louis Board of Aldermen to approve a series of regulatory ordinances targeting different polluters. The final success came in April 1940, after several years of especially heavy smoke palls, when the board of aldermen approved an ordinance requiring domestic users to use clean fuel or smokeless combustion technology. The result was a series of smokeless days that city officials claimed was the result of the smoke ordinance.[71]

Pittsburgh officials and representatives of smoke control groups visited St. Louis and returned with glowing reports. The press demanded that the city council pass similar legislation. In response, Mayor Cornelius Scully appointed a Mayor's Commission for the Elimination of Smoke with wide representation from steel and coal corporations, labor unions, the health professions, the media, and civic groups. A technical advisory subcommittee provided needed advice and information about fuel availability and combustion equipment. The commission recommended a procedure similar to St. Louis's, requiring the use of smokeless fuels or the utilization of smokeless mechanical equipment. While the act contained provisions for regulating railroad and industrial smoke, the focus of the ordinance was on domestic coal consumers.

Figure 1.4. Women campaigning for clean air, 1941.
(Used with the permission of the Western Pennsylvania Historical Society.)

The Pittsburgh City Council approved the smoke control ordinance by an 8 to 1 vote in early July, but its full implementation would have to wait until the end of World War II. The willingness of the representatives of the coal corporations and of the mine workers union on the mayor's commission to support the ordinance can be explained by the belief that bituminous coal could serve as the basis for new industries processing local coal for smokeless fuel, coal tar, and for coal gasification. Many of the smaller mine operators, however, were skeptical and warned that creation of an industry to manufacture processed coal was "wishful thinking." Instead, they pushed for the use of better combustion technology to reduce smoke emissions from bituminous consumption. Their doubts were to be borne out in the postwar period when inadequate supplies of smokeless coal threatened the success of the ordinance.[72]

Implementation of the Smoke Control Act in the postwar period focused first on domestic users, considered to be inefficient energy consumers whose emissions were harder to disperse because they were closer to ground level. Individual householders could choose among several alternatives in order to meet the law's requirements. One was to purchase smokeless combustion equipment that would permit the continued use of high-volatile coal. A second was to install combustion equipment using alternative fuels such as natural gas or oil or to retrofit existing equipment to burn these fuels. A third alternative was to use smokeless coal in existing stoves and furnaces. Because the last option required the least capital outlay, most working-class households preferred it. From the perspective of enforcement, however, this alternative posed the most difficulties.

There were approximately 100,000 homes using hand-fired coal-burning stoves and furnaces in Pittsburgh, including about 69,000 that depended on coal stoves for cooking as well as heating. To enforce against the individual domestic consumer was an impossible task for the Bureau of Smoke Prevention's twelve inspectors. The bureau solved the enforcement problem by focusing on the coal distribution yards (approximately thirty) and the coal truckers, forbidding the yards and independent truckers from selling high-volatile coal for use in hand-fired equipment. Truckers hauling coal for consumption in the city had to be licensed and to have license numbers painted on the side of the trucks for easy identification. Those caught hauling illegal high-volatile (or bootleg) coal were subject to fines, as were dealers who made illegal sales.[73]

Popular Pittsburgh history attributes the success of the city in cleaning up its air to the Smoke Control Ordinance and its implementation. Smoke control is pictured as the important first step in the Pittsburgh Renaissance, brought about by a public-private partnership spearheaded by Mayor David Lawrence, banker Richard King Mellon, and the nonprofit Allegheny County Conference on Community Development composed of the city's major CEOs. Undoubtedly the Smoke Control Ordinance and the reduction of air contamination was a major step in reforming the industrial city. As has been discussed, however, some decision makers chose to believe that clean air could be achieved without major harm to the coal and related industries. Coal had been the major element in the rise of Pittsburgh to industrial prominence and many of the city's industrialists expected that it would remain as a major factor in the city's and region's economy.

The Smoke Control Ordinance was implemented at a time when the nation was already in the middle of an energy transition. Coal had suffered during the 1920s and the Great Depression as the competing fuels of oil and natural gas cut into its markets. Rising incomes after the Depression caused sales of improved home-heating equipment and consumer appliances to rise, as natural gas and fuel oil companies greatly expanded their marketing campaigns. Household oil burners increased in sales more than sixfold between 1929 and 1941 and the sale of gas home-heating units and other gas appliances greatly increased in locations where pipelines made natural gas newly available. Even Pittsburgh's local gas utilities saw increases in demand from domestic and industrial users.[74]

Confronted by their vulnerable market position, the Pittsburgh regional coal companies fought hard to protect their interests. There were limits to their influence. Probably the key reason for their inability to weaken stringent municipal smoke control regulations was the influence of Richard King Mellon, who exercised enormous economic power through his leadership of the Mellon banking interests and linked energy corporations such as the Pittsburgh Coal Company, Koppers Coke, and Gulf Oil. Mellon had a keen interest in smoke control both because his father had sponsored the original Mellon Smoke Investigation and because he believed that cleaner air was vital to Pittsburgh's economic future. In 1945 he helped engineer a merger between the United Smoke Council, the leading antismoke organization, and the Allegheny County Conference on Community Development, the pivotal group that brought about the Pittsburgh Renaissance.[75]

Control of smoke, however, in the eyes of corporate decision makers and politicians, did not necessarily mean the end of coal's dominance in Pittsburgh for domestic heating, industrial uses, and railroad fuel. Because of issues surrounding the uncertain availability of competing fuels, especially natural gas, they were confident that treated coal and mechanical stokers would sustain coal production as well as stimulate new industries producing smokeless coal, stoves, and furnaces after the war. In 1943, for instance, the U.S. Bureau of Mines noted, "Lower cost to consumers and availability at all times . . . are the principal factors favoring the use of coal."[76] In addition, restrictions on the availability of oil and natural gas and limited production of new furnaces during wartime suggested that coal would continue to dominate. The coal industry therefore looked forward with some confidence to continued control of markets in the postwar period.[77]

Significant developments in the area of natural gas availability, however, boded ill for the future of coal especially in Pittsburgh domestic markets. In 1945 the Texas Eastern Transmission Corporation purchased the twenty-four-inch "big inch" and the twenty-inch "little inch" pipelines—constructed by the federal government during the war to transport oil from the Southwest to industrial markets—and converted them to natural gas. In 1947 these pipelines reached the Pittsburgh region and supplied clean, cheap natural gas to local gas utilities.[78] A second major factor was the expansion of underground storage pools for natural gas that allowed the gas industry to transport and store gas from the Southwest in the summer and withdraw it during the winter to satisfy residential needs, thereby eliminating a serious problem that had led to winter gas shortages particularly for industrial users. Between 1944 and 1954, the capacity of national underground storage pools increased from 135 billion to 1,859 bcf (billion cubic feet). In Pittsburgh, the Equitable Gas Company, the largest Pittsburgh based natural gas utility, increased its storage from 8,943 mcf (million cubic feet) in 1947–48 to 25,064 in 1952–53. Peoples Natural Gas Company, another major Pittsburgh utility, expanded its storage inventory from 4,600 mcf in 1945 to 13,920 in 1950.[79]

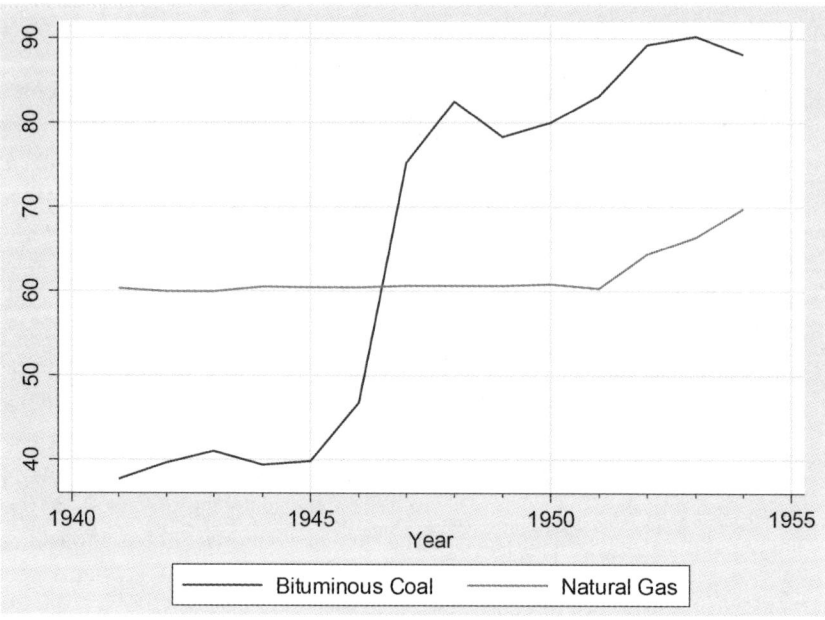

Figure 1.5. Coal and gas prices in Pittsburgh, 1941–54.
Source: American Gas Association, *Historical Statistics of the American Gas Association, 1966–1975* (Arlington, Va.: AGA, 1977), 231, table 3.

Figure 1.5 shows that the retail prices of coal increased in the immediate post-war years, while natural gas prices in Pittsburgh remained stable. In Pittsburgh, as well as other cities such as Chicago and Cleveland, gas became cheaper than bituminous coal in 1947, and customers began to switch to the cleaner fuel. Discussions of heating fuel and furnaces appeared in newspapers and popular magazines, as did advertisements for gas and conversion units because utility companies launched extensive sales campaigns. A series of strikes by coal miners, particularly in 1946 and 1949–50, undoubtedly accelerated the pace of change as the price of coal increased in response to the work stoppages.[80]

In Pittsburgh the high price of coal and the advantages of natural gas combined with the requirements of the Smoke Control Ordinance to encourage customers to switch fuels and purchase new combustion equipment. The increased use of gas is reflected in changes in domestic heating apparatus. In 1940 the U.S. Census reported that 81 percent of Pittsburgh households burned coal and 17.4 percent natural gas; by 1950, the numbers were 31.6 percent for coal and 66 percent for natural gas.[81] Large numbers of homeowners and commercial users installed either new gas boilers or conversion units for gas in their coal-burning boilers. Some also installed mechanical stokers. In the same period Equitable Gas and Peoples Natural

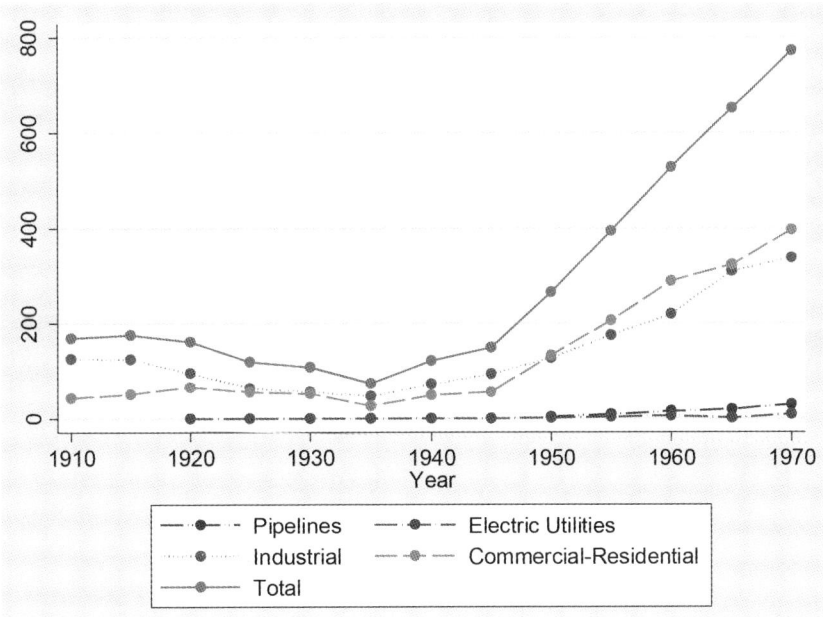

Figure 1.6. Natural gas consumption by type, 1910–70.

Source: G. Gschwandtner, K. Gschwandtner, and K. Eldridge, "Historical Emissions of Sulfur and Nitrogen Oxides in the United States from 1900–1980," EPA report, October 1983 (Washington, DC: EPA).

Gas reported that their residential and commercial sales had doubled while their industrial sales had increased over a third.[82]

The winter of 1947–48 was critical for the new statute. Some of the more dire predictions of the ordinance's critics were borne out. Supplies of smokeless fuels proved to be inadequate and prices were inflated. In 1948, bituminous coal was $10.70 a ton while "Disco," or treated coal, was $17.10. The ordinance fell hardest on coal-burning low-income families because fuel costs composed a larger percentage of their budget than for higher income groups. They often purchased their coal by the week in bushel lots from itinerant truckers, since they had neither the cash nor the storage space to buy larger amounts. Because of a lack of familiarity with the characteristics of low-volatile coal, as well as the sale of low-grade mixtures, many Pittsburghers had serious problems obtaining sufficient heat. Complaints about the regulation poured into city hall, newspaper offices, and radio stations, and several city councilmen attempted to have the regulation suspended. "Undoubtedly," an important city official later admitted, "some very real difficulties were imposed on many people."[83]

In spite of the many difficulties with fuel supply, the heating season of 1947–48 showed a considerable improvement in air quality compared with previous years. An unusually mild winter aided in reducing the smoke palls. "PITTSBURGH IS CLEANER" reported the *Pittsburgh Press*. The worst smogs were gone, homes were cleaner, and white shirts did not develop black rings around the collars. The United Smoke Council boasted in a publication titled "The New Look in Pittsburgh" that it was losing its reputation as "the Smoky City."[84] Soaring natural gas adoption for residential purposes in the coming years, as shown in figure 1.6, ensured that the smoky city was gone.

Another major fuel change that reduced Pittsburgh's smoke burden further in the 1950s was railroad conversion from coal-burning steam locomotives to diesel-electrics. Smoke investigators often pinpointed locomotives as offensive smoke generators, responsible for a large fraction of the city's smoke burden; trains were considered especially polluting because of their mobile nature. Railroads were concerned with coal not only because of fuel availability and costs but also because it was their largest single item of revenue tonnage.[85] The Pittsburgh Smoke Control Ordinance of 1941 was the nation's strictest, and in 1947 the Pennsylvania House of Representatives approved a bill giving the Allegheny County Commissioners jurisdiction over railroad smoke but coal-burning locomotives remained a problem.[86]

Improvement would come with a major change in technology—from coal-burning to diesel-electric locomotives. In the 1930s railroads had used diesels as switching engines in some cities and substituted them for steam locomotives on long-distance passenger runs, but railroads had resisted the capital costs involved in adopting the new technology while locomotive manufacturers largely remained wedded to the steam locomotive. The coming of World War II further retarded the transition, and in 1946 steam locomotives in service numbered 39,592 compared with 5,008 diesels. The change to diesel locomotives, however, occurred with great swiftness after this date. By 1951 there were almost as many diesels as steam locomotives in the nation and by 1960 there were 30,340 diesels and only 374 steam locomotives.[87] Pittsburgh railroads were part of this transition. By 1952, the Pennsylvania Railroad, which operated the city's largest number of passenger and freight trains in the city, had replaced almost all its steam locomotives with diesel electrics. The same pattern was true for Pittsburgh's other roads—in its 1955 report the Bureau of Smoke Prevention observed, "The Diesel locomotive has solved the smoke problem of the railroads."[88]

Effects of the Energy Transition

The shift from coal to natural gas as a fuel for domestic and commercial heating and the replacement of the steam locomotive with the diesel electric had significant effects on environmental quality, on the health of Pittsburghers, and on Pittsburgh's civic ambience. The most evident environmental improvement was the

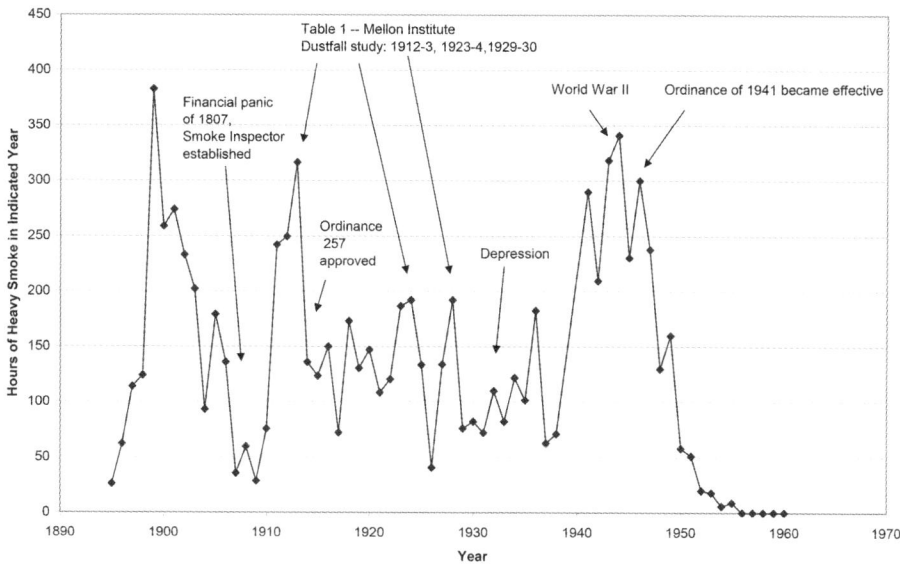

Figure 1.7. Hours of heavy smoke per year in Pittsburgh.

Sources: Cliff I. Davidson, "Air Pollution in Pittsburgh: A Historical Perspective," *APCA Journal 29* (October 1979): 1035–41; and Cliff I. Davidson and Debra Davis, "A Chronology of Airborne Particulate Matter in Pittsburgh," in *History and Reviews of Aerosol Science*, ed. G. J. Sem, D. Boulard, P. Brimblecombe, D. S. Ensor, J. W. Gentry, J. C. M. Marijnissen, and O. Preining (Laurel, N.J.: American Association for Aerosol Research, 2005), 347–70.

reduction in the number of hours of heavy smoke that began in the winter of 1947–48, as is reflected in figure 1.7. Some of the reduction is attributable to declines in industrial production and the implementation of more efficient combustion technologies in the immediate postwar period, but the magnitude and timing of the change suggests that the shifts in heating and railroad fuels also played key roles.

In regard to health, a large epidemiological literature links exposure to air pollution to higher morbidity and mortality. Pittsburgh was particularly subject to particle pollution that is associated with higher winter mortality. Such pollution was especially generated by coal burned for home heating and discharged from chimneys at relatively low levels. Throughout the twentieth century public health investigators had sought to link up coal use with specific medical problems but with limited success. The 1909–14 Mellon Smoke Investigation, for instance, produced a series of papers on the influence of smoke on health that were inconclusive. Later studies of the relationship between smoke and pneumonia by the Pittsburgh Department of Public Health and the Air Hygiene Foundation of the Mellon Institute for Industrial Research also failed to establish a strong causal connection.[89]

The availability of stronger data and means of analysis today makes it possible to shed better light on the relationship between coal burning, health, and the positive effects of the transition to natural gas for home heating. In a national study of the effects of retail coal use on mortality, results indicate the monthly mortality rate is related to retail sales of coal. A one percent increase in retail coal sales caused a 0.06–0.10 percent increase in the overall monthly mortality. Other studies have found that declines in total suspended particles (TSP) cause declines in infant mortality.[90]

Many of the benefits of smoke control in terms of improved air quality, increased sunshine, and health were difficult to quantify, but the Bureau of Smoke Prevention did attempt to put a dollar figure on savings resulting from "greater cleanliness." In 1950 the bureau calculated the total savings as $26,808,000, or $41.00 per capita per year. This included savings on cleaning costs, building depreciation, laundry bills, injury to vegetation, and fuel costs. These calculations do not include the fuel cost savings and improvements in industrial efficiency that took place in industries that substituted natural gas for coal.[91]

From the perspective of the civic culture, smoke control played a critical role in the Pittsburgh Renaissance, the first attempt by a major industrial city to renew itself. A 1949 *National Geographic* article on the city's renewal noted that "[s]moke elimination has become a dramatic symbol of the whole powerful impetus toward civic improvement." A *Newsweek* article in the same year claimed that smoke control had cut the smoke pall by half and that the "result has been a revelation to the city's dwellers" because it had convinced Pittsburghers that "better things, somehow, can get done," giving them faith in the whole Renaissance redevelopment program. There is no doubt that without smoke control the Pittsburgh Renaissance would never have been accomplished.[92] The city's boosters now heralded it as a Renaissance city no longer dependent on coal and its noxious by-products, smoke and dust fall.

Although Pittsburgh is famous within the annals of energy history for the fortunes derived from exploitation of the Pittsburgh coal seam and for its "smoky city" nickname, it has actually experienced several energy transitions. This chapter has attempted to provide insights into the most significant of these transitions— those involving coal and natural gas. The first occurred in the late nineteenth century and the second took place in the decades after World War II. Issues relating to supply and demand, fuel prices, labor issues, and technology shaped these transitions, as did forecasting failures, unexpected events, and public policies on smoke control.

The discovery of natural gas in the Pittsburgh region was an unexpected event although gas was known to exist in the oil fields in the northwest corner of the state. The first transition period was marked by exaggerated forecasts in regard to the extent of the local natural gas supply. By approximately 1890 supplies began to fail, resulting in a return to coal for many domestic and industrial users. The second and longer transition was driven by the arrival in the region of extensive gas supplies from the Southwest provided by the inch pipelines. Gas use had unanticipated elements as well; authorities had thought the transition would be driven by cleaner processed coal and not a different fuel. Public authorities also failed to anticipate the rapidity of the replacement of coal-burning locomotives with diesel electrics in the 1950s, even though the latter technology had been available for some years.

Each of these transitions had significant effects on the environment, on concepts of civic improvement, on public health, on the economy, and on labor. In the first transition, natural gas availability stimulated the growth of the iron, steel, and glass industries. The city's downtown experienced a commercial building boom as industries and banks constructed new headquarters. Production in the mining areas, however, was reduced and large numbers of miners were put out of work. When natural gas supplies later declined, industries and households returned to coal, bringing back heavy smoke palls. In turn, this happening stimulated a movement toward clean air through regulation. It would be approximately fifty years before success was achieved and it would, at best, be only partially attributable to public policy. Finally, the transition from coal to natural gas in both periods resulted in improvements to the public health as the smoke burden was reduced.

2

The Energy Capital of the World?

Oil-Led Development in
Twentieth-Century Houston

Martin V. Melosi and Joseph A. Pratt

Although incorporated in 1836, modern Houston is the product of oil-led development in the twentieth century, when the southeast Texas town grew
into a full-scale metropolis. Houston is currently the fourth largest American city
in population and the largest in area, and it sits at the center of the tenth largest
metropolitan area (in population) in the United States. Its self-proclaimed status as
the "Energy Capital of the World" is more than a hollow brag. In an age when oil
and natural gas still dominate the world's energy supply, the Houston region contains the oldest, largest, and most diverse complex of oil- and natural gas–related
activities in the world. This dynamic economic development played a central role
in the city's growth, form, and culture. The case study of Houston, therefore, is a
logical starting point for the analysis of the emergence, evolution, persistence, and
decline of energy capitals.

The Physical Setting

Geology and geography shaped the setting and the context for the emergence
of Houston as an energy capital, but human activities determined the pace and
direction of industrialization and urbanization. Houston is a product of the Tex-

as Coastal Zone, an area of approximately 20,000 square miles with about 2,100 square miles of bays and estuaries, 375 miles of coastline, and 1,425 miles of shoreline.[1] Texas has two shorelines—one along the Gulf of Mexico and another along the bays. Bolivar Peninsula, Galveston Island, and Follets Island are grass-covered barrier flats and sandy beaches that separate the bay areas from the Gulf of Mexico.[2] Nature's design provides a measure of protection from hurricanes for the industrial sites and refinery towns that grew along the shores of Galveston Bay and of the Houston Ship Channel that connect the bay to the city of Houston.

The entire Galveston Bay drainage basin covers 33,000 square miles of land and water from the Dallas–Fort Worth Metroplex to the Texas coast.[3] Above the bays two major river streams cut into the coastal plain. The Brazos River and Oyster Creek flow through the western portion of the Texas Coastal Zone.[4] From these rivers and bayous came the abundant freshwater that played an important role after 1920 as cooling water for the region's growing refining complex.

Within the Houston-Galveston area is approximately 2,300 square miles of land—a broad area of flat coastal plain between marshes and the areas of pine and hardwood forests along the Trinity River and north of present-day downtown Houston.[5] Situated approximately forty-nine feet above sea level on prairie some fifty miles from the Gulf of Mexico, Houston is linked geographically, geologically, and climatically to the Texas coast. The city sits on a coastal plain that comprises gently dipping layers of sand and clay, favorable for artesian water. It historically has drawn water from the Chicot and Evangeline aquifers running through the city. Extraction of the groundwater has led to serious land subsidence and saltwater intrusion when water is drawn out too aggressively.[6] In the twentieth century, the extraction of oil and gas also contributed to subsidence.

Surface water from Houston drains into the Gulf of Mexico via an elaborate network of bayous. With an average yearly rainfall of forty-two to forty-six inches, the area frequently floods. Since urbanization removes much of the filtering capacity of the soil, the severity of flooding increased as the city expanded. The region also is susceptible to hurricanes and tornadoes. Water and the pollutants it carries have been the greatest natural threat to Houston and its neighbors.[7] Besides its extensive waterways, Houston also is a heavily vegetated city. Large portions of the region are forested, with substantial tree growth along the bayous. To the south, the area is covered with a combination of prairie, marsh, forest, and abandoned agricultural lands; to the north, the pine forests of East Texas extend down into the region. As the Houston metropolitan area spread under the impetus of oil-led development, the ground and canopy cover diminished markedly. Concrete replaced much marshland and pine forest, producing a pronounced "heat island" effect in the area.[8]

Houston's climate is a defining characteristic of the region. It is subtropical and

humid, with prevailing winds bringing heat from the deserts of Mexico and mois-
ture from the Gulf. The region's topography is also distinctive. The flat, boggy
coastal plain gives the region a physical blandness; throw in stifling heat, and living
in this environment can be downright oppressive. The use of air conditioning on
a massive scale after World War II made the region's climate more bearable while
shaping patterns of life and work.[9] Before or after the coming of air conditioning,
however, few have come to Houston to enjoy its mild climate, its beautiful mountains, or the crystal clear water and white sands of its beaches. They have come
instead for economic opportunities in the form of jobs produced largely by the region's abundant resources and the energy and innovativeness of its oil-led economy.

Close proximity to the Gulf of Mexico for shipping, ample intracoastal waterways, good water supplies, and large tracts of arable land have been valuable assets.
In addition to oil and natural gas, the region has a variety of other exploitable resources, including timber, sulfur and salt, sand and gravel, shells for lime, abundant wildlife and sea life, and petroleum reserves.[10]

Exploitation of these resources, however, has come at a high price.[11] As one geologic study noted, "The attributes that make the Texas Coastal Zone attractive for
industrialization and development also make it particularly susceptible to a variety
of environmental problems."[12]

Extensive dredging of channels and passes resulted in discharge of sediment
into bays, ultimately modifying circulation patterns, and affecting water quality
and estuarine plants and animals. Because of increased cultivation, the construction of irrigation and drainage canals and urban paving result in streams accelerating the transport of sediment into bays, as well as increasing nonpoint pollutants
including pesticides and herbicides. The straightening and lining of several major
bayous in the region with concrete has encouraged flash flooding. Thermal effluent from manufacturing processes and power generation can be lethal to fish. Discharge of organic materials and trace metals too numerous to mention—including
oil production, pipelines, spills, and chemical production—adds significantly to the
pollution load of all watercourses. Other actions also have had significant impacts
on oyster reefs, submerged aquatic vegetation, intertidal marsh vegetation and animal life, and freshwater wetlands.[13]

A final natural endowment of the region, vast deposits of oil and natural gas in
this region and in sections of the southwestern United States that could reach the
Gulf Coast via pipelines, shaped both the environmental costs and the economic benefits of development. Petroleum propelled regional growth throughout the
century, giving the Houston area its national and international economic identity.
The economic benefits were great, as were the environmental costs and the uncertainties caused by often severe swings in oil prices. For most of the twentieth
century, the major oil-related companies and their supporters in government made

the key calculation of the balance between benefits and costs, usually tipping on the side of benefits. On a more personal level, hundreds of thousands of individuals made their own calculations, choosing to migrate to the Houston area from farming regions in the surrounding area and throughout the nation in search of opportunities to improve their lot in life.

Houston and Oil-Led Development: An Overview

The age of oil arrived dramatically in Houston with the epoch-defining Spindletop oil strike ninety miles east of Houston near Beaumont, Texas, in January 1901. This spectacular gusher grabbed the attention of the world and launched an oil boom on the Texas–Louisiana Gulf Coast. The importance of Spindletop to the region and the nation is captured in the engraving on the monument near the discovery well: "On this spot on the tenth day of the twentieth century, a new era in civilization began."[14] In this "new era," oil and natural gas emerged as the preferred energy source of the twentieth century, gradually displacing coal in a variety of markets while encouraging fundamental changes in transportation, manufacturing, and even geopolitics. The southwestern oil boom sparked by the Spindletop discovery altered the competitive balance in the oil industry by giving rise to a new group of "local" rivals that included Gulf Oil, The Texas Company (Texaco), and Sun Oil. These companies challenged Standard Oil's control of the industry. The giant reserves subsequently found on the Texas Gulf Coast and throughout the southwestern United States greatly expanded the supply of oil, encouraging the adoption of refined petroleum for fuel and paving the way for the rise of the gasoline-powered automobile.[15] The new era in oil brought marked departures in the location, scale, and technical complexity of the petroleum industry, and Houston was well positioned to exploit them.

The timing of the discovery of vast supplies of oil near Houston was critical in the city's subsequent rise as an energy capital. Modern Houston was born at the same time and place as the modern petroleum industry. Before 1901, oil remained primarily a source of illumination; after that date, it quickly became a fast-growing new source of energy.[16] The city grew up with the modern oil industry, and many oil-related companies that began as local concerns ultimately grew into national and international concerns. In this sense, the region enjoyed what business strategists describe as a "first-mover advantage" in many oil-related industries. Indeed, one reason for the rapid rise of oil as an energy source was that local industries and individuals along the Gulf Coast and throughout the southwestern United States quickly switched from expensive imported coal to inexpensive local oil as a primary energy source. The city took advantage of its head start and grew steadily over a long period into an important and well-established part of what became a global system of oil and gas production, refining, transportation, and use. Capitalizing on a combination of its abundant natural resources and the risk taking, innovativeness, and work ethic of its people, Houston became synonymous with oil in the way that Pittsburgh meant coal and steel and Detroit meant automobiles.

The location of Houston and the newly discovered oil fields on the Texas–Louisiana Gulf Coast also proved important for the region's persistence as an energy capital. A black gold rush to the region after 1901 led to the discovery of other large oil fields throughout the upper Texas–Louisiana Gulf Coast and then in surrounding states. The southwestern United States quickly emerged as the center of oil production in the world, a position it held into the 1950s. Houston and its surrounding region remained an important center of oil production for much of the twentieth century, but pipelines increased the region's reach by tying its deepwater port and its refineries to many of the major fields of the Southwest. This assured the region's sustained growth as an oil capital even as oil boomtowns in Texas and surrounding states regularly rose and fell with the depletion of nearby oil fields. By the 1940s, half of the world's production of oil was located within 600 miles of Houston, and the region surrounding the city could boast of 4,200 miles of pipeline reaching out to hundreds of oil fields in Texas, Louisiana, Oklahoma, and New Mexico.[17] Pipelines and refineries could not be moved after construction, and the Gulf Coast refining region from New Orleans to Corpus Christi benefited from investments in such facilities during decade after decade of oil-led development.

The persistence of oil-related investments in this broad region throughout the twentieth century produced an unusually large cluster of oil-related activities that built on the existing momentum, drawing more and more petroleum specialists to the Houston area. During four eras of oil-led development, the petroleum-related core of industries remained dynamic and grew increasingly diverse. The distinguishing characteristic of Houston as an energy capital became the size and diversity of this core—for lack of a better word, its thickness.

The Region before Oil

In the 1850s, Frederick Law Olmsted, the American landscape architect who designed New York City's Central Park, traveled through Texas. His impression of the region east of Houston was as follows: "Upon the whole, this is not the spot in which I should prefer to come to light, burn, and expire; in fact, if the nether regions . . . be a boggy country, the avernal entrance might, I should think, with good probabilities, be looked for in this region."[18] Yet in the last half of the nineteenth century, the population of the region that became the Houston Metropolitan Area grew from less than 15,000 to almost 125,000. By the late nineteenth century, Houston was well prepared to absorb the benefits of massive oil-led development in the southwestern United States. At the turn of the twentieth century, Houston already was a city on the rise. It served as the center of trade, railroad transportation, law, and banking in a growing region with good prospects for future expansion. It had strong ties to the national economy through its exports of cotton and timber, the close collaboration of its law firms with intercontinental railroads, and the well-established ties of its banks with money center banks and other outside investors.[19] These strengths gave it advantages that the smaller city of Beaumont could not match. Houston also had an aggressive civic/business elite and a habit of openness to ambitious outsiders that was not matched by its local rival, Galveston, which was also more vulnerable to hurricanes than its sister city forty miles to the north.

Throughout the late nineteenth century, migrants from smaller towns in Houston's hinterland brought their ambitions and talents to the larger stage represented by Houston, which they identified as a city on the move.[20] Hardworking farmers and sharecroppers in the agricultural regions outside of Houston proved eager to find pathways out of rural poverty, and they had begun to migrate to the cities on the coast even before the oil boom. Also important in the region's preparation to emerge as an energy capital was a stable political environment conducive to economic growth yet with ample powers to regulate the societal costs of growth. In short, when the opportunity for oil-led development knocked, Houston was ready and willing to open wide its doors.

The Oil Boom, 1901–20

Tens of thousands of migrants came to the oil fields of the Texas Gulf Coast in response to the Spindletop discovery, which transformed Beaumont into the temporary center of the world oil industry. Almost overnight, this thriving town of about ten thousand became a boomtown, a hothouse of capitalist development where more than fifty thousand sought oil wealth. Substantial capital flowed into the region from well-heeled investors from the established industrial cities of the northeastern United States. The Mellon family of Pittsburgh provided much of the financing for the newly created Gulf Oil, which initially was the largest oil company at Spindletop. New York investors backed two of the other largest concerns active in Beaumont and the surrounding region—The Texas Company (Texaco) and a refining company that became Magnolia before later being absorbed into Standard Oil of New York (Mobil).[21] These three companies took strong roles in developing Spindletop and in finding and developing other fields in the region, and they also built the first substantial refineries on the Gulf Coast in Beaumont and Port Arthur.

Local entrepreneurs and capitalists participated widely in the oil boom. Houstonians, as well as many people who migrated to the region to cash in on the oil boom and then stayed, helped organize and finance oil companies and an array of oil tool and supply companies that grew to serve the needs of the regional petroleum industry. Because of the great distance between Houston and the existing centers of oil production in Pennsylvania, West Virginia, and Ohio, few "foreign" companies (Texans then used "foreigners" to identify those from outside the state) could quickly set up shop in the region. This left plenty of competitive space for local entrepreneurs to move into many of the new industries needed to supply specialized equipment and services to the booming Texas oil industry. Several of these companies developed products for regional markets and then grew into national and international concerns. Companies such as Hughes Tool and Cameron Iron Works were born at Spindletop and gradually grew into giant global competitors headquartered in the region.[22]

To prevent the takeover of the newly discovered oil fields by the most feared and dominant "foreign" oil company, John D. Rockefeller's Standard Oil, the Texas legislature passed laws that Texas courts used to limit Standard's early involvement in the regional oil industry. These state laws helped prevent Standard Oil from controlling the new oil field, enabling Gulf Oil and Texaco to expand and become large enough to compete with Standard Oil.[23] Over time, these local companies became major international companies, and they retained strong ties in the region through regional headquarters, oil production, and the expansion of what remained of their largest refineries.

Migrants to the region included experienced oil men like Joseph Cullinan, a longtime employee of Standard Oil who came to the region and became the first president of Texaco.[24] Others followed Cullinan's path from Spindletop to Houston. The founders of the Humble Oil & Refining Company, which became an important part of Standard Oil of New Jersey (now ExxonMobil) through acquisition in 1919, chased the boom in Beaumont before discovering the large Humble field north of Houston and moving their headquarters to the city.[25] Many more people from small Texas towns joined the migration to the big city of Houston in search of their main chance for wealth or at least for expanded opportunities.

Workers by the thousands came to the coast from the interiors of Texas and Louisiana, looking to work in the oil industry in search of a better life than that available to family farmers and sharecroppers.[26] Some took jobs in the oil fields; others chose the factory work offered by companies such as Hughes Tool, whose large plant to manufacture drill bits just outside downtown Houston quickly became a major employer in the region. So many people rode trains to Houston seeking work at Hughes Tool that conductors would at times simply call out "Hughes Tool" instead of "Houston" at the main railroad station in the city.[27] The growing petroleum refineries—and associated manufacturing—attracted wave after wave of "country boys" to the region, first to the Beaumont–Port Arthur area and then to the refineries that sprang up along the Houston Ship Channel after World War I.

As the boom subsided in Beaumont and the discovery of new fields moved westward, Houston captured many of the permanent economic benefits of the oil boom that had gripped the Gulf Coast in the first two decades of the twentieth century. Houston was the early choice as headquarters city for Texaco and other oil companies and a variety of important companies in oil-related businesses. Specialized oil-tool manufacturers, service companies, and oil supply companies established factories and administrative headquarters in the city. Among other things, they manufactured drill bits, oil storage tanks, equipment for managing the control of oil from oil wells, and instruments needed to find and produce oil. Specialized companies also grew to construct pipelines and, later, refineries. In the two decades after the Spindletop discovery, the region from Beaumont to Houston became an important national center of oil and oil-related industries. By the end of the boom era, the region had developed the essential industrial base and infrastructure needed to support the operations of vertically integrated oil companies, as well as independent producing and refining companies.

The opening of the Houston Ship Channel (HSC) in 1914 cemented the region's growing leadership in all things oil. Extensive dredging transformed an existing stream, Buffalo Bayou, and a channel through Galveston Bay into a deepwater channel to the Gulf of Mexico.[28] Before the discovery of oil, cotton, timber, and other commodities had at times been able to make their way down Buffalo Bayou and through Galveston Bay to the port of Galveston, forty miles south of Houston.[29] Railroads provided more dependable transportation than the shallow bayou. The growing need to transport crude oil and petroleum products required additional waterway improvements. These went forward in the early years after the discovery of oil, when Houston's civic leaders struck a deal with the U.S. Army Corps of Engineers to share the cost of this badly needed transportation improvement. This act of political entrepreneurship became one of the most used symbols of the "can-do" spirit of Houston. Public-private cooperation in deepening the Houston Ship Channel demonstrated the willingness of the public sector to promote oil and economic development. A city open to outsiders and fervent in its worship of the civic religion of economic growth had laid the foundation for sustained oil-led development in the first two decades after the Spindletop discovery in the region.[30]

The Growth of a Refining Region, 1920–45

From World War I forward, a giant complex of petroleum refineries, an equally large collection of petrochemical plants, and many other oil-related businesses grew along the banks of the HSC.[31] In the 1920s and 1930s, spurred by the impetus of the opening of the ship channel, oil surged past cotton as the engine of growth for the Houston economy. In 1935 almost half of all Texas oil was shipped through the growing Port of Houston. *Fortune* magazine asserted that "without oil Houston would have been just another cotton town."[32] Today, petroleum remains the top import and export, along with other petroleum products, crude fertilizers and minerals, and organic chemicals. This is true of other ports in the broad coastal region stretching more than 200 miles to the southwest from Houston to Corpus Christi and 350 miles to the east to New Orleans, with Baton Rouge and Lake Charles, Louisiana, and Beaumont and Port Arthur, Texas, in between.

Although all phases of the industry were important, refining left its mark most prominently on the coastal region from Houston to Beaumont. From the early days of Texas oil forward, this area had a national identity as a major refining center. The giant refineries in the region were powerful magnets attracting people and capital to the area. Each of the largest plants has employed thousands of workers from the 1920s through the present. Many other people in the region took jobs supplying the specialized services and goods needed by the refineries and the needs of their workforces.

The Gulf Coast refining region's historic and geographic center was the 100-mile coastline from Houston to Beaumont–Port Arthur, which from the 1920s forward remained the largest center of refining in the nation. By 1927, less than a decade after the opening of the Houston Ship Channel, eight refineries with a capacity of approximately 125,000 barrels of crude a day already operated along the channel. During the heyday of regional refining from World War I into the 1970s, refining and then chemical production continued to expand in the broad section of Texas and Louisiana from New Orleans to Corpus Christi, which continues to the present to be the site of one of the largest concentration of refinery-chemical plants in the world. Most of the large, integrated oil companies—many of which had regional offices in Houston—chose this prime location for their major American refineries.[33] In 2008, the Texas Gulf Coast alone housed a refining capacity of over four million barrels a day and produced about 25 percent of the nation's refined goods.[34] These manufacturing plants added economic value to the crude oil that passed through them. The jobs and capital accumulated from turning crude oil into sophisticated products accrued to the refining region, not to the oil-producing regions that fed crude into the refineries through pipelines.

Over time, the steady growth of manufacturing jobs in the refineries and chemical plants gave the region a large and distinctive industrial working class. Houston was a city of opportunity for working people. Poor white and black tenant farmers and family farmers from East Texas and western Louisiana fled depressed agricultural conditions in the interior and moved to the factories along the ship channel. They were joined in growing numbers during and after World War II by Mexican nationals and Mexican Americans who also made the journey from rural to urban lives. Cajuns from South Louisiana moved to the urban-industrial centers to the west at about the same time, and many crossed the border into Texas to take industrial jobs in the area from Houston to Beaumont.[35] Many of these migrants had learned basic mechanical skills working on farm equipment and early automobiles. Accustomed to hard work, they responded to the lure of factory jobs that offered steady pay, paid vacations, health insurance, and the promise of upward mobility across generations. The often successful quests of individual workers for better lives for themselves and their offspring gave the region an optimistic tone; the collective impact of hundreds of thousands of industrial jobs in oil- and chemical-related manufacturing drove economic growth in the region for much of the twentieth century, playing a vital part in the expansion of Houston into a major city.

Large, permanent investments in giant refining complexes, pipelines, and modern oil shipping facilities differentiated the Houston area from many regions that remained primarily centers of oil production. Oil fields might come and go in boomtowns throughout Texas and Oklahoma, but the refineries remained on the Gulf Coast, bringing to the region billions of dollars in investments and tens of thousands of jobs for more than a hundred years. In addition to the workforces required to operate major plants, large numbers of people over time worked in specialized construction companies that built and expanded and maintained these plants. Refineries and chemical plants, as well as specialized service and supply companies, employed a growing body of professionals who developed the new technical processes for the refineries and developed new products for the petrochemical plants. The region also attracted highly educated engineers, accountants, and managers. Over time, the number of unskilled workers and craftsmen needed in the refineries dwindled as the plants became more automated; at the same time, skilled and technical jobs grew, changing the mix of workers in the refineries and gradually in the city as a whole.[36] For about fifty years after World War II, the region could be aptly described as the refining and chemical capital of America, but other important related activities also grew steadily in these decades.

New Layers of Oil-Related Industries: Petrochemicals, Natural Gas, Offshore Oil, 1940–1970s

As Houston continued to benefit from the expansion of the global oil industry after World War II, the chemical and natural gas industries—through rapid and sustained growth—added two new dynamic businesses to the regional economy. Close ties through technical processes, hydrocarbon chemistry, and ownership linked these businesses to the traditional production and refining of oil. As oil production in the region and throughout the southwestern United States began to decline in the postwar years, these two new businesses provided several strong bursts of growth and a measure of diversification to the oil-related core of manufacturing.

The emergence along the Houston Ship Channel of a major petrochemical complex during the late 1930s and World War II reinforced the economic impact of oil on the region. This dynamic new industry made extensive use of oil and natural gas in its manufacturing processes, and its plants were often constructed near oil refineries and owned by oil companies. World War II brought significant changes in the industry, which grew in response to the growing need for aviation fuel, synthetic rubber, and other petroleum-based products. The large investments in war-related manufacturing came from a combination of funds from business and government, and many of these facilities were absorbed into the private sector after the war. Much of the synthetic rubber used in the war, for example, came from new plants built along the Texas Gulf Coast, and this vital new industry remained concentrated there after the war.[37]

During the postwar boom, the region from Houston to Beaumont became the largest center of chemical manufacturing in the world, adding a new layer of diversity to the region's oil-related core. By 1950 there were twenty-seven chemical plants along the Houston Ship Channel, and chemicals remained for several decades one of the fastest-growing industries in the nation and in the region. By the 1980s, the Houston-Beaumont area had more than half of the petrochemical capacity in the country. This growth continued, and by 2008 the Houston Ship Channel alone was the site of a $15 billion petrochemical complex that was the largest in the nation. At that time, some four hundred chemical plants in the region employed more than 33,000 people.[38] These plants had close ties to regional refineries, which were the source of much of the "feedstock" used to make the many new chemical products created in the postwar era.

Natural gas was the key feedstock for the production of basic chemicals, and it also provided the basic fuel for the refineries, chemical plants, power plants, and much of the remainder of the region's industrial economy. From the early twentieth century, natural gas, which companies often found while exploring for oil, played an important role in the regional economy as both industrial and domestic fuel. This high quality and—at least until the 1970s—inexpensive fuel helped attract manufacturing to the region and to other parts of the Sunbelt blessed with large supplies of this relatively clean-burning fossil fuel. Natural gas also supplied the fuel for the generation of much of the electricity that powered Houston's sustained expansion. The availability of natural gas for domestic and industrial fuel gave the region an important competitive advantage over regions still dependent on coal and other fuels.

In the postwar years, the impact of natural gas on the regional economy increased sharply with the opening of much broader national markets for natural gas. From the 1930s forward, technical advances in long-distance gas transmission connected the vast gas supplies of the southwestern United States with the vast demands for fuel of the northeastern United States. Houston quickly emerged as the center of the nation's booming natural gas transmission industry. Most of the major transmission companies had headquarters in Houston, including national leaders such as Tenneco, Transco, Texas Eastern, and Panhandle Eastern.[39] These companies built a new generation of skyscrapers in Houston, where they housed large numbers of administrative and technical jobs. By 2008, fifteen of the nation's twenty largest natural gas transmission companies had corporate or divisional headquarters in Houston, where they controlled about 57 percent of the total natural gas shipped in the United States.[40] Several companies, led by Tenneco, became leading corporate citizens that supported major civic, cultural, and educational initiatives. Since the 1970s, many in the region have referred to Houston as the nation's "oil capital," but it has an even stronger historical claim as the "natural gas capital" of the United States. As the role of natural gas has grown compared with that of oil in the nation's energy mix since the late twentieth century, this distinction has become increasingly important for the region's future.

The expansion of offshore production in the Gulf of Mexico after World War II added still another source of growth to the regional economy, as Houston and New Orleans both provided critical services to this technologically intensive segment of the petroleum industry. Specialized constructions firms such as Houston-based Brown & Root became prominent in the design and building of giant offshore platforms, and a variety of new service and supply companies rose to meet the needs of offshore drilling, production, and transportation.[41] As offshore exploration and production in the Gulf of Mexico moved into deeper waters over time, this source of oil and natural gas became increasingly important in the overall energy supply of the United States. Both Houston and New Orleans added substantial administrative and technical jobs, and numerous smaller cities along the Texas–Louisiana Gulf Coast grew into significant manufacturing and supply centers for the offshore industry. An important symbol of Houston's leadership in the development of the advanced offshore technologies has been the Offshore Technology Conference, still the largest international conference on offshore technology, which has been held annually in Houston since 1969.[42]

As these new layers of industrial activities expanded in the region, its oil-related industrial core became thicker and more diverse, surpassing the concentration of similar activities in other energy capitals. This core grew steadily in the quarter of a century after World War II. It then accelerated frantically in the boom of the 1970s and the early 1980s, when two large and unpredicted increases in oil prices, brought by the energy crises of 1973–74 and the Iranian Revolution of the late 1970s, drove Houston skyline and its oil economy as a whole to new heights.[43] During these heady boom days, Houston was the place to be for companies in industries related even indirectly to oil. Led by Shell Oil in 1971, numerous companies established larger regional or national headquarters in the region. More of the technical work of the industry came to the region, along with specialized legal and financial services, technical training, and administrative services. The new wealth that flowed into Houston transformed a provincial oil town into a maturing metropolitan area, complete with high culture, good restaurants, and major league sports.

Drilling companies—which, at least in retrospect, should have known better—greatly expanded their activities in response to price surges after the Iranian Revolution in the late 1970s that seemed to herald a new era of ever-rising oil prices. The city hummed with unprecedented oil-related activity and optimism in the brief period of very high oil prices, from about 1978 through 1983. For a moment in time, the boom seemed destined to go forward for decades as oil prices continued to surge. The irrational exuberance of this unsustainable super boom became evident only in retrospect, when individuals and businesses throughout the regional economy calculated the damage from the great oil depression of the late 1980s.

Concentration in Houston after the Oil Price Bust of the 1980s

The bubble burst in the mid-1980s when a sharp drop in global oil prices devastated the global petroleum industry and, along with it, the Houston region's economy. A hard truth seldom acknowledged during the long oil boom of the twentieth century now confronted Houston: an oil capital remained vulnerable to major problems within its leading industry. Despite the past success of the oil-related core in generating jobs and despite the complexity and depth of the oil-related core of industry, all were, after all, vulnerable to sharp declines in the oil and natural gas industries. The key lesson learned was simple: the region needed to diversify its economy into areas not directly affected by oil prices.[44]

Much of the new, "non-oil" economy that grew in Houston before and after the price bust had taken root during the long oil boom after World War II. Two of the major departures outside the oil industry that added economic diversity were space and medicine; both had strong indirect ties to the oil industry. Also significant was the impact of the oil industry on higher education and real estate development. Construction of the Johnson Space Center in Clear Lake (twenty miles south of Houston) had been orchestrated in the early 1960s by an alliance of George Brown (a founder of Brown & Root and Texas Eastern Gas Transmission Company), Morgan Davis (president of Humble Oil), and Vice President Lyndon Johnson (who enjoyed the strong support of the oil industry in the 1950s).[45] The philanthropic support of many Houstonians who had made their original fortunes in oil fostered the growth of another important engine of growth, the Texas Medical Center, which became one of the major employers in Houston in the second half of the twentieth century. Many of these same individuals and their philanthropic foundations supported the development of private universities such as Rice University and state universities such as the University of Houston; strong institutions of higher learning facilitated economic diversification. Even the real estate development industry that built the Houston suburbs had strong ties to the oil industry. For example, Friendswood Development, a major developer that built much within the suburban Houston developments of Clear Lake and Kingwood, originally was affiliated with Exxon and named after one of its important regional oil fields. The Woodlands, a planned community about thirty miles north of the city, was developed by oilman George Mitchell.[46] If the oil bust of the 1980s showed the need for economic diversification, the response to the bust revealed the deep economic strengths from more than eighty years of oil-led development that could be drawn on to build a more balanced regional economy.

After three or four shaky years of oil depression, the region rebounded, growing stronger and more diverse with new economic development not only in space, medicine, and real estate, but also in high-tech enterprises and an expanding service market. Many energy companies responded to the oil bust by consolidating operations or by merging with other companies. Time after time, the consolidation of companies resulted in the movement of positions and even headquarters to Houston, making the city a net gainer of jobs as the industry shed positions. This was clearest at the turn of the twenty-first century, when cutbacks in the offshore industry in the Gulf of Mexico led to substantial relocations of employees from New Orleans to Houston. In this same era, the merger of Conoco and Phillips petroleum companies resulted in the relocation of many functions historically done in Phillips's headquarters in Bartlesville, Oklahoma, to Houston.[47] Similarly, Chevron's merger with Texaco gave the combined company a larger presence in the region.

After more than one hundred years of oil- and gas-led development, the city has few rivals as an oil and gas capital in the early twenty-first century. Other regions can boast of significant oil-production or refining. None matches the Houston region's concentration of oil refining and production, petrochemical production, oil and natural gas transportation, administrative functions, petroleum-related research, and specialized services and supply companies. The Houston area provides a case study of a region well prepared to absorb the economic benefits of the development of a new source of energy and blessed by the timing and location of its discovery. The region's steady, persistent evolution into a fully developed energy capital produced a thick and diverse oil-related economic core that included well-developed supporting institutions such as universities and philanthropic foundations. The case of Houston shows that under the right circumstances, a region can build sufficient scale and scope in energy-related activities to become the location of choice for a wide variety of energy companies. In the twentieth century, Houston developed the oldest, largest, strongest, and deepest cluster of oil- and natural gas–related expertise in the world.

Table 2.1. Population, Houston Metropolitan Statistical Area (Houston-Baytown-Sugarland, TX, MSA), 1837–2005

Year	Population
1837	1,500
1850	14,773
1860	29,801
1870	42,962
1880	63,729
1890	76,959
1900	122,785
1910	176,589
1920	256,023
1930	439,226
1940	627,311
1950	908,822
1960	1,364,569
1970	1,903,192
1980	2,753,155
1990	3,342,247
2000	4,715,407
2005	5,280,077

Sources: Proximity, "Resources to Create and Apply Insight," http://www.proximityone .com/metros.htm#top10; Greater Houston Partnership Research Department, "Houston Facts, 2000," http://www.houston.org/tophoustonfacts/houstonfacts2000; Real Estate Center, Texas A&M University, "Houston, Tex.: Metropolitan (MSA) Population and Components of Change," http://recenter.tamu.edu/Data/popm/pm3360.htm; David McComb, "Houston," Handbook of Texas Online, http://www.tsha.utexas.edu/ handbook/online.

Energy and Urban Infrastructure

A long-term record of the impact of Houston's sustained economic development on the region's total population and on the physical expansion of the city of Houston can be seen in tables 2.1 and 2.2. However, sustained spatial and population growth in Houston brought a range of challenges. Cities and regions that reap the long-term economic benefits of energy production have been physically transformed by the burgeoning energy industries, requiring the construction of extensive new infrastructure to meet the demands of both the growing energy-related

Table 2.2. Area of Houston City Limits, 1910–2000

Year	Square Miles
1910	17.4
1920	36.5
1930	71.8
1940	72.8
1950	160.0
1960	328.1
1970	433.9
1980	556.4
1990	539.9
2000	617.0

Source: U.S. Bureau of the Census, "Population of the 100 Largest Urban Places,"
1910–90, http://www.census.gov/population/www.documentation/twps0026.html;
Greater Houston Partnership Research Department, "Houston Facts, 2000," http://
www.houston.org/tophoustonfacts/houstonfacts2000.

sector and the surging population. Demands for water systems, transportation and
communication networks and facilities, sources of power (particularly electricity),
and educational institutions put pressure on cities to expand their infrastructure,
frequently in cooperation with the business community benefiting from that expansion. In some cases, however, energy industries compete directly with municipal infrastructure needs. The case studies of water, transportation, and education
in Houston illustrate the complex reality of the shared or contested infrastructure
required for the growth of its energy industries and its population as a whole.

Water

The transformation of Houston's water supply system from one largely dependent on groundwater to one increasingly dependent on surface water is a good
example of how energy-led industrial development influenced urban infrastructure
and urban services.[48] Indeed, the dramatic shift from dependence on groundwater
to surface water is intimately connected to the needs of the oil and petrochemical industries and other enterprises, especially along the Houston Ship Channel.
Houston developed its first public water-supply system in 1876. By national standards, the Bayou City's commitment to a centralized water source and distribution network was typical for a city of its age, size, and location. The system, however, was unique insofar as the city relied on groundwater from countless wells
from 1887 until the 1940s.[49] Until it sought to develop surface water prior to World

War II, Houston was the largest city in North America to rely exclusively on well water.

In the nineteenth and early twentieth centuries Houstonians believed that their source of water was clean, abundant, and never-ending.[50] Theoretically, the underground supply was abundant for many future generations; realistically, demand on the system increased so rapidly that productive wells were depleted or ceased to be free flowing. Population growth accounted for much of the escalating demand, but agricultural and industrial uses also were important. After the opening of the Houston Ship Channel, pressure on the aquifer from industrial use rose dramatically.[51] The immediate response to growing demand was to sink new wells, to add pumps to existing wells, to build new pumping plants, and to extend distribution lines.[52]

Contentment with the water supply tended to mask structural deficiencies in the system. By the mid-1930s, the water system was essentially a collection of water plants and distribution mains pieced together by expansion into subdivisions along the fringes of the city and through modest annexation.[53] The challenges to the water system eventually eroded the city's confidence in the groundwater supply. Artesian well pressure had begun to decline as early as 1910, and for several years the water level dropped by an average of five feet annually. With increased industrial pumping along the ship channel for cooling and other purposes, the depth to retrieve water increased. The deterioration of the wells—plus the lack of a wholesale rate for water—caused many industries and owners of commercial buildings to drill more of their own wells. By 1941 the public supply satisfied less than 40 percent of the total demand of the metropolitan area.[54]

The decline in well productivity—and growing independent action of commercial and industrial enterprises—led several experts to view the problem as nearing a critical stage. In the 1930s and early 1940s, almost forty reports on the water supply's condition were written but they were often contradictory or inconclusive.[55] Alvord, Burdick & Howson—a Chicago engineering firm retained by the city— issued a report in February 1938 that favored the use of the San Jacinto River as a single, inexpensive, and reliable water source. The report came at a time when groundwater sources were becoming increasingly costly and difficult to obtain. The San Jacinto was the nearest surface supply with a large drainage area above the ship channel. A reservoir site was available only fifteen miles from the city's industrial district.[56] The study painted a poor picture of the existing system, adding that it "is virtually a group of small town supplies without the distribution facilities or interconnections essential to the delivery of water for either fire or domestic use." In addition, the city's investment in the system was less than one-third that of the average city of its size.[57]

In May 1937 (before the Alvord, Burdick & Howson report was made public) the engineering staff recommended filing an application with the State Board of Water Engineers to appropriate water from the San Jacinto River to complement the groundwater withdrawal. G. L. Fugate, chief engineer of the Water Department, pursued the combined groundwater/surface water program into the 1940s, viewing the damming of the San Jacinto as a source principally to supply industrial demand.[58] The onset of World War II finally pushed Houston toward surface supply. On September 10, 1941, the city filed an application with the Federal Works Administration for financial support to improve the water-supply system and to obtain a supplemental supply from the San Jacinto River. Wartime exigencies directed the federal government's interest in the project to the eastern portion of the city around the ship channel, designated as a "defense area." The industries there would employ an estimated 90,000 workers during the war.[59] In July 1942 the War Production Board authorized the San Jacinto River Conservation and Reclamation District to build a dam and other facilities on the San Jacinto to supply water for war industries along the ship channel and in the Baytown area.[60]

Because of the urgent need for the water, the War Production Board declared in August 1942 that the agency best showing the ability to deliver water to industries in the Baytown area would be favored. Neither the city nor the water district would permit a grant to the other, and the federal government announced its intention of constructing the facility itself. Ironically, since the city's preliminary plans and surveys were well advanced, the Federal Works Agency adopted Houston's program and, in November 1942, employed the city as its architect-engineer with Fugate as contract engineer.

Actual construction of the dam began in December 1942 by Brown & Root, for delivery of water to industries at Baytown and in the Pasadena area in 1943. Two open canals from the river were constructed to serve the ship channel: the West Canal leading to Pasadena and the East Canal terminating at the Humble Refinery in Baytown. Despite the increase in available supply from the San Jacinto, distribution facilities still did not reach remote sites along the ship channel, and as a result wells served the increasing demand from new industries. In June 1944, Houstonians voted for a $14 million bond issue to increase the amount of groundwater supply and for more mains, and also to buy the West Canal from the federal government, to build another dam across the San Jacinto River, and to construct a filtration plant. Because of the need for additional funding, Lake Houston Dam on the San Jacinto was not placed into operation until 1954. The new public water supply provided water for the city of Houston and the industrial complex from Houston to Baytown, and also supported local irrigation for various products including rice.[61]

The debate over water in the 1930s, coupled with the acquisition of wartime industries along the ship channel, began Houston's transition from a city totally dependent on groundwater to one eventually dependent on a dual supply, and ultimately to virtual dependence on surface water. The ability of the new surface supply to save the city from its faith in the shrinking resource of groundwater did not grow out of some enlightened city policy. Indeed, the requirements of the refining and petrochemical industries in particular dictated the change. And for a time, the new surface supply benefited industry much more than the city. Without industrial demand for water, Houston's transition to a surface supply may have been much further in the future than World War II.

Transportation and Urban Sprawl

Transportation was another key issue facing the booming region. A prototypical "Sunbelt city," Houston grew up with the automobile. With no widespread investment in public transit before the coming of oil in 1901 and the opening of the Houston Ship Channel, the city expanded rapidly just as cars came into general use. It is symbolically fitting that the decade of the city's fastest growth in the twentieth century was the 1920s, when auto use took off in the region and around the nation. In that decade, basic changes in refining technology to produce more and better gasoline combined with state gasoline taxes designated for highway construction to encourage increased automobile use. Despite periodic efforts to introduce public transit in the form of streetcars, an expanded bus system, and a light rail system, Houston has remained a consummate automobile city into the twenty-first century. According to the U.S. Census, the city had 1,053,788 cars in 2000, which amounted to 1.3 vehicles for every commuting person. In fact, 90 percent of all commuters in the city ride in cars, with 82 percent of the commuters as lone drivers.[62]

The same combination of changes in refining technology and gasoline taxes continued to spur road construction throughout the remainder of the century, with federal gasoline taxes contributing to the building of the massive interstate highway system after the 1950s. Highway construction remained a major regional employer throughout these years. As a well-developed system of highways steadily expanded to serve the needs of the growing region, it shaped the pace and direction of Houston's growth. Intensified suburbanization began in the 1920s and 1930s, when both jobs and roads connected the central city to the refinery towns that emerged east of the city along the industrial corridor, which grew on the banks of the Houston Ship Channel and the banks of Galveston Bay—Pasadena, Galena Park, Baytown, Deer Park, and Texas City. Most of these towns remained independent of the city of Houston, with taxes and payments in lieu of taxes from the refineries helping to fund city services.

In subsequent decades, the city sprawled outward in every direction. Favorable state laws, such as the Municipal Annexation Act (1963), allowed the city to aggressively annex adjoining areas and thus further enlarge its territory. By 1999, this generous annexation policy had allowed Houston to reserve approximately 1,289 square miles for future annexation.[63] The availability of inexpensive land surrounding the city kept housing prices down, providing inexpensive housing for the millions of migrants to the region. Cheap gasoline provided the fuel needed to commute longer and longer distances. Inexpensive electricity produced primarily by abundant and low-cost natural gas allowed for the air conditioning that made the city livable during its long, harsh summers. Local, state, and federal governments responded to the demand of citizens to build more and more roads reaching farther and farther out from the city. For much of the twentieth century, the region as a whole accepted a "mass transit" strategy of more highways filled by more cars, often with one driver per car.

The basic highway system for the Houston Metropolitan Area at the turn of the twenty-first century consisted of Interstate 45 providing the main north–south axis, Interstate 10 the primary east–west axis, and Interstate 610 looping around downtown Houston to relieve the pressure of through traffic on the central city.[64] Well-developed state roads created spokes in the two wheels of highways around the city: Loop 610 and the more recently constructed Beltway 8 farther outside the city's borders. Good highways extending out into the city's hinterland for hundreds of miles facilitated migration by making it easy to move to the coast and to return to the interior for visits. The flat coastal plain thus shaped both the region's highway system and the resulting patterns of urbanization and suburbanization.

By the turn of the twenty-first century, "Houston" (as defined by economic and commuting ties, not by political boundaries) had become one of the world's largest cities in area, stretching thirty or forty miles from downtown Houston in every direction over a broad area of the Texas Gulf Coast. Much of this land had been farmland in the early twentieth century, but by 2000, from Katy to Conroe to Baytown to Galveston to Sugar Land, each freeway exit looked much the same, the homes in each subdivision merged into several generic floor plans, with all roads leading into and out of Houston. The city exhibited the worse kind of urban sprawl—patternless, unplanned (Houston is the largest city in the United States without zoning), and highly decentralized. In the postwar years, Houston spilled out into ten counties and had half the population density of Los Angeles. Miles outside the central city, a series of mini-downtowns and ever-expanding suburban developments/towns grew along the region's well-developed highway system.[65]

Oil provided an economic thread that loosely bound together the sprawling Houston metropolis. In the decades after World War II, the oil core directly or indirectly generated a majority of the jobs in the region. The public highway system thus can be seen as transportation infrastructure built and maintained in large part to meet the needs of those working in the oil industry. The industry also made extensive use of these highways to distribute its products in large trucks to service stations throughout the region. The vast improvements made in the region's waterways had similar dual uses. All shippers could take advantage of the ship channel, the intracoastal canal through the region, and the Port of Houston, but for much of the century, shipments from oil and related industries dominated the use of these facilities.

Private, not public, investment built much of the transportation infrastructure required by modern oil operations specifically for the uses of the industry. An extensive pipeline system bound together the Gulf Coast refining/petrochemical region into a distinctive "spaghetti bowl" of interconnected plants; the tankers that plied the Gulf of Mexico and came up into the ship channel were privately owned, as were the docks at the plants on the ship channel.

All in all, the combination of public and private investment financed the transportation infrastructure that allowed the oil industry to grow and enabled its employees to take advantage of inexpensive housing by commuting long distances to work in private automobiles.

Education

Although educational institutions generally are not treated as "infrastructure" in the same sense as water and transportation systems, they nonetheless played important roles in the evolution of the Houston region as an energy capital. Historically, the state of Texas has been a southern state in terms of its public schools, with low levels of funding and a low rank in comparative quality. This reflected both the impact of Jim Crow segregation in the schools and the low cultural value placed on formal education in the poor sections of rural Texas, which fed the migration of workers to the Gulf Coast refining region's growth. Many of these migrants, however, did not compare Houston's schools with those in New York or California cities; instead, they viewed the region's public schools as a giant leap forward for their children when compared with the public schools available in rural East Texas or south Louisiana or northern Mexico.

The schools in the towns around the major refineries generally were superior to those in the rural areas from which the refinery workers migrated. The presence of the plants influenced the quality of the public schools in "refinery towns" such as Pasadena, Galena Park, Deer Park, Baytown, and Texas City. Baytown, one of the first new towns to emerge after the opening of the ship channel, was founded in the early 1920s as a company town for workers at the new Humble Oil (now Exxon-Mobil) refinery. Like many other refinery towns, it remained an independent city with its own school districts, and the refineries and chemical plants gave Baytown and other refinery towns distinctive student bodies and healthy tax bases to support their schools. In the late twentieth century, many of the companies on the ship channel also became partners with the school districts in supplying computers and other equipment and providing volunteers to work in the schools. The educational and social life of these schools created a common ground for the families of those who worked along the ship channel.[66] One common goal of many students in these schools was to find choices other than work in unskilled and semiskilled jobs in the refineries, and education could open such choices.

As time passed, regional community colleges and colleges and universities trained tens of thousands of people from within the region for white-collar professional positions inside and outside the oil and gas industries. The original endowment of the most prominent regional university, Rice University, came from a local merchant before the discovery of oil in the region; but generous support and leadership followed from the founders of Humble Oil and Texaco, as well as the later contributions from a wide array of oil-related companies such as Brown & Root.[67] Rice developed outstanding engineering programs that equipped students to succeed in the oil industry, but its training could also be put to good use in many other industries.

The public universities that served the region included the University of Houston and Texas Southern University in Houston and other nearby state universities such as Sam Houston State in Huntsville and Lamar University in Beaumont. After its creation in 1927, the largest of these, the University of Houston, expanded and grew in quality under the impetus of major contributions from Houston independent oilman Hugh Roy Cullen and others from the oil and gas industries.[68] Flagship state universities outside of the region, led by the University of Texas in Austin and the Texas A & M University in College Station, also trained many students who subsequently worked in Houston's oil cluster. For much of the twentieth century, these state universities offered good educations at very low costs. Systems of community colleges grew to provide vocational training and to feed transfer students into the four-year universities. These well-developed institutions of higher learning offered specialized technical degrees in great demand in the oil and petrochemical industries. Graduates—at times the children or grandchildren of refinery workers—often found engineering and technical jobs in the oil-related complex. Although this "educational infrastructure" made excellent use of sustained philanthropic support from individuals, companies, and foundations with ties into the oil-related core, it was, of course, available for those who sought careers in other fields.

The region's philanthropic foundations also contributed to regional improvements far beyond the specific needs of the energy core. Numerous individuals who had built fortunes in oil-related enterprises, or in business, legal, or financial services provided to these companies, created foundations that transferred the funds generated by their original endowments across generations and through time.

Many of Houston's civic and business leaders had migrated in their youth to the region, leaving small towns in search of opportunity in a growing city. Those who succeeded often felt a special attachment to their adopted city, and they proved generous in their efforts to help improve the quality of life in Houston.[69] These men and many of their wives took the lead in organizing and financing civic initiatives. They helped bring high culture to the region, supported the improvement of universities, and brought major league sports to the city. Their foundations continued to support these and other similar commitments after their deaths. With the passage of time, many of these foundations also launched new initiatives designed to improve the quality of life in the region by addressing some of the problems associated with oil-led development.

Environmental Costs of Oil-Led Development

The most visible problems have been environmental and health risks. Houston and other energy capitals have absorbed most of the environmental costs of producing energy for national and international markets that are far removed from pollution generated during the production and processing of oil and coal. Effective regulation of pollution was slow in coming, in part because the technology of energy production had been the historical focus of investment by the energy industries, while investment in the technology of pollution control lagged. The political and legal processes for negotiating societal solutions to environmental problems have shaped regional responses to social and physical costs, and such processes have differed sharply across time and place.[70] Because petroleum has been both the major industry and the major fuel for modern Houston, this self-proclaimed "energy capital" has also been the de facto "oil pollution capital of America."[71] The scale of the regional oil industry and its long years of operations in and around Houston created distinctive levels and forms of air and water pollution, as well as toxic wastes and brownfields.

The timing of oil-led development affected regional responses to pollution. Almost from the moment the Houston Ship Channel opened, it was severely polluted by oil run-offs from refineries and tankers and oil fields. During the first national oil pollution crisis after World War I, a national survey of coastal waters identified the ship channel area and Sabine Lake, which served as the outlet to the sea for refineries in nearby Beaumont and Port Arthur, as two of the most polluted sections of the coastal waters of the United States. The regulatory response to the severe oil pollution in the 1920s promised much, but delivered little. The U.S. Congress debated the passage of strict pollution controls before passing the Oil Pollution Act of 1924, which did little except threaten future actions.[72] The oil industry as a whole responded by cleaning up the most visible pollution around facilities and taking measures to reduce future emissions.

For the next thirty years, the industry retained effective control over the issue through a form of self regulation that kept oil pollution out of the public eye and off of the political agenda.

Such measures dampened political pressures for stricter controls over oil pollution through the late 1920s, when other more pressing issues rose to dominate political discourse. During the Great Depression of the 1930s, economic recovery trumped other issues, including pollution control. World War II then became the key concern for Americans. Mobilization for war brought far-reaching changes to the Houston region, especially in its oil-related core of industries. A combination of business and government investments built new synthetic rubber plants, expanded the production of 100-octane aviation fuel in regional refineries, and constructed massive new pipeline systems to assure the supply of crude oil and refined oil products. While contributing mightily to the Allied victory, the rapid industrial expansion of the war years also added to the backlog of environmental problems that would have to be faced at some time in the future.

Patterns of living set in the postwar economic boom in Houston made these problems even more difficult to address. The metropolitan area sprawled outward as the availability jobs, cheap land for houses, and cheap energy for transportation and air conditioning made the region a magnet for workers in search of better opportunities. As the population spread farther and farther out into the suburbs, so did regional pollution. By the 1950s, emissions from the plants along the ship channel and from automobiles made Houston's air among the worst in the nation. Emissions from the region's booming cluster of petrochemical plants were largely unregulated in the decades after World War II, which meant that several generations of workers in the years before the enforcement of strong federal regulations of safety, health, and the environment in the 1970s were exposed to substances now known to be harmful. With little concern for the long-term implications for public health, toxic wastes were dumped in the central city itself and also on the far outskirts of the city, there to become the next generation's problem when suburban residential areas later spread farther out of town.[73]

When the early wave of stricter environmental regulations took effect in the 1960s and 1970s, Houston faced something akin to affirmative action in dealing with environmental problems. Not only did officials have to address current sources of pollution, but they also had to take care of a backlog of problems that had accumulated over more than half a century of largely unregulated growth. Fifty years of accumulated water pollution by oil had to be cleaned up in the region's waters. Serious air pollution intensified as more cars and trucks used regional highways and the refining complex expanded. By the turn of the twenty-first century, regulators scrambled to respond as Houston briefly passed Los Angeles as the region with the nation's worst ozone problems. Health officials had to confront an epidemic of cancer as they sought to understand the health hazards of various emissions from refineries and petrochemical plants. The remediation of toxic waste sites required officials at times to try to locate responsible parties no longer active at the site.[74] "Heat island" effects increased steadily from the paving of acre after acre of the region's sprawling suburbs and roadways, revealing the symbiosis of Houston's intense climate and the results of rapid expansion. Intense flooding, related to extensive nonpermeable surfaces and substantial annual rainfall, exacerbated problems from nonpoint pollution by washing all kinds of toxic materials from lawn fertilizers to heavy metals into the city's extensive network of bayous that drain the city and ultimately spill into the Gulf of Mexico.[75] One estimate suggests that of the 16,000 miles of streams and shorelines in the thirteen-county region encompassing Houston and Galveston, 66 percent do not meet water quality standards.[76]

Air pollution possibly has been the most widespread pollution issue that is connected, in part at least, to the oil and gas industry in Houston. As one writer for NASA noted, "Houston has a serious air quality problem. Since 1999, the Texas city has exchanged titles with Los Angeles as having the most polluted air in the United States defined by the number of days each city violates federal smog standards."[77] The problem, of course, began much earlier than 1999, and is linked very closely to Houston's weather and location as a perfect environment for the formation of ozone. Ozone, a form of smog, is a hazardous combination of nitrogen oxides and volatile organic compounds in the air that are intensely heated by the sun. Sources of these chemicals are particularly rich at petrochemical refineries and plants, at chemical manufacturing facilities (more than four hundred in Houston), and in tailpipe emissions from cars, trucks, and buses. Other chemical sources also can be found where construction equipment is run, at power plants, and in small businesses like gas stations and dry cleaning establishments that use chemicals. Even plants and animals emit some of these chemicals.[78] *A Report of the Mayor's Task Force on the Health Effects of Air Pollution* (2007) found levels of air pollution in the city "unacceptable," and not only generated by ozone, but by a variety of hazardous air pollutants (HAPs) possibly exceeding 170—with 12 in particular presenting "a definite risk." The report also found that while air pollution was ubiquitous in Houston, the substances identified as "definite risks" were found "in greater numbers in several East Houston neighborhoods adjacent to the Houston Ship Channel."[79]

Oil had been the lifeblood of the regional economy throughout the twentieth century, making Houston a city of opportunity for wildcatters and refinery workers alike. But after the 1970s, the societal bill came due for cleaning up the severe environmental problems associated with the region's production, processing, transporting, and use of oil. Confronting these problems required the enforcement of strict new laws often strongly opposed by those in charge of businesses in the oil-related core as well as by many of those who worked for them.[80] The region had not previously made a sustained effort to balance the economic benefits of oil with its environmental costs. Part of the reason for this neglect was the widespread political sentiment that economic growth trumped environmental quality; part was a political system that traditionally promoted the oil industry instead of regulating its activities. The intervention of the federal government, which represented interests broader than regional interests alone, ultimately proved decisive in ushering in a new era of stricter regulation on environment, health, and safety.

A political system that accommodated the interests of the oil industry made it difficult to define a long-term balance between the benefits of oil-led development and pollution control. For most of the region's history, a broadly shared societal consensus among a majority of the population, rich and poor, favored oil development largely unrestrained by pollution controls. "Opportunity" and "economic growth" were the twin tenets of local boosterism. Those who called for stricter controls of pollution had to overcome more than regional attitudes favoring growth. At the local level, business interests had long dominated both politics and civic leadership. Historically, their idea of a "healthy business climate" included low taxes, weak unions, and very limited regulation. Scholars have used the descriptive label "free enterprise city" to describe the dominance of Houston's political and civic cultures by conservative businessmen, many of whom ran oil- and gas-related companies.[81]

Political realities also included the entrenched power in the governance process at the state and federal levels of the well-organized interests of the major industries that produced the bulk of the region's industrial pollution. Many of the basic decisions about Houston's oil-related development were made by private corporations in the global energy economy. In Houston, as around the world, price dictated the key decisions on energy use and, to an extent, the approach to pollution control. But the political process played pivotal roles in channeling government promotion and blocking government regulation.[82] The economic importance of oil in the state of Texas and in the city of Houston, in particular, skewed political decisions toward policies that promoted the oil industry and away from policies that constrained the industry.

At the same time, understaffed and underfunded state regulatory agencies could do little to identify the sources of pollution, much less to reduce it. In a state committed to limited government power and spending, a fragmented collection of state agencies with only limited authority over pollution was no match for the mounting problems with oil pollution in and around Houston. This was true at least until the federal government preempted much of the traditional authority of states over pollution control after the 1960s.[83] The state's one-party political system through the 1960s also proved to be a barrier to change because those who advocated state's rights in defense of Jim Crow had ample reason to support the "state's rights" arguments of those who fought against federal government involvement in pollution control.

An important departure in the economic/environmental history of the region has been the recent emergence of efforts to create more effective pollution controls while also encouraging continued economic growth. In the late twentieth century forward, a growing number of civic leaders and grassroots organizations in Houston recognized that high levels of pollution could become a barrier to economic growth and the creation of jobs in the region. For much of Houston's previous history, the oil industry in the region enjoyed the fruits of a strong tradition of business–government relations that stressed the need for low taxes, inexpensive nonunion labor, and very limited regulation to sustain a healthy business climate that encouraged profits, economic growth, and job creation. Was it possible that cleaner air and water—along with other quality-of-life issues—could be added to a new revised list of attributes that defined a healthy business climate?[84]

The Future of Houston as an Oil Capital

In the region's oil-related core, far-reaching changes in the late twentieth and early twenty-first centuries have called into question the continued growth of the region as an oil capital. Most obvious is the decline of oil production in the region, in Texas, and throughout the nation. As domestic oil production has declined, exploration and production has moved on to other parts of the world or to much more expensive and risky places in the United States such as deepwater Gulf of Mexico. The wildcatters and independent oil producers who became the symbol of Houston's emergence as an oil capital in mid-century have declined as a major force in the regional economy, which has increased the dominance of major international companies in oil-related industries in the Houston economy. The other pillar of Houston's century-long expansion as an oil capital has been the thousands of factory workers in its giant refineries and chemical plants. The lack of construction of new refineries and the sharp reduction in the jobs in existing plants, as they continue to automate production processes, present a troubling prospect: the decline of the upward mobility that has made Houston a city of opportunity for so many migrants over the last century. Here again, some compensation for this loss has come from the growth of white-collar jobs that design and manage the automated system being put in place. But can oil-led regional growth continue without increases in the domestic production of oil?

Yes, it can, although probably not at the same high rate attained for most of the last century. The world demand for oil and natural gas continues to rise, and the demand for cleaner-burning gasoline and diesel fuel continues to grow. The world-scale, technically advanced refineries and chemical plants on the Gulf Coast will continue to play important roles in the production of refined goods and of a changing array of chemical products for the world's largest market for such products: the United States. Indeed, one place to stretch supplies of future oil will be these plants, which will expand their historical quests to search for processes to produce more products from each barrel of crude oil, to produce cleaner products, and to use hydrocarbon chemistry and refining expertise to produce useful products from "unconventional" fuels such as oil sands and shale gas. More important for the region will be the substitution of natural gas production and processing for oil production and refining. New technologies such as hydraulic fracturing have begun to increase the supply of domestic natural gas, and Houston—the natural gas capital of the nation in terms of administration, pipeline location, and technology—will reap many of the benefits of this growth no matter where the new gas is produced.

In the early twenty-first century, growing concerns about the impact of fossil fuels on climate change have called into the question the future of the oil industry. On a more local level, pollution generated from production and consumption of fossils fuels and its various products also raise questions as to whether Houston may be some sort of sacrifice zone for other parts of the world seeking its vital energy and chemical products. It seems clear, however, that oil will continue to play important roles in the economies of the United States and the Houston region for decades, if not generations. During the time required for other energy sources to become economically competitive with petroleum, the thickness and diversity of the region's oil-related core will prove difficult for other regions to match. One key factor in the region's long-term competitive advantage in oil will be the leading multinational oil and oil supply and service companies with long, strong ties to Houston. Another will be human resources rich in technical, financial, and managerial talents currently housed in the region. No matter the fate of oil, however, natural gas and chemicals will continue to boost Houston's economy well into the future. The city should also benefit from the infrastructure and the accumulated resources fostered by a century of oil-led development. Particularly important in Houston's ongoing efforts to diversify the regional economy will be the educational institutions that historically have been shaped by oil development. In a sense, higher education in the region has been a part of its oil-related core, and this core has shown and should continue to show the capacity to diversify, both inside of and outside of the petroleum industry. Sometime in the future, Houston will cease to be the energy capital of the world, but it is well equipped by history to remain the heart of a dynamic region. The city also has been permanently imprinted— physically and culturally—with its historic relationship to the oil and gas industry.

3 Making a Lemon Out of Lemonade

Louisiana's Petrochemical Corridor

Craig E. Colten

A blended agricultural and industrial landscape dominated the lower Mississippi River floodplain in the late nineteenth century. Between Louisiana's Gothic political capitol in Baton Rouge and the vibrant economic and social capital in New Orleans, sugar planters oversaw the cultivation of thousands of acres of sugarcane and managed the grinding mills that carried out the initial processing. Small-scale, dispersed industrial activity was a fundamental component of a largely rural agrarian economy hugging the banks of the massive river.

New Orleans in 1900 was the twelfth-largest city in the country and functioned as a classic mercantile city. Bankers and cotton factors guided the movement of agricultural commodities, such as cotton and sugar, to distant manufacturing centers. The port handled outbound commodities and received manufactured goods from distant ports. A limited industrial base refined raw sugar and pressed cotton for export. About a hundred miles upstream in Baton Rouge, state politics dominated the much smaller local economy, supplemented by modest cottonseed- and timber-processing industries.

By the late twentieth century, the fundamental character of the riverside landscape had changed dramatically. Sugarcane had declined in prominence, although it had not disappeared. Farmers had converted some cane fields to pasture, and urban sprawl encroached on other former plantation lands. The most dramatic change came when corporations acquired thousands of acres of floodplain property where they built sprawling petrochemical processing plants situated safely behind the massive federal levee system. Most sugar mills have closed as transportation and economies of scale enabled the consolidation of these primary processors. New Orleans has declined precipitously in its national ranking. Condominiums and hotels now occupy New Orleans's former cotton presses and Baton Rouge's cottonseed mills have been dismantled and riverboat casinos dominate much of its riverfront. Many of the remaining plantation houses serve as tourist destinations. Tourism has risen in importance, particularly in the cities, but is dwarfed along the river road by the riparian petrochemical complex.

Several factors enabled the conversion of the sugarcane fields along the river to a dynamic and expansive, yet conflict-ridden, petrochemical complex. The prevailing narrative is that the mere presence of natural resources and a viable waterway led to an almost inevitable industrial development of the lower Mississippi. Federal investments in petrochemical production during World War II accelerated the expansion of industrial activity and transformed the floodplain into a manufacturing complex that stretches for over one hundred miles along the river.[1] While the accounts of locational allurements and the war-related growth are valid, the development of the Louisiana petrochemical complex is more nuanced in both its chronology and geography. Without the federally financed levees, built initially to protect plantation lands and New Orleans, it is unlikely industry would have invested so heavily along the lower river. Key technologies, such as pipelines in the early twentieth century and later off-shore drilling platforms, enabled the efficient exploitation and transport of crude oil. Changes in the market for Louisiana sugar and also in cane grinding technologies decreased the value of agricultural property and enhanced opportunities for new industries to dislodge an older one. And finally, favorable state policies that offered incentives to, and tolerated offensive practices by, industry contributed to the evolving landscape along the lower river. These locally distinct developments significantly contributed to the emerging energy-oriented regional economy.

Despite celebrations by economic development boosters at each factory opening, the arrival of petrochemical firms to the lower river valley was not universally welcomed. Any account of the petrochemical complex demands more than an exposition of its economic benefits. Industry's reliance on the Mississippi River as a bottomless sink for its effluent eventually overwhelmed even the mighty river and tainted public water supplies. Explosions, chemical releases, toxic waste dumps, NIMBY battles, and labor conflict engendered public apprehension toward and direct opposition to the petrochemical facilities. Beginning with the 1980s oil bust and continuing into the twenty-first-century plant downsizing and closures, the sweet sugar landscape of the nineteenth century has turned into what some see as a lemon—a potent, yet sour to some, ingredient in the state's economy.

Making the Floodplain Safe for Industry

The Mississippi River provides a prominent resource for economic development in southeastern Louisiana. Water that drains from over 1.2 million square miles of the North American interior maintains a channel that can accommodate ocean-going ships as far upstream as Baton Rouge, and provides a means for the inexpensive waterborne movement of bulky cargo such as grain, coal, and petroleum products to and from distant interior locations. Its navigational advantages prompted the Louisiana Purchase in 1803 and helped New Orleans attain a prominent position among the country's cities during the antebellum period. More than any other river system in the country, the Mississippi proved ideal for the nineteenth-century commerce carried by steamboats, and with federal investments in navigational improvements during the twentieth century, it has remained a major transportation corridor for barge fleets.[2]

While it served as a transportation resource, the river also posed a hazard to riparian land uses. This gargantuan fluvial system, that carried the runoff of much of the U.S. interior, delivered regular floods. When copious spring rains and heavy snowmelts simultaneously coursed through the Ohio and upper Mississippi River basins, high water could threaten activities along the banks of the lower river. The floodplain was literally constructed by repeated inundations of the relatively flat expanse of land on either side of the river. High water to some extent benefited agriculture with the delivery of nutrients and land-building sediment, although it frequently disrupted cultivation. Baton Rouge sits atop an ancient terrace (about fifty feet above sea level) and during the post-Pleistocene land-building phase was largely immune to flooding. Downstream from Baton Rouge, the terrace veers eastward away from the watercourse, and the river was largely responsible for shaping the natural topography all the way to the Gulf of Mexico. Immediately adjacent to both banks of the river are shoulders of slightly higher ground known as natural levees. These features are modest berms of relatively well-drained soils that are highest near the river and slope gradually downward as one moves away from the river. Soils grade from tillable sediments near the river's edge to fine clays over a distance of about two miles, and natural vegetation correspondingly shifts from willow to oak forests to cypress swamp. The natural levees, which flooded last and drained first, provided the best sites for human occupation for multiple reasons. As a consequence, unlike many other river basins, initial settlement was attracted to the flanks of the waterway not only for the transportation advantages, but because mucky back-swamp soils deterred agriculture, urbanization, and other land uses

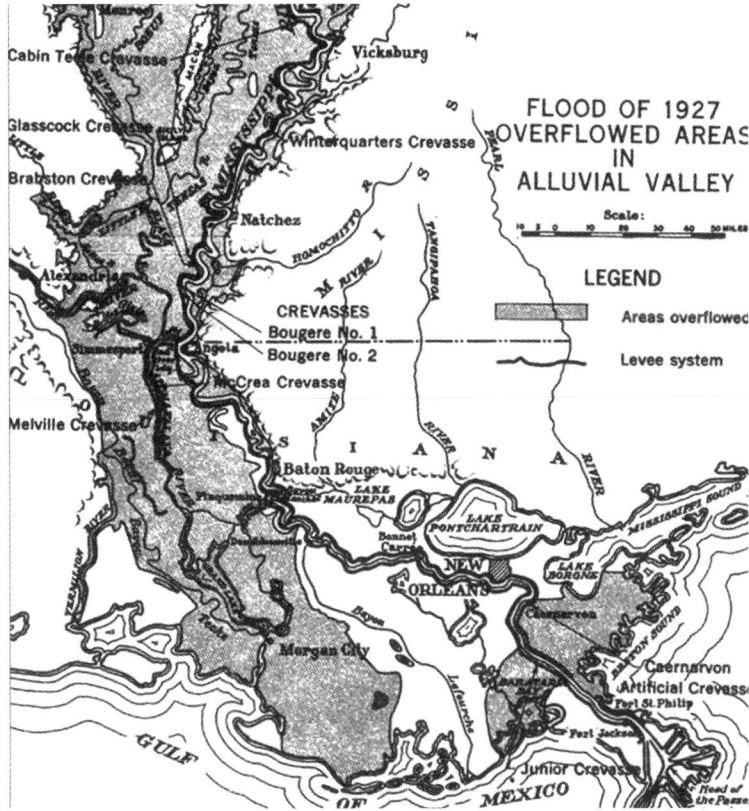

Figure 3.1. Flooding of the Mississippi Valley in 1927, illustrating how effective the levees were in what became Louisiana's petrochemical corridor.

Source: U.S. Army Corps of Engineers (New Orleans: U.S. Army Corps of Engineers).

beyond the slightly elevated shoulders. Topography defined the optimal urban, agricultural, and industrial zone.[3]

Despite the modest protection offered by the natural levees, floods were a threat to the lower river inhabitants. Upon planting the colonial capital at New Orleans, the French began work to fend off high water, and by 1726 laborers had erected an earthen levee between the incipient city and the ancient river. Colonial law required landowners to construct levees along their riverfront plantations, and thereby gradually extended the barrier both up- and downstream. The policy's intent was to protect both agricultural land and the emergent metropolis of the lower valley. By the time of statehood in 1812, levees extended upstream to Baton Rouge on the east bank, nearly to the Red River on the west bank, and some distance downstream from New Orleans on both sides of the river. Floods continued to overwhelm these human-made structures during the nineteenth century, and in 1879, Congress created the Mississippi River Commission (MRC) to take on the job of constructing a more consistent levee system, ostensibly for navigation purposes.[4]

After numerous studies, debates about levee-building strategies, and struggles to obtain funding, the MRC gradually oversaw the construction of an improved levee system by 1926. The record flood of 1927 tore gaping holes in what had been considered an adequate protection system, inundated over twenty thousand square miles of land, forced the evacuation of over six hundred thousand valley residents, and caused an estimated $363 million in direct property damage.[5] While damage was massive between Memphis and Baton Rouge, the area that later would become Louisiana's chemical corridor largely escaped the high water (see figure 3.1). Land along the east bank between Baton Rouge and New Orleans had some of the oldest and strongest levees and withstood the flood. There was flooding on the west bank, but between Donaldsonville and New Orleans riverfront sugarcane lands remained safe behind their levees. Both intentional and river-caused breaches below New Orleans allowed inundation across much of the delta. In terms of the region's formal flood protection strategy, this event forced the abandonment of the levees-only policy. Yet at the same time it demonstrated that levees, despite massive failures upstream, could protect much of the floodplain accessible to oceangoing ships—particularly if modifications to the protective system were implemented.

Congress, the MRC, and the Corps of Engineers responded to the overwhelming flood of 1927 with numerous adjustments to the flood control system, many of which directly benefited the lower river corridor. Plans called for higher levees for much of the upstream system, plus two outlets in Louisiana to divert huge quantities of water through the Atchafalaya River basin and through Lake Pontchartrain. The changes represented a shift to the "levees and outlets" approach. The enlarged levees, augmented with one of the two outlets, prevented unwanted crevasses after 1937. By 1960, over $1.5 billion had been invested in these alternations. According to Corps of Engineers' estimates, fending off the floods between 1927 and 1960 prevented some $5 billion in property damage.[6] The major flood in 1973 provided the greatest post-1927 test of the completed outlets. It prompted the lone opening of both floodways which effectively rerouted water out of the main stem of the Mississippi and prevented serious flooding in the industrial corridor.[7] While there was minor flooding in the lower river corridor in 1973, mainstem levee failures were not the cause and the federal system protected the major industries.[8] Opening the floodways allowed extensive inundation of the backwater "reservoirs." Damage to oilfield service facilities at Venice and shipyards at Morgan City (near the mouth of the Atchafalaya River) were the principal direct impacts to industry, and there was extensive flooding of sugarcane and other agricultural lands.[9] Subsequent floods in 1975, 1979, 1983, 1997, and 2008 prompted the Corps to open the Bonnet Carré Spillway near New Orleans, but these events barely threatened the protective structures in Louisiana's energy corridor. High water can, and does, disrupt river traffic and thereby interferes with waterborne movement of raw materials and feedstock among the various processors. During the major recent floods, the Coast Guard has either restricted navigation by cautioning shippers to moderate their speed to prevent levee damage due to wake or by closing the river temporarily.[10] While disruptive, river floods have not directly inundated any major industrial operations since 1973—hurricane-induced flooding has damaged refineries below New Orleans, however.

The huge federal investment in this flood protection system has been essential to the petrochemical development of this region. Recognized as an appealing element in industrial site selection, but seldom touted as a federal subsidy to the region's manufacturing development, the levees have proven vital to protecting all the plants downstream from the ExxonMobil plant atop the terrace at Baton Rouge. Yet, the construction and maintenance of the levee system was an indirect subsidy. The petrochemical producers inherited a valuable infrastructure, largely designed and built before their arrival with the intent to protect planters and New Orleans.

Sugarcane Landscapes in Transition

The arrival of the Standard Oil refinery (now ExxonMobil) to the bluff at Baton Rouge in 1909 began a century-long transformation of the Mississippi River's riparian landscape. Standard acquired 213 acres of riverfront property and built its refinery at Baton Rouge. The Baton Rouge agglomeration of primary and secondary refining that grew over the next seventy-five years intruded on cotton, not cane, fields and was more urban in its setting than more recent industrial clusters. Its position north of Baton Rouge minimized offensive odors and dangerous emissions reaching the city's main population centers when southerly winds prevailed. A millgate community arose nearby, and adjacent neighborhoods became a white, working-class district where families could live close to the dominant employers. Just upstream from New Orleans, the New Orleans Refining Company (later acquired by Shell) began developing a refinery and an adjoining company town, Norco, in 1918. Downstream from New Orleans, two refineries were operating on the urban margins in the late 1920s—Sinclair in Meraux and Pelican in Chalmette. The New Orleans area refineries all occupied former agricultural lands. This trio of early refining complexes grew as extensions to the two largest cities in the lower river, displaced some agriculture, and established a pattern of widely dispersed, but urban-oriented, industrial clusters that persist to the present.

Changes in the sugar economy in the twentieth century contributed to the availability of land for industry. In 1900 sugarcane was the dominant land use in the river corridor between Baton Rouge and New Orleans. Tariff protection in the postbellum period had enabled planters to re-establish their plantations, substituting wage for enslaved labor, but they retained most other elements of the traditional plantation system—grinding mills and on-site residences for laborers in particular.[11] Sugar cultivation in the parishes along the river between Baton Rouge and the river's mouth occupied over 96,000 acres in 1910, but declined to about 72,000 acres in the corridor by 1949. By mid-century, there was less demand for land for sugarcane. In addition, geographer John Rehder calculates that the sugar economy supported, at its nineteenth-century peak, 1,495 grinding mills in Louisiana—most along the Mississippi, but with many along bayous Lafourche and Teche to the west. By 1900, that number stood at only 300. Changing technologies and consolidation of land holdings contributed to the declining number of mills. Moving into the early twentieth century, steam replaced animal power and greatly increased mill capacity. Eventually, motorized transportation enabled the movement of larger loads of cut cane from the field to the mill and accelerated mill consolidation. Increased mill capacity during the last third of the twentieth century, decisions to close rather than update obsolete units, and global competition in the sugar market further eroded the number of mills in Louisiana from 112 in 1922 to 43 in 1970, 20 by 1995, and only 18 by 2003.[12] That number has continued to fall slowly. With each mill closure, a potential petrochemical industrial site becomes available.

Elizabeth Vaughan suggests that the U.S. price support system has enabled Louisiana farmers to continue cane cultivation, inhibiting more economically rational land-use adjustments—that is urban or industrial.[13] Despite consolidation of sugar processing, in some river parishes cane acreage increased during the last half of the twentieth century. Ascension and Iberville parishes are just downstream from Baton Rouge and witnessed considerable industrial development, yet both also experienced expansion in cane acreage between 1949 and 2007. Conversely, suburban St. Charles and West Baton Rouge experienced considerable sugarcane acreage decline. As industry swallowed up acreage, it did not completely displace sugarcane production. Land planted in cane rose from 72,000 acres in 1949 to a crest, driven by rising sugar prices, of 106,000 acres in 2002, before falling to 94,000 acres in 2007. By the mid-1970s manufacturing occupied approximately 73,000 acres in the river corridor. While industry obviously uprooted some cane fields, it has not squeezed agriculture off the floodplain. Nonetheless, the total acreage in cultivation of all crops in the river parishes dropped by about half between 1949 and 2002. Furthermore, where manufacturing operations have occupied prime floodplain lands, they diverted cane cultivation to less productive, often ill-drained soils in the backswamps—some of the areas inundated by the 1973 flood.[14] Industry has obviously pushed a considerable amount of agriculture either off the floodplain or onto more marginal soils.

Although cane fields have succumbed to encroachment by industry, suburbs, and grazing, this process might have been more extensive without government supports for sugarcane. Even with the persistence of a traditional agricultural economic activity, land prices in the lower river have not been too high to discourage industrial development.[15] And, both population and industrial growth has slowed in Louisiana since the 1980s, thereby minimizing encroachment pressures on agricultural land uses. Sugarcane cultivation continues in the shadow of the petrochemical complex, and during the global energy price spike of the early twenty-first century, hopes rose that cane would play a role in an emerging biofuel industry. The coexistence of cane and petroleum landscapes will continue and perhaps become even more intertwined.

Erecting an Energy Economy

Upon deciding to build a refinery in Louisiana, Standard Oil selected a strategic site atop the flood-proof terrace just north of Baton Rouge and at the head of navigation for oceangoing transport ships. Strict application of antitrust laws barred Standard from operating in Texas, but the 1909 construction of a Louisiana refinery gave the corporation access to oil in nearby fields and water transport advantages that were comparable, if not superior, to those enjoyed by the Texas Gulf Coast refinery district. As part of its Louisiana plans, Standard quickly constructed a pipeline to transport its Oklahoma crude to the new refinery that began receiving crude in 1910. A pipeline from the north Louisiana fields near Shreveport soon added to the flow of petroleum, and additional lines eventually linked the refinery to oil fields in Louisiana's coastal parishes.[16] Subterranean transport provided a cost-effective means for delivering raw materials and became a fundamental technology for the movement of crude and refined products in Louisiana.

Standard briefly operated as the sole Louisiana refinery, and provided a node for the gradual expansion of related petrochemical operations over the ensuing years. Facilities for producing chemicals for aviation fuel, synthetic rubber, and ordnance for the military during World War II expanded the petrochemical presence on the Baton Rouge bluff.[17] By 1945, Ethyl Corporation (antiknock compounds), Allied Chemical (soda ash and chlorine), and Copolymer (synthetic rubber) shared the terrace location with Standard Oil.[18] Downstream, New Orleans Refining Company (later Shell) also contributed to the war effort by producing aviation fuel.[19] A pair of refineries built during the interwar years downstream from New Orleans provided additional capacity.[20] By the end of World War II, Louisiana refineries could produce 366,000 barrels of refined oil annually—over a third of total U.S. capacity. Without question, the war effort and federally supported expansion provided a critical boost to the Louisiana petrochemical complex and the economic growth of the river corridor.[21] Beyond the three dominant refinery nodes and an

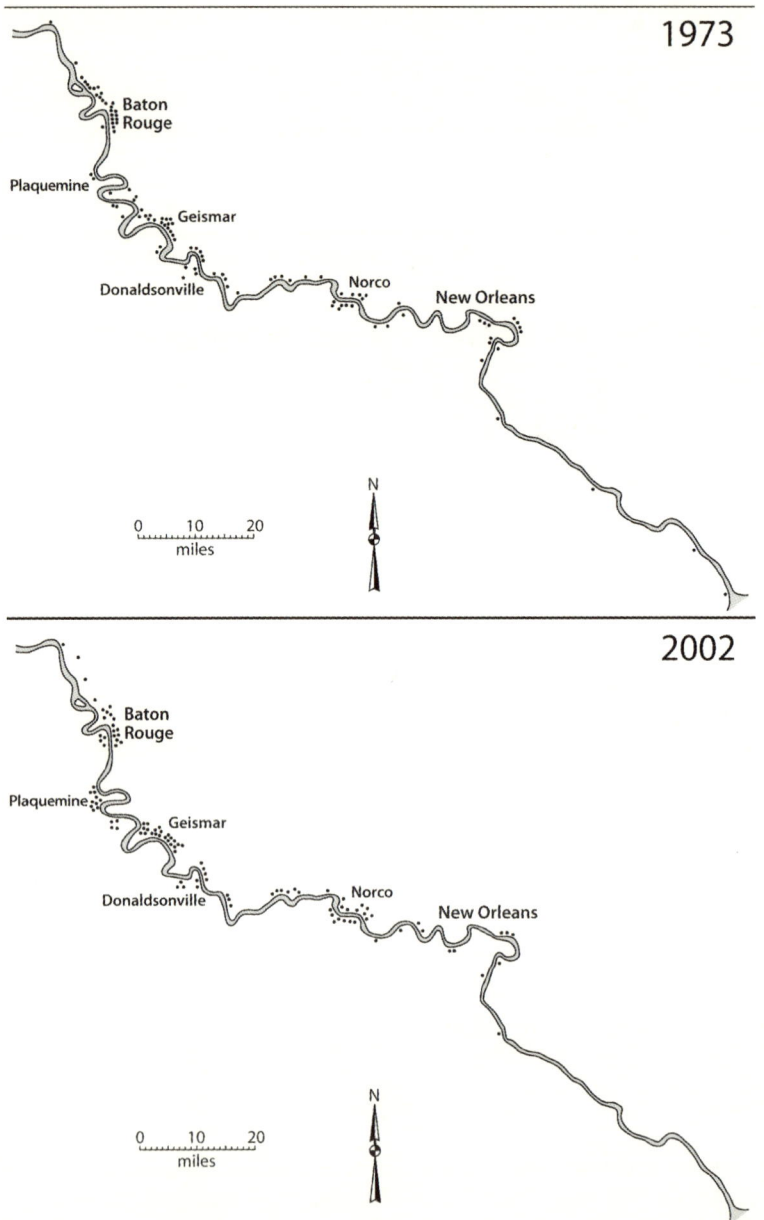

Figure 3.2. The Mississippi River Petrochemical Complex, 1973 and 2002. Cartography by Clifford Duplechin. Original cartography after Yodis and Colten.

Sources: Illinois Central Railroad 1974; and Elaine G. Yodis and Craig E. Colten, *Geography of Louisiana* (Boston: McGraw-Hill, 2007).

incipient petrochemical complex at Baton Rouge, the other principal industries on the floodplain by war's end were two sugar refineries and prominent shipbuilding activities in New Orleans. Sugar, nonetheless, still dominated the rural riparian landscape in the late 1940s.[22]

The insertion of petrochemical and other industrial operations into a functioning, albeit changing agricultural setting continued in earnest after World War II. Several related factors enabled the expansion of the petrochemical complex and also reflected its growing economic and political influence. The state resumed its generous ten-year tax exemption for plants and their equipment in 1946. Petrochemical and power-generating industries took advantage of this incentive more than other manufacturers. By 1974, 83 percent of exempted investment was in these two categories, with petrochemical firms accounting for 47 percent of the total; much of this development took place in the river parishes.[23] Abundant water for processing, cooling, and transport also lured industry to the banks of the Mississippi River. In response to the expanding industrial presence, the state invested in major infrastructure improvements. It built new highway bridges in Baton Rouge and New Orleans by 1950, and added a third near Donaldsonville in 1963. These structures relieved the reliance on ferries, diminished the function of the river as a barrier to land transportation, and reflected a commitment to entice more industry to the state. New oil exploration techniques and portable drilling rigs facilitated the development of rich oil fields and natural gas fields in the southeast Louisiana wetlands and nearby coastal waters, thereby increasing nearby raw material sources. Crude production in Louisiana climbed from about 250 million barrels in 1950 to its peak of nearly 550 million around 1970. Louisiana's petrochemical complex had risen to second place nationally in terms of refining capacity by 1979, and it had become a force in state taxes and expenditures and also in private-sector employment along the river.[24]

Louisiana's environmental policies were also extremely accommodating. The state's water pollution authority largely rubberstamped applications for permits for industries seeking to discharge waste into the Mississippi River. In terms of enforcement actions, of 163 pollution abatement orders during the 1950s, the state issued only 1 to an oil refinery, while frequently targeting sugar mills.[25] Both Louisiana's policy and its natural resource base suited the needs of a growing number of manufacturers that relied on the river for waste removal.

Several things distinguish the Louisiana energy-oriented agglomeration. Much of the post-1945 development was rural—with the exception of urban refining complexes adjacent to Baton Rouge and New Orleans (figure 3.2).[26] Almost all the facilities constructed during the second half of the twentieth century occupied sites along the river and some distance from the state's two largest cities. While small agricultural towns welcomed the industrial facilities and the opportunities they provided, labor was inadequate for both plant construction and operation. Consequently, a mobile workforce, drawing from the two major cities and also the entire river road corridor, commuted considerable distances from home to work. Corporations even encouraged their managers to buy houses in Baton Rouge or New Orleans to facilitate their resale in the event of eventual transfers.[27] Industry's presence has contributed in innumerable ways to growth and economic development along the river during the postwar years.

Additionally, the impact of the oil economy spread far beyond the river parishes and contributed to the state's distinct petrochemical economy. Onshore and, increasingly, offshore exploration and drilling transformed the state's coastal parishes and they became very much oriented to extraction and oilfield services. Offshore drilling began in earnest in the 1930s and came to dominate state production by the late 1970s. Lafayette and Morgan City became centers for local oil-related companies and offshore drilling platform fabrication. Throughout the state's Acadian heartland—a second distinct energy-oriented region—opportunities in the oil fields and associated activities provided a considerable income boost to those willing to shift from natural-resource collecting or agricultural pursuits to oil exploration and exploitation.[28]

Between 1940 and 1984, the energy economy drove a dramatic shift in the per capita income within the state. In the petrochemical refining district along the river, six parishes were among the state's twenty top income-earning parishes in 1940. By 1980, all but two of the state's top twenty per capita–earning parishes were in the state's southern section. This included all but one of the petrochemical corridor parishes and those in Acadiana that participated in the energy economy.[29] The shift to deepwater drilling since the 1980s has reinforced the coastal orientation of the industry and its economic impacts.[30]

Throughout the last quarter of the twentieth century, Louisiana grappled with its growing dependence on a petrochemical economy, particularly during dramatic fluctuations in the global oil economy. When oil prices rose sharply following the OPEC oil embargo in 1973, oil production and per capita incomes in Louisiana's coastal and river corridor parishes followed. For the state as a whole, employment in the oil and gas industry doubled between 1973 and 1981. When energy prices peaked in 1980 and slid into an eventual collapse in 1985, oil and gas production and per capita incomes dropped in Louisiana's two energy regions and the state lost over half its oil and gas production jobs. During the recovery from 1986 to 1990, the energy-driven parishes experienced a slow return to parity with the state's noncoastal parishes that had not suffered as severely during the oil bust. Economists noted that the state's "undiversified" economy made it exceptionally vulnerable to volatile changes in the international oil economy.[31] While the entire U.S. economy suffered due to the embargo and subsequent energy price fluctuations, Louisiana's lack of diversification produced particularly disruptive economic repercussions. The fact that high crude prices drive exploration and production while driving down demand for refined product distributes the impacts of the economic cycles on a different schedule and geography. This phenomenon lessens, to a degree, the negative impacts, but unleashes negative consequences with both petroleum price increases and declines.

Fundamental changes throughout southeast Louisiana have accompanied each economic oscillation. During the 1980s oil bust and in the period since the 2005 hurricanes, the energy industry has consolidated some of its Louisiana-based exploration and research and development activities to Houston, thus reducing the number of high-skill jobs once abundant in the coastal parishes and along the river.[32] The impacts of these adjustments were prominent both in New Orleans and places like Houma and Lafayette. Price spikes brought on by the double onslaught of Hurricanes Katrina and Rita in 2005 had a mixed impact on the Louisiana energy regions. The damage wrought by the hurricanes to offshore rigs disrupted production and thereby interrupted the state's vulnerable energy-extraction economy. But the price spike that followed temporarily enhanced state revenues—in terms of production—and somewhat offset the disruption of refining activity. The hurricanes also prompted Congress to authorize additional revenue sharing with Louisiana and other coastal states from Outer Continental Shelf extraction activities. Conversely, the sharp rise in natural gas prices prompted the closure of several manufacturing firms using natural gas as a primary feedstock.[33] As the nation entered an economic depression in 2008, Louisiana enjoyed a brief spike in revenue due to the temporarily inflated oil prices. Ultimately, Louisiana, and particularly its petrochemical corridor and coastal parishes, experience dramatic economic turbulence tied to oil price fluctuations. Commonly the ups and downs are not synchronized with national economic trends, and Louisianians consider themselves "recession proof." But the dependence on the energy economy makes the state especially susceptible to energy recessions. Despite efforts to diversify, the state and especially the chemical corridor and coastal parishes remain heavily dependent on the energy economy.[34]

Conflicts in the Corridor

Energy-related activities have direct ties to the natural setting where they take place, and many impacts from Louisiana's industries have proven disruptive. Water and air pollution flow from production and processing facilities. Fires, explosions, and other turbulent events can threaten human and biotic communities. It is these undesirable, but inevitable, outcomes of the energy economy that counterbalance the benefits produced by high-paying jobs and demand difficult social decisions about the management and regulation of the industry.

Oil and gas extraction can wreak havoc on small waterways in producing areas. Brine in particular has fouled streams and bayous and prompted some of the earliest pollution control efforts in the state. Louisiana's first explicit water pollution law in 1910 sought to prevent the release of brine in the state's rice-growing region during the season when farmers diverted river water to their fields. This legislation sought to protect traditional agricultural pursuits over the upstart oil production activities.[35] As the processing industry grew, state policies became more accommodating. The state allowed the largest petroleum trade association to help pay the salary of the staff developing state pollution definitions and policy in the 1940s.[36] In the petrochemical corridor, industry boosters proclaimed that the large volume of the Mississippi River enabled it to adequately handle refinery effluent. Yet, as early as 1942, New Orleans industry advocates cautioned that site selection must consider the potential impact of refineries on public water supplies.[37] This consideration had not deterred construction of the Shell Refinery at Norco, not far upstream from New Orleans's water intakes. The refinery's discharges to the river in the 1940s fouled the urban water supply and prompted complaints. In response, the state sought a cooperative, rather than confrontational, solution. Government officials convinced Shell to divert its effluent to a bayou draining away from the Mississippi to protect New Orleans's drinking water supply. The state also worked with Standard Oil in Baton Rouge to prevent it from releasing some oily wastes into the Mississippi that were causing downstream taste concerns.[38] These efforts to redirect effluent from the river during the 1940s represent public recognition that even the ultimate diluting machine had limits.

A series of water pollution studies over the next two decades registered increasing concern with the impact of petrochemical releases into the river. In 1951 the U.S. Public Health Service (USPHS) reported that stream pollution was only a problem in a few local areas where industrial wastes caused problems—such as the state's largest city.[39] By 1956, the U.S. Geological Survey (USGS) observed that industrial pollution was an "ever present danger" to New Orleans's drinking-water supply. Over the next several years the state's Stream Control Commission investigated the situation and reported that refineries and chemical plants were releasing objectionable wastes. By 1960 the commission prodded nearly two dozen manufacturers to modify their waste-release practices to reduce undesirable impacts.[40]

These modest efforts were more than offset by the overall increase in volume produced by a burgeoning industrial presence and accidental chemical spills, dramatic fish kills, and reports of cancer-causing chemicals in the river. One of the early turning points in public tolerance for using the river as a sink was a major industrial release near Baton Rouge in 1960 that fouled downstream public water supplies. In response, the state mandated that industrial plants had to report spills to a central government office that would notify downstream water-supply operators to close their intakes. This warning system did not penalize industry, but placed the responsibility for preventing impacts on the state and municipal water suppliers.[41] Over the next decade, however, state and federal agencies incrementally shifted greater responsibilities for protection to industry.

After a massive fish kill in Louisiana during the winter of 1963–64, the USPHS determined that wastes from an agricultural chemical manufacturer hundreds of miles upstream in Memphis, Tennessee, were responsible. The federal agency mandated that Velsicol Chemical send its wastes to a land disposal site far removed from the river.[42] This event moved the social response to wastes through several critical thresholds: Government agencies and society at large (1) acknowledged that small quantities of industrial wastes, and not just agricultural chemical run-off, posed a serious environmental risk; (2) recognized that the impacts of industrial wastes could be felt well beyond the local scale, and (3) repositioned chemical wastes as a direct hazard to public water supplies and not just a matter of taste.[43]

In 1974, an exposé of high cancer rates in communities along the lower Mississippi furthered the reaction to petrochemical pollution of the Mississippi.[44] Additionally, investigations of organic chemical contaminants in the early 1970s underscored the public health concerns of the time and eventually provided a basis to expand government regulation and granting permits of industrial wastes.[45] The outcome of shifting attention to chemical wastes during the period between 1960 and 1974 was heightened public apprehension about consuming river water, which propelled an increasing reliance on bottled water in the New Orleans area. This change also contributed to a fundamental shift from solid support for the petrochemical industry to a growing opposition to the energy economy that had overextended its unquestioned welcome. Concern with long-term exposure to minute quantities of toxic ingredients that could result in a deadly disease provided the basis for an entirely new regional moniker: Cancer Alley.

If water pollution was not enough to blemish the reputation of the region's manufacturing giants, a series of environmental mishaps added to the public's growing apprehension. During Hurricane Betsy in 1965, a barge loaded with highly toxic chlorine broke free from its mooring, disappeared as it drifted downstream in the storm's chaos, and eventually sank near Baton Rouge. This accident prompted a feverish search and rescue mission to locate and raise the barge before its contents leaked.[46] Over the next few decades, earthshaking explosions rocked the Norco Shell facility in 1979 and 1988. These blasts caused fatalities, spewed hazardous debris and chemical emissions, shrouded the communities with ominous smoke clouds, and heightened anxiety among those living nearby. A 1989 Christmas Eve explosion at the Exxon refinery in Baton Rouge produced one fatality along with numerous injuries and sparked massive fires in storage facilities. This event reemphasized the risk of refineries at the upper end of the corridor.[47] A national survey found the Baton Rouge–New Orleans corridor had one of the highest rates of chemical plant explosions in the country, which validated the burgeoning local anxieties.[48] A high national ranking for explosions was not the only undesirable distinction earned by the chemical corridor. In both 1998 and 2000, Louisiana ranked second among states nationally for the volume of on-site releases according to the U.S. Environmental Protection Agency's (USEPA) Toxics Release Inventory. Six of the top-ten Louisiana emitters and six of the leading ten parishes also were in the chemical corridor.[49] A litany of other environmental issues—such as Superfund and brownfield sites, groundwater contamination, conflict over high ozone levels in Baton Rouge, coastal wetland damage, and lax enforcement by Louisiana agencies —galvanized citizen opposition to the region's powerful economic engine.[50]

Environmental Justice in the Chemical Corridor

In a state noted for its accommodating stance toward polluting industries, Louisiana citizens have frequently taken an adversarial stance against both companies that degrade the "sportsman's paradise" and the government that recruits them.[51] Indeed, citizen petitions declared as early as the 1920s that the new refineries downstream from New Orleans were public nuisances.[52] While not an environmental justice action, this early resistance to industry reveals a public that was skeptical of larger industrial impacts. According to William Fontenot, longtime community liaison with the Louisiana Attorney General's office, some of the earliest environmental justice activities in the country took place in direct opposition to threatened industrial pollution in the chemical corridor.[53] Activism is also borne of the particular residential, racial, and employment geography of the chemical corridor.

Many of the chemical plants acquired portions of former sugarcane plantations—generally former grinding mill sites—where they encroached on existing communities. Sugar plantations had retained more of their resident labor force after the Civil War than cotton plantations. Heavy demand for skilled labor gave African American workers a modest degree of leverage in terms of securing better wages and benefits, and owners commonly maintained living quarters on-site for these employees.[54] In addition, throughout the lower Mississippi River valley there are small, linear African American residential hamlets. Relicts of lands given to freed slaves after the Civil War, these clusters have often become neighbors to sprawling petrochemical complexes, when agricultural landowners sold tracts to industrial purchasers. These hamlets often housed impoverished African American populations who had little means to relocate. All the while, many plantation owners vacated their former mansions and moved to town. Those making factory site selections observed that it was hazardous to place two plants immediately adjacent to one another, but they apparently gave little thought to their proximity to preexisting communities. Despite being adjacent to the new industries, few

African Americans were able to gain employment in the plants during the manu-facturing expansion of the 1960s. Racism blocked their entry into the better-paying positions and economic circumstances prevented their exodus. Thus, they experi-enced the disamenities without the benefits of high-wage jobs. In contrast, many white employees traveled from the larger cities or more distant towns and thus did not endure exposure to chemical releases during their nonworking hours.[55] Con-sequently, into the 1980s most workers tended to side with their employers and defend environmental practices as a small price to pay for good jobs. There were white working-class neighborhoods adjacent to refineries in Chalmette, Narco, and Baton Rouge, so not all industrial districts created the potential for race-based conflicts.[56] Nonetheless, the racial and class geography of the 1960s created the po-tential for more broadly based environmental justice conflicts that emerged in the 1980s.

Some of the most vigorous environmental justice confrontations have arisen from the "fence-row" neighborhoods. Representing one of the first such efforts, citizens in Alsen, a small African American community north of Baton Rouge and its industrial complex, mobilized and successfully opposed the operation of a toxic waste incinerator, industrial emissions, and uncontrolled industrial waste dispos-al. Before the term "environmental justice" had gained widespread use, this early 1980s effort represents a watershed in community opposition toward the heavy re-liance on using the local environment as a sink for hazardous wastes.[57]

Public opposition to industry gained momentum and forged new alliances with the lockout of employees at the Geismar BASF plant in 1982. Labor activists, seek-ing to restore union members to jobs and taking advantage of widespread appre-hension of chemical plant safety sparked by the Bhopal, India, disaster at a similar type of facility, argued that replacement workers could not be relied on to operate the plant safely. Although unions were politically weak at the time, the local affili-ate pressed the timely environmental issue. Public concern with a potential chemi-cal accident at BASF galvanized an unprecedented alliance among union chemical plant workers, local and national environmental groups, and African Americans who lived nearby. In the wake of events at Love Canal in upstate New York in the late 1970s, even the local press began questioning the environmental performance of the petrochemical industry. As part of its long-running battle with BASF, the union erected a billboard that asked if BASF was the gateway to "Cancer Alley." This sign, erected in 1987, echoed growing concern with high cancer rates in the chemical corridor and introduced for the region a label in opposition to the pro-development terminology. While this conflict was not born of an environmental justice issue, the alliances developed during this five-year labor struggle strength-ened the environmental justice movement in Louisiana.[58]

Those alliances—which included the community liaison in the attorney general's office and statewide organizations such as the Louisiana Environmental Action Network (LEAN) that provided guidance and continuity; academics at Tulane, Xavier, and LSU who provided legal and technical expertise; national and international environmental and labor organizations that could mobilize wider public attention; and the press which increasingly gave sympathetic coverage to the state's environmental blemishes—continued to impact environmental justice activities. During the 1990s community activists countered a state-supported effort to permit the Japanese chemical producer Shintech to build a vinyl chloride plant near the town of Convent. Central to the opponent's case was a strategy that reversed the way communities looked at pollution sources. As part of the permitting process, the USEPA procedures focus on the population in proximity to a particular facility when considering potential impacts; industries are the center of the spatial analysis and oftentimes their cartographic depictions fail to depict people and the institutions where they spend time. Organizations in Louisiana developed maps that plotted schools and residential areas as the center of concern and dramatically illustrated the looming presence of the proposed facility. This cartographic reversal helped disrupt the permitting process. While the persistent opposition prompted Shintech to withdraw its initial plan in 1998, it nonetheless successfully constructed the plant upriver at a more isolated site.[59] In another case, near the Norco Shell chemical plant, residents of the African American hamlet known as Diamond waged a long-term struggle to secure a buyout from Shell so that they could move out from under the shadow of the production facility.[60]

Seeking to avoid the risks associated with toxic releases in the event of explosions and also to avoid conflicts with neighbors, several chemical producers have purchased entire fencerow settlements. Upon acquiring the property, the new owners remove the houses and use the newly opened spaces as "greenbelts" or safety buffers.[61] Dow's Plaquemine facility reported spending $14 million acquiring the community of Morrisonville and relocating its approximately 300 residents.[62] Georgia Pacific purchased the community of Reveille not many miles downstream. In urban Baton Rouge, Exxon created what appears to be a sprawling park east of its refinery when it bought out tens of square blocks of residential homeowners. Yet, it is not a public space; rather, it is a buffer to minimize risk of lawsuits and personal exposure. These relocation efforts, desired by the residents of Diamond, have also become the object of criticism by environmental justice advocates who charge that uprooted residents suffer a loss of community.

Events in the wake of Hurricane Katrina (2005) have further heightened public antipathy toward the petrochemical industries. Closure of some plants resulted in job losses and widened the base of opposition toward former employees. A sizable spill from a storage facility in Meraux, caused by Hurricane Katrina, tainted flooded homes and spawned a lawsuit for damages. When a barge collided with a tanker on the Mississippi in 2008, it produced a sizable spill and closed the river below New Orleans for days, further emphasizing concerns over safety and economic dependence on the industry.[63] Despite nearly thirty years of conflict, controversy, and uncertainty, Louisiana will not let go of this economic mainstay, nor will the industry and its environmental impacts disappear any time soon. Environmental justice efforts have forced some cleanup activities, deterred the opening of some petrochemical-related facilities, and heightened public awareness, but the state still actively courts the petrochemical industry and its inevitable consequences. While the struggles of Louisiana residents had produced uneven results, it has become a key case study for the emerging global environmental justice movement.

Writing for an industry trade group, economist Loren Scott repeatedly proclaims that Louisiana was fortunate to have a natural endowment of sizable oil and natural gas deposits.[64] Such enthusiastic observations echo early twentieth-century boosters who did not foresee the eventual environmental, public health, and economic impacts that dependence on the energy industry would bring to the state. The broader view in the twenty-first century is that the state has enjoyed considerable economic benefits, but the costs have been high. As in other zones of industrial expansion, tens of thousands of acres of prime agricultural land have been converted to industrial uses—some irreversibly. What is unique to the Louisiana energy region is that both the nineteenth-century agricultural land use and the twentieth-century conversion were dependent on federal levees, which amounted to a sizable, if indirect, subsidy. Funding this massive technological system was a critical factor in the development of the lower-river industrial complex. And although industry has encroached on agriculture, it has not driven sugarcane off the floodplain. Indeed, agricultural acreage expanded during the same time that industry was expanding most rapidly. Consequently, sugarcane acreage is still interspersed among the refineries, but energy dominates in terms of its economic value. If energy prices soar again, biofuels may produce a synergy between the two floodplain economies.

While we may never know the full public health impacts of the petrochemical complex, emissions to the air, land, and water have fouled Louisiana's environment and entered the bodies of those who have been exposed to them. The state's economic incentives to industry have been the object of criticism for their inefficiency in producing jobs, their use to attract the more polluting industries, and their denial of much-needed state tax revenue. Public opposition to environmental hazards emerged full force in the 1980s and revealed serious political and social strains. Despite considerable resistance to industrial abuses, the environmental justice movement has experienced modest and mixed successes. Accidents, spills, and sizable emissions continue to characterize a troubled industry that has experienced cyclic downturns followed by rejuvenation with new crude and natural gas discoveries. New technologies have enabled both the discovery and exploitation of these previously undetectable and unattainable resources. Fluctuations in the global price of oil send the Louisiana economy to dizzying heights followed by dramatic plunges. A local economy tied to the fortunes of the international energy economy, unavoidably, follows its turbulent path. And in early 2010 forecasters began touting a rebound of the petrochemical industry, as the global economy began to slowly emerge from the economic collapse of 2008. Louisiana's chemical corridor is in many respects subsidiary to Houston and the larger global energy economy, but it will remain a dominant force in the state's economy for the near future.

4

Los Angeles, the Energy Capital of Southern California

Sarah S. Elkind

Los Angeles has some competition for the title of Energy Capital of Southern California. Southern Kern County saw oil development before Los Angeles did; oil companies pioneered offshore oil drilling in Santa Barbara County. Opposition to oil drilling, too, is more associated with Santa Barbara than Los Angeles because of the massive 1969 blowout in Santa Barbara Channel, which many historians credit with energizing the 1970s environmental movement, at least in California. Cases could be made, too, for naming Hoover Dam, or perhaps the Diablo Canyon Nuclear Power Plant, as California energy capitals because of the controversies that surrounded their construction, and their importance in California's electricity grid. Nonetheless, I think Los Angeles will do as an energy capital. Oil profoundly shaped the landscape, economy, and culture of the region. The inherent conflicts that arose because of the presence of oil deposits under residential sections of the city, the early adoption of zoning, a wildly speculative real estate market, LA's carefully constructed image as a bucolic respite from the dirty, industrial East, and the relationships between oil production and consumption in the quintessential automobile city raise a number of issues central to the history of energy development and its impacts on cities and society.

History of Oil in Los Angeles

Los Angeles and coastal Orange County remain important areas of oil production, refining, and transportation today. The region boasts a third of California's refineries; the Los Angeles and Long Beach Harbors host a fifth of the state's petroleum terminals. Additional oil fields, refineries, and processing facilities are located north of Los Angeles in Ventura, Santa Barbara, and Kern Counties.[1] All of this began with oil prospecting in the wake of geological surveys of California in the 1860s, which were in turn fueled by the invention of processes that distilled kerosene, a lamp fuel, from asphalt or crude oil. Kerosene production in California began before this period with raw materials imported from the East Coast. But oil seeped from the ground in many places in California. Railroad surveys of the 1850s noted oil and gas seeps from the Central Coast of California south to Los Angeles.[2] The most famous of these seeps is the La Brea tar pits in downtown Los Angeles; in fact, tar still oozes onto walkways in the park, site of both the George C. Page Museum that houses the paleontological finds from the La Brea tar pits, and the Los Angeles County Art Museums. In 1859, two of San Francisco's kerosene merchants moved south to produce kerosene from the La Brea tar pits and from bitumen deposits in Carpenteria, north of Los Angeles. These ventures lost money and lasted only a few years.[3] Other early oil ventures were even more speculative; drilling firms incorporated to acquire water rights, deal in real estate, and to "improve, develop and cultivate lands whether for mineral or agricultural" purposes.[4] This first oil boom also soon faded, in part because California's heavy crudes were not well suited to kerosene production.

Thus, Edward Doheny's subsequent "discovery" of oil in Los Angeles hardly amounted to a surprise. Los Angeles's oil boom began a little later, in 1892, when Doheny drilled a shallow, but productive well about a mile from city hall, near the La Brea tar pits. Within a decade of Doheny's initial success, over a thousand other wells had sprung up in and near downtown Los Angeles.[5] In 1911, Standard Oil laid pipelines from the downtown Los Angeles oil field to the coast, built its second California oil refinery, and founded the city of El Segundo.[6] Other oil refineries soon followed, eventually making Los Angeles a center for oil processing as well as extraction. Back in the oil fields, however, the oil rush sparked by Doheny's find seemed to play out quickly. Then, between 1917 and 1926, prospectors found extraordinarily rich oil deposits dotted elsewhere around the Los Angeles basin. Just three of these—Huntington Beach, Signal Hill, and Santa Fe Springs—produced more than three quarters of California's total oil output. These new oil fields swamped the market, causing oil prices to drop by two-thirds in eighteen months.[7]

Residential land development created a host of unusual problems in these new oil fields. Much of Huntington Beach, Signal Hill, Long Beach, Venice, and Playa

Del Rey had already been developed as residential communities or subdivided into small house lots before oil was discovered. Where subdivided lands had already been sold, oil drilling had to take place on leases with boundaries that conformed to these residential plots, and oil companies had to negotiate drilling leases with each property owner. Because subdivision prevented a single oil company from leasing large sections of any given field, the collision between oil drilling and residential real estate development generated a peculiar and problematic style of oil development characterized by closely packed derricks operated by dozens of companies, each competing to pump oil out of the ground as fast as possible. The result was both wasteful and dangerous. Oil companies spent more money on drilling and derricks than they would have in a field with larger leases. Rapid drilling and pumping depleted the oil pressure necessary to move oil out of its underground deposits. Moreover, in these crowded oil fields fire spread with devastating ferocity. Those living in the shadow of this black gold rush faced natural gas explosions, fires, runaway wells that sprayed oil around for hundreds of feet, the constant clack and clatter of drills and pumps, and spilled oil that befouled the entire community.[8] As one veteran of the chaotic Huntington Beach oil rush complained: "These oil men have taken everything except the food in the icebox. . . . My back yard is an oil well, a sump hole. My fence is gone and the inside of my house is a mess."[9]

Drilling in the City and on the Beaches

Oil strikes in and around Los Angeles's residential neighborhoods posed a real quandary for residents. On the one hand, oil promised riches, or at least regular income from leases and royalties, and clearly fueled substantial suburban growth.[10] On the other, the mess and danger that followed in the wake of oil development threatened to make city fields unlivable and dramatically reduced the value of properties not leased for oil development. As a result, residents and city officials struggled for decades to find the right balance between accommodating and restricting oil development. Efforts to contain oil development evolved relatively quickly; the Los Angeles City Council enacted the first drilling restrictions in 1897, just five years after Doheny opened his first well. This ordinance prohibited drilling within 800 feet of Elysian Park or Echo Park—two parks located relatively near the downtown Los Angeles City field—and within 1,800 feet of Los Angeles' other city parks—which happened to be farther away from known oil deposits.[11] In 1908, Los Angeles implemented one of the first zoning laws in the nation. By the 1920s, zoning restricted drilling in commercial and residential sections of the city. In spite of these and other regulations, conflict over oil drilling remained a constant in Los Angeles land-use politics right through to the 1950s.

Although drilling in areas zoned for residential development would eventually prove an important test of Angelenos' attitudes toward property rights and regulation, one of the first disputes over drilling involved recreational rather than residential lands. In 1898, California oil companies operating in Summerland, north of Los Angeles, began erecting oil derricks on piers extending out from the shore.[12] Los Angeles oil companies soon copied them, so that by the 1920s, derricks, drilling piers, fences, pipes, and equipment proliferated on Los Angeles's beaches. In the furious competition of the times, some oil companies bought or fenced land along the shore, more to block their competitors' access to drilling sites than to protect or develop their own leases.[13] Simply walking along the beach through this maze of equipment became increasingly difficult. Spilled oil, derrick fires, noise, and fumes from the oil-drilling operations transformed the beaches from recreational to industrial landscapes, befouled the beaches, and threatened tourism and growth in shoreline communities.

To some extent, the chaos that accompanied oil piers grew from the legal vacuum in which so-called tidelands drilling occurred. In this era, California and most other states recognized oil rights in two ways. First, oil companies owned all the oil they pumped out of the ground, but not oil still in the ground; ownership of oil is governed by the right of capture. In other words, the mineral rights attached to a particular piece of property conveyed merely the right to extract oil from that land, not ownership of a specific quantity of oil. Second, when the owner of a parcel also owned the mineral rights underneath that parcel, that owner could sell or lease those mineral rights to an oil company. The state, not individuals, owned submerged lands and their mineral rights in California, but the earliest pier drilling took place before the 1921 Submerged Land Leasing Act established a clear tidelands leasing process.[14] Although the Submerged Land Leasing Act brought some order to coastal oil development, this legislation promoted oil development of the beaches and permitted California to collect revenues from tidelands oil wells. It did not address the crescendoing complaints about industrial exploitation of Los Angeles's beaches.

As the 1920s proceeded, a public campaign to remove oil wells from the beaches mounted. This was fueled by a gathering movement to expand public ownership of beaches nationwide. In Los Angeles, advocates of public beaches traded on the idea of beaches as a uniquely public resource that ought not be exploited for private gain. In letters to the county supervisors, Angelenos demanded new legislation to "clearly define the public's right to full enjoyment of the beach, without regard to the plans or desires of private interests."[15] Where this impulse manifested as calls for city ordinances to ban oil wells within a quarter- or half-mile of the beach, the oil companies fought back. Standard Oil and others of California's largest oil companies supported statewide legislation to ban new drilling piers in 1928. They preferred state to local legislation and hoped that moving oil wells from the beach itself to solid land behind the beach would reduce opposition to shoreline drilling.[16]

Not incidentally, they saw this legislation as a means to reduce competition in the oil fields. Moreover, by this time the oil companies had discovered how to whipstock, or drill diagonally, from dry land into the oil deposits they had previously reached from piers. When the pier drilling ban passed in 1928, however, the California governor celebrated it as a victory for the public in a long battle to protect the priceless California coast from the private exploitation and ruin.[17] He did not draw attention to oil company support for the measure, or to the way the bill might affect California's small, independent oil companies.

The pier drilling ban clarified drilling rights but did not assist local governments in the task of regulating the impact of oil drilling on their communities. Mayors, city councils, and county supervisors did not have adequate regulatory tools for this task. Zoning, their best tool for constraining land use within city limits, could be changed either through wholesale rezoning or piecemeal zoning variances. The existence of dozens of independent municipalities in the Los Angeles area also interfered with municipal efforts to protect residential property by limiting the reach of drilling restrictions. Then major oil companies used state and national legislation to trump local regulations. The 1936 ballot proposition that legalized whipstocking into California's coastal oil pools illustrates the oil companies' ability to bypass local restrictions. Standard Oil wrote the proposition. Los Angeles voters and leaders opposed it because they feared, with some justification, that it would open new areas of the city to oil wells. The proposition passed over Los Angeles's objections, largely because it earmarked oil royalties from tidelands oil deposits for public park acquisition throughout California.[18] In other cases, national priorities overshadowed local concerns or redirected public priorities away from restricting oil drilling.

Protecting the beaches remained the one consensus that restricted drilling activities through the first half of the twentieth century. From the 1920s through the 1940s, Angelenos demanded that local officials ban oil wells within 2,500 feet of the shore on the grounds that oil companies exploited the beach "to the detriment of the community as a whole."[19] These proposals either failed outright or were never enforced.[20] Even if city or county governments had prohibited drilling near the beaches, these rules would have applied only to new wells; oil production along the Southern California coast from Santa Barbara to Orange County continues to this day.

Regulating new oil development in residential areas proved even more difficult than protecting the beaches. In large part, this was because of property owners' direct financial interest in oil development. Specifically, property owners sought to profit from oil extracted from their own lands, but fought to prevent nearby drilling that might reduce their property values. For the next decades, debates over urban oil wells revolved around precisely these questions of property rights and the distribution of the costs and benefits of oil drilling. This was particularly true from the late 1920s through the early 1950s, as new oil discoveries repeatedly reintroduced this same regulatory quandary, and as the crisis mentality of the Great Depression, World War II, and, finally, the Cold War largely reversed any momentum to regulate drilling that the fights over the beaches had generated.

The first test of zoning restrictions in this time of crisis began with the discovery of the Venice–Del Rey field in 1929. Los Angeles annexed Venice in 1926, just three years before the strike, so Los Angeles's relatively strict rules about drilling in residential neighborhoods clearly applied to Venice. However, Venice residents fought against the implementation of these protective codes.[21] The first protests came at a public hearing in January 1930, when five thousand Del Rey and Venice residents demanded unrestricting drilling. Meanwhile, thousands of landowners signed petitions urging the City Planning Commission to permit drilling in the Venice field. A former city attorney for Venice called drilling "possibly one of the most popular issues with the greatest unanimity of opinion in the area as has ever been heard of."[22] The fact that he sought to lease his own land in Venice to an oil company may have shaped his sense of the thing. The Los Angeles City Council followed public opinion and allowed drilling in residential Venice.

As derricks, tanks, and pumps sprouted across the beach town, the dramatic social and economic consequences of unrestricted drilling quickly shifted public opinion. The "fiasco at Venice" became an object lesson for subsequent debates over the merits and dangers of residential oil development.[23] In 1931, for example, public opposition mounted quickly to a proposal to drill a test well by whipstocking from a residential neighborhood into the strata under Elysian Park. Initially, owners of property abutting the proposed well supported the idea, so the City Planning Committee voted to allow the company to drill.[24] But the opposition, including the Los Angeles Chamber of Commerce, the Municipal League, the League of Women Voters, and numerous neighborhood associations, demanded that the planning committee reverse itself in order to protect the city park. The Los Angeles Times called the oil well "adverse to the public interest and the welfare of the city, though pleasing to particular, though short-sighted, property-owners."[25] Nine hundred people attended a raucous public hearing in January 1932. The proponents argued that their land ownership included the freedom to drill for oil; opponents argued oil drilling was an "invasion of property rights" and a threat to property values.[26] Opponents also cited the damage that oil development had done in places like Venice: "I . . . saw beautiful sections of this city ruined and I hope that such will never come again. Human greed and avarice, unless restrained, would destroy our beautiful beaches, our residential areas, without any compunction."[27] The shrill protests moved one city councilor to propose an outright ban on drilling in residential neighborhoods.[28] Although this ordinance failed, opponents of the

Elysian Park well ultimately triumphed on the strength of their assertions of the rights of abutters and of the adverse impact of drilling on real estate values in Los Angeles generally. Real estate, after all, was one of Los Angeles's most important industries, even in 1931.

The drilling calculus changed again with the outbreak of World War II. Wartime military buildup affected all sectors of the American economy, but Southern California oil wells received particular attention because they were a main source of fuel for the Pacific Fleet. Moreover, even though there was no national oil shortage, the Petroleum Administration for the War advocated suspending all local oil regulations to increase the production of aviation fuels. California oil companies complied eagerly, opening one hundred new wells in just two months in 1942, and increasing the state's oil output by six percent.[29] One company invested so much money in drilling and exploration that it could not pay its employees.[30]

Wartime pressures and opportunities inspired firms to renew proposals for drilling in residential areas ruled off-limits to oil wells in the previous decades. So, for example, in January 1942, J. E. Elliot and Shell Oil proposed a deep well in an unincorporated corner of land called the Gilmore Island, between the La Brea tar pits and West Hollywood.[31] Even though this well would have stood outside city limits, it lay so close to residential areas in Los Angeles proper that the Los Angeles municipal officials had to approve the project before Shell could proceed. Residents around Gilmore Island rejected the wells as "nothing more . . . than a wildcat oil scheme."[32] Over two hundred of them protested Shell's project at a public hearing in June 1942. In response, Shell defended the project as necessary for the war effort, and Secretary of the Navy Frank Knox and the federal Office of Production Management urged city officials to approve both city drilling generally, and the Gilmore Island project in particular.[33] Mayor Fletcher Bowron sided with the abutters, vetoing ordinances to allow Shell to proceed on the grounds that oil wells so close to residential areas threatened the rights of the city and the interests of property owners.[34]

The city council overrode Bowron and, in so doing, unleashed a small flood of applications for new wells in residential areas where city officials had previously, and sometimes repeatedly, prohibited oil development. The rush of wartime drilling proposals included a new bid to drill under Elysian Park, a plan almost identical to the plan rejected in 1932. As in 1931, abutters supported the project; some two thousand signed leases to allow the oil company to drill on their land.[35] The Los Angeles City Park Commission also gave the oil company permission to drill under the park, citing the war emergency. The City Planning Commission approved many more variances on the grounds that "the Federal government had stated that the oil industry must produce great quantities of oil for war purposes at the earliest possible date."[36]

The city did finally permit the Elysian Park oil well, reversing its 1932 denial, out of deference to the war emergency. But opposition to residential drilling continued throughout the war. Mayor Bowron led the campaign for local control of oil regulations. He was outraged that the oil companies used the war emergency to circumvent hard-won regulations, and he was frustrated that federal officials backed the oil companies in what should have been a local affair. Many of his constituents greeted proposals for wartime drilling with equal skepticism. They sent letters to the federal officials asking, "The Government took my son and can take my oil—but why should I ruin my home to make an oil company richer?"[37] They insisted that the oil companies used the war as an excuse "to enrich themselves" at public expense.[38] If the military needed oil, they argued, the federal government should extract it, or shut down these wells after the war.[39] When accused of being unpatriotic, as they routinely were, opponents of drilling in the city fields retorted that the oil companies' efforts to bypass local oil regulations were far more unpatriotic.[40]

Two things lend credibility to the opposition. First, even as the Petroleum Administration for the War and Secretary Knox pushed Los Angeles to permit unprecedented drilling in the city, they refused to open the Elk Hills Naval Petroleum Reserve to increased drilling, and reduced oil output from rural areas.[41] Increased production from either of these two sources would have yielded urgently needed oil much faster than prospecting for new deposits in Los Angeles's residential neighborhoods.[42] Second, oil companies refused to promise to close wartime oil wells after the war ended. Bowron proposed this as something of a compromise, arguing that if oil companies drilled new wells out of wartime necessity and true patriotism, they should be willing to close them when the emergency passed.[43] The industry rejected this on the grounds that they could not afford to drill unless they were allowed to operate productive wells for many years. To be fair, drilling exploratory wells is expensive, but the oil companies' economic arguments confirmed for many Angelenos that the oil industry was "a selfish type of enterprise"[44] that operated "*not* in the best interests of the majority of the citizens," or, as manager of the Beverly Hills Chamber of Commerce said, "a group hiding behind the American flag, working for a selfish interest."[45]

Ultimately, Secretary Knox's call for more oil "to win the battle of the Pacific" gave the petroleum industry a critical military justification for expanded oil production.[46] Harold Ickes and the Petroleum Administration for the War not only rejected Bowron's proposals to limit the operating life of oil rigs in residential areas, but also sought to increase Los Angeles oil production by streamlining new drilling permits and decentralizing regulatory oversight.[47] The director of the Los Angeles Petroleum Administration for the War announced that he wanted "every field and pool in the state, including those within the city of Los Angeles" developed as swiftly as possible.[48] Bowron and his allies in the campaign to limit the impact of oil on residential Los Angeles were accused of "using the oil issue as a political football" and of interfering with the war in the Pacific.[49]

Some Angelenos remained adamant that sacrificing everything, including their homes and communities, to defeat the Axis was not worth it; others were willing to accept oil wells if oil companies also sacrificed. This comment letter to the *Los Angeles Times* is typical: "If the government really needs the oil, I will let them drill in my front yard and turn the proceeds over to the Red Cross, if they will agree to stop when the war is over."[50] But even those willing to accept oil wells as an emergency measure could not countenance "wrapping the flag" around urban drilling for profit. Others believed that the war did justify expanded oil production: 90 percent of the letters sent to the *Los Angeles Times* in February 1944 favored urban drilling for the war effort. This number may have been inflated by ads placed by a local oil company urging Angelenos to show their support for oil wells. Even so, the letters were enthusiastic about the war, but less sanguine about the oil wells.[51] In the most poignant of these, Ora E. Knight argued: "There are so many things we have had to put up with and endure because of the war effort that it seems positively infantile nonsense to make such a fuss over aesthetics. If I could have a seat on the bus every day, for instance, I would be happy to sit in the shadow of an oil well. If I could only have my friends and relatives back from the service an oil well even within a few hundred yards would not make me unhappy."[52]

Pressure to permit drilling in Los Angeles's residential neighborhoods continued after World War II. When oil companies found oil near the Los Angeles airport and in Boyle Heights, they argued for zoning variances on the grounds that the city's and state's future depended upon "oil, oil and more oil."[53] The calculus justifying oil drilling changed significantly at the end of World War II, although economic growth and the danger of dependence upon foreign oil remained powerful arguments for relaxing zoning and drilling regulations. The fact that Los Angeles residents and elected officials resisted residential drilling even during World War II reflects both the importance of property values in American cities and the difficulty of balancing various rights associated with property ownership. But clearly, the local government authority was insufficient to defend residential property rights in the context of a national crisis. The presence of oil in residential areas undermined urban planning, land-use regulation, and other local authority. All of this left many in Los Angeles suspicious toward the federal government and the oil industry.

The cumulative effect of this pressure on local regulation can be interpreted as an emergent government culture of accommodating oil development within the city. The noise, pollution, odors, and traffic associated with new oil development had considerable impact on oil-producing neighborhoods. These impacts made it extremely difficult for city officials to justify denying those near oil fields the right to drill their own lands; the cumulative effect was the constant erosion of no-drill areas. Each new strike seemed to inspire enthusiasm and public pressure to permit drilling. But the expansion of drilling areas was nearly always closely followed by reaction and public protests. This left city officials with few options, and so they turned to technological fixes to reduce oil nuisances where they could not ban drilling altogether.[54]

Social Impacts

Oil development in the first half of the twentieth century reshaped Los Angeles's society as much, or more, than it did the city's landscape. Most historians of Los Angeles peg the region's industrial growth to two factors: first, federal construction of the port at San Pedro, which finally gave Los Angeles a deepwater harbor and, not incidentally, extended the reach of Los Angeles's energy and industrial firms far beyond the local sphere; and second, the discovery of oil, which created a local demand for machinery and provided local companies with ample, cheap fuel. Beginning in 1918, the Los Angeles Area Chamber of Commerce Industrial Department boosted their region as the place where nature uniquely benefited industry, and where eastern and midwestern firms would find favorable land prices, labor rates, and tax structures. Los Angeles's open shop labor market was a major selling point, but so was its inexpensive hydroelectricity and the ready availability of petroleum-based fuels.[55] The city's efforts met with outstanding success through the middle of the twentieth century, as Los Angeles transformed from an agricultural to largely industrial economy, and as industrial suburbs sprang up around the Los Angeles Basin.[56] The construction of the port may have ended Los Angeles's geographic isolation but the oil industry served a pivotal role in Los Angeles's industrial growth.[57] When Los Angeles's industrial sector did take off in the late 1910s and 1920s, it brought enormous population growth, rampant real estate speculation, and significant cultural change. Los Angeles's African American community felt these changes particularly keenly.

Until the 1920s, African Americans regarded the American West as a haven from the racism and limited economic opportunities of the South. The black population was small and cohesive and regarded the racial climate of Los Angeles as unusually open, even for the American West.[58] Black Angelenos owned homes in larger numbers and in a wider range of neighborhoods than in other northern or western American cities. This openness was relative, of course. In Los Angeles, blacks found integrated schools and public facilities and, for as long as Harrison Gray Otis ran it, a very sympathetic *Los Angeles Times*, but they still had to contend with employment and housing discrimination and segregated public beaches.[59] In the first decades of the twentieth century, however, the hardening of segregation and racial tension across the country began to reshape race relations in Los Angeles, too. Longtime African American residents started to report a change in racial attitudes as early as 1913, as public theaters and other establishments seemed to suddenly deny access to black patrons. Titus Alexander, a prominent black attorney, conducted his own survey in 1913 and found that only three of Los Angeles's many saloons would serve blacks, and some theaters discriminated by charging African American patrons far more—a dollar instead of a nickel—than whites.[60]

Racial exclusion grew more pronounced in the 1920s. Segregation officially ended at the public beaches in 1927, but the city parks department suddenly segregated swimming pools, and more neighborhoods adopted restrictive covenants that prohibited the sale of homes to African American buyers. Los Angeles's black community blamed the change on the arrival of white migrants from the American South, and most notably workers from the Texas and Oklahoma oil fields who came west with the oil boom.[61] Indeed, that migration was substantial; Huntington Beach's population exploded, increasing sevenfold in 1920 and 1921.[62] Los Angeles County's population nearly doubled between 1910 and 1920, and then doubled again by 1930; residential suburbs around the oil fields grew even faster than the city as a whole.[63] As far as the African American population was concerned, oil transformed what had been a relatively open racial environment into an increasingly closed one.

The oil industry, indeed most of Los Angeles's industries, offered few opportunities to blacks. Most of Los Angeles's industrial development took place in newly incorporated industrial suburbs; workers lived in nearby residential suburbs. So, for example, Long Beach grew initially as a bedroom community for the San Pedro port and oil industries; workers in the Standard Oil Company refineries in El Segundo lived in neighboring Vernon and Huntington Park.[64] These suburbs were almost all restricted, off-limits to blacks. This meant that the distinctive suburban industrialization developed by the Los Angeles Chamber of Commerce and fed by the oil industry itself contributed to the hardening of race relations which blacks blamed on Southern migrants.

Oil undergirded Los Angeles's growth in the twentieth century. It brought a new population to the city, created the foundation for industrial development, and added to pressure to build what has become one of the busiest ports in North America. This growth, and indeed, the oil boom itself could not have taken place without critical technological innovations, including those which permitted near-shore oil derricks and whipstocking once the drilling piers were outlawed. But the iterative discoveries of new oil deposits beneath the residential sections of Los Angeles left a profound legacy. In some places, residential development coexisted with oil—particularly once innovations in sound- and fire-proofing drilling rigs reduced the danger of oil development. In others, such as Boyle Heights and, at least temporarily, Signal Hill and Venice, oil discoveries resulted in the removal of residential neighborhoods or halted residential development in progress.

The impact of oil drilling on Los Angeles also left profound political legacies. The first surfaced in the 1920s, as oil companies sought to reshape public outrage about beachside and pier drilling into state regulations that enhanced major oil companies' control of Los Angeles resources. In the 1930s, the competition between major and independent oil firms led to national debates over oil conservation and rights to the tidelands. The key problem here, and the root of the particularly chaotic drilling practices in the city fields going back to Doheny's first well, lay in California's approach to property rights and oil. California used the rule of capture for petroleum rights, which awarded ownership of oil to whoever pumps oil out of the ground and into a pipeline or barrel. This nearly always sparked an oil rush, as nearby property owners dashed to drill and pump, too, lest they lose out. Eventually, most states replaced the law of capture with "correlative rights," a system of rights that encourages unitization and cooperative oilfield development.[65] In Los Angeles, the law of capture not only led to the kind of oilfield development most incompatible with residential or commercial land use, but also increased the bitter competition between major oil companies and their smaller, independent rivals. The major oil companies used the specter of oil fires, chaotic development, and unclear mineral rights to promote legislation that slowed oil drilling and gradually increased minimum distances between oil wells. These proposals had great public benefits, as the regulation of shoreline drilling in Los Angeles suggests, but they also contributed to concentration, if not monopoly, in the oil industry.

One side-effect of this, in Los Angeles, was a challenge to state management of offshore oil reserves. Beginning in 1924, independent oil companies began to apply to the Department of the Interior for near-shore "Tidelands" leases under the Federal Mineral Leasing Act.[66] These smaller firms looked to the federal government because California's slant- and pier-drilling laws now restricted their access to their state leases, and because they believed that the major oil firms had manipulated the California legislature to exclude them from tidelands drilling.[67] The Department of Interior ignored these applications for many years, until Senator Nye and Harold Ickes both began to pursue federal control over all underwater minerals in the United States, including tidelands oil.[68] The creation of a federal naval oil reserve in the tidelands did not ultimately succeed, but it did spark immediate controversy. The House Judiciary Committee called the proposal "the entering wedge" of federal claims to state resources.[69] The state of California passed the California State Lands Act, asserting state control of mineral leasing. The outbreak of World War II halted any real action on the tidelands, but the issue reemerged in 1945 when the Senate and House failed to override Truman's veto of a quitclaim bill that would have renounced federal claims to offshore oil in Texas, Louisiana, and California.[70] The issue reappeared regularly in Congress and even in the Supreme Court until 1953, when President Eisenhower finally put an end to the debate with an executive order that ceded to the states all offshore resources within three miles of their coasts. By this time, the independents were very small players indeed, and the issue of tidelands oil was no longer cast in terms of monopoly at all.

Although the tidelands battle moved away from Los Angeles pretty quickly, it reflected lingering distrust of the oil industry, one legacy of the Southern California boom. From 1920 to 1950, oil companies embodied corporate malfeasance much as railroads and utilities had in the Progressive Era. The rhetoric of private exploitation that echoed through accusations that oil wells violated the public's rights to the beach expressed profound distrust of the oil industry. This distrust reappeared during World War II, in public outrage that the oil companies refused to end production in residential areas at the end of the war. Skepticism and frustration turned to scapegoating when air pollution emerged as a major regional problem. In the mid-1940s, acrid clouds of industrial fumes and automobile exhaust first coalesced into smog in downtown Los Angeles. The Los Angeles Chamber of Commerce and elected officials immediately sought to reduce emissions from burning garbage dumps, orchard smudging, and backyard incinerators. The Los Angeles Chamber of Commerce, in particular, mobilized a voluntary and unusually effective effort to reduce smoke from Los Angeles's many factories. But the public blamed the oil refineries, and continued to do so for years after chemists proved that automobiles were to blame.[71] This persistent distrust, too, was a legacy of the problems and presence of the oil industry in Los Angeles's neighborhoods.

PART II   Distant yet Central?

Perth, Calgary, and Stravanger

The U.S. energy capitals discussed in the prior section emerged as longstanding centers of production because of their proximity to the natural resources and because they possessed transportation networks, adequate capital sources, and the necessary business and political elite that allowed them to control the harvesting, processing, and distribution of those resources. Another group of cities, such as Calgary (Canada), Stavanger (Norway), and Perth (Australia) emerged as energy capitals despite their distance from initial refining locations, their late entry in the industry, the absence of significant manufacturing within their boundaries, and their remoteness from later oil discoveries. Instead, their business and governmental leaders situated their cities to capture the financial and administrative functions of the oil and gas industry in their respective countries when new resources were discovered, giving these metropolises new economic and political power and earning them the title of energy capital.

In discussing Calgary, Matthew Eisler distinguishes energy capitals from other industrial cities by recognizing "hosting economies that are dependent in varying degrees on the extraction and export of energy." Despite its distance from the most productive natural resource fields in the Province of Alberta and despite the fact that it lacks significant petroleum processing, manufacturing, or transshipment facilities, Calgary qualifies as an energy capital.

Like other contributors to this volume, Eisler illustrates the importance of the political culture in the emergence of the metropolis. In the 1950s and 1960s, the city's conservative political and business elites worried about the social implications of forming a traditional industrial base; they feared working-class politics that might result from such economic activity. Instead, they chose to maintain the status quo and recruited major multinational corporations to develop Alberta's petroleum resources, including a few deposits near Calgary. After these companies had harvested the most easily recovered resources, Calgary-based independent companies emerged in the mid-1970s to extract the remaining marginal reserves. To do so, they relied upon financial support from the state and the development of new, science-based technologies. Other state support included the building of "a large post-secondary research complex specializing in geological and health sciences and introducing a favorable royalty regime that incentivized marginal resource extraction." Thus, while Calgary entrepreneurs kept the more environmentally destructive aspect of energy development distant from their city (many of the refineries are located in the provincial capital of Edmonton), they fostered the growth of their metropolis through research and development and the influx of a well-educated professional class.

Gunnar Nerheim introduces us to Stavanger, Norway, a city that experienced early economic and population growth in the late nineteenth and early twentieth century with its booming sardine canneries, an industry that declined after World War I. The local shipyard saw a revival after World War II with the construction of large oil tankers, but the growth of Japanese shipyards brought stagnation in the 1960s when a new opportunity arose for the city. Various multinational oil corporations chose Stavanger as a convenient location to center their offshore operations on the Norwegian Continental Shelf. Geological explorations of the North Sea revealed the presence of oil fields and the British and Norwegian governments began to award blocks for drilling. Lacking experience in the oil industry, Norway initially relied on the expertise of established corporations to locate and develop its offshore resources. By 1972, however, Norwegians became increasingly concerned about foreign control and the Parliament established a state-owned oil company, Statoil, and the Norwegian Petroleum Directorate to oversee petroleum development in Norwegian waters. The government headquartered both entities in Stavanger just as the OPEC embargo sent oil prices soaring. As oil and gas revenue became an important part of Norway's gross domestic product, oil and service companies built their facilities in Stavanger providing a twelvefold job increase, although many positions went to foreign skilled workers in the early days. As Statoil matured into a dominant company in the Norwegian sector of the North Sea and a major international oil company, however, more and more Norwegians found positions in the industry.

When the impact of the global decline in oil prices hit Norway in the late 1980s, the nation and Stavanger learned the negative consequences of their dependence on oil revenues; Stavanger's fortunes were now closely intertwined with international political events and a volatile global commodities market. In the end, Nerheim observes, despite periodic ebbs in oil prices that result in temporary economic downturns, there is no incentive for oil network cities such as Stavanger, or North American energy capitals such as Calgary or Houston, to move away from the industry while oil remains the most important raw material on the planet.

Like Calgary and Stavanger, Perth earned its status as an energy capital relatively late in the global game. The city experienced steady if unspectacular growth from the late nineteenth century as the economic and political center of the state of Western Australia and through the development of the state's gold deposits. As a more traditional resource capital, Perth experienced typical boom and bust cycles. Since the 1980s, however, the city's population soared exponentially as Perth dominated the mining of iron ore, the development of offshore oil and gas fields on the northwest shelf of Australia thousands of kilometers away, and the export of such resources to large, fast-growing Asian markets. The absence of any other significant metropolis in Western Australia allowed Perth-based corporations to consolidate their capital and expand their influence. In addition to corporate investments, the state government of Western Australia has funded pipelines and other essential infrastructure, particularly for liquid natural gas used domestically for industrial purposes and power generation. Within Perth's more immediate hinterland, in the suburb of Kwinana just thirty-five kilometers to the south, the state government struck a deal with Anglo-Iranian Oil (the predecessor to British Petroleum) in the 1950s to facilitate the construction of a refinery there. As the industry expanded in Kwinana, feeding the corporate coffers in Perth and contributing to a high-rise skyline, however, rural communities adjacent to Kwinana experienced air quality problems that compromised the long-term health of many residents. As with all energy capitals, the costs and benefits of development were not equitably distributed.

Calgary, Stavanger, and Perth all entered the energy industry at a time when multinational corporations had already consolidated their interests and dominated the extraction, processing, and distribution of resources around the world, yet each city was able to maintain substantial control over much of the revenues generated from its energy hinterlands. Why, given this late date of entry, did they succeed? First, they carved for themselves more limited roles than their U.S. counterparts had enjoyed. These cities did not need to construct the physical infrastructure essential to petrochemical refining and manufacturing. Second, they possessed entrepreneurs and government officials capable of developing local and national investment capital and crafting policies that ensured that most of the revenues generated in their energy hinterlands remained in Canada, Norway, or Australia, and that much of it returned to their energy capitals. Although somewhat distant from the location of the key minerals, these three energy capitals experienced economic development and desirable population growth while often avoiding the pollution surrounding production.

Third, the cases of Canada, Norway, and Australia all highlight the continued interplay between industry and government in the creation of energy industries, and suggest that the greater roles played by their centralized governments in protecting national interests were essential to their success, given the late entry date. Fourth, Calgary, Stavanger, and Perth all developed their energy roles at a time when consuming countries were trying to counteract the growing power of the OPEC imperium.[1] Finally, at the time these three energy capitals emerged, Canada, Norway, and Australia enjoyed advantages that Mexico and Gabon, discussed in the next section, did not when they entered global competition: mature, stable governments; relatively diversified national economies; and relatively homogeneous and well-educated populations. Nonetheless, as Nerheim reveals, their status as energy capitals still left their economies subject to the price swings in the international oil marketplace.

Scoping Perth as an Energy Capital

Jenny Gregory

Perth is a city that owes its prosperity to mining. It has witnessed successive mining booms—first in the 1890s, then the 1930s, the 1960s, and the 1980s, followed by the long boom since the turn of the twentieth-first century that appears to have saved the nation from the worst of the global financial crisis of 2009. Each boom has left its mark on the city.

This chapter first provides a brief overview of the history of mining booms in Western Australia and their impact on the state capital, Perth; second, outlines the history of resource industries—coal, gas, and oil—in Western Australia; and third, examines the development and impact of the regional center of Kwinana as a case study of the influence of energy resources on the city. Kwinana is to the south of Perth and was established as a new town in conjunction with the establishment of the British Petroleum Oil Refinery in 1954. Within decades it had been subsumed by the metropolitan area of Perth. The proximity of Kwinana to the refinery subsequently resulted in environmental problems. Last, this chapter poses questions about the nature of Perth as an energy capital.

Mining Booms in Western Australia and the City

Founded in 1829 as the capital of the Swan River Colony—a private British settlement—Perth initially languished. The introduction of male convicts by the British government in 1850 provided a much needed free labor force to build an infrastructure of roads and public buildings, and British funds to pump prime the economy.[1] But even in 1883, when the population of Perth had crept up to just over six thousand, a visitor could comment "you feel yourself more out of the world in Perth than in Siberia."[2]

The promise of mineral treasures in the Colony became apparent in the late 1880s, and with early discoveries of gold came self-government in 1890. The big finds of 1892 established the Eastern Goldfields, one thousand miles inland from Perth, as one of the world's major goldfields. Both gold production and the quality of gold mined peaked in about 1903 but, although that first rush gradually lost its intensity, substantial levels of gold mining continued.[3]

The gold rush and its aftermath had a spectacular effect. Western Australia's population rose from nearly 50,000 in 1891 to over 184,000 in 1901; and Perth's population, which reached 70,700 in 1901, nearly doubled to 111,400 within another decade. Perth had become a modern commercial city. A host of magnificent new public buildings had been completed, and private investment resulted in the construction of many banks, insurance, commercial, and retail buildings, as well as numerous hotels and theaters. The American City Beautiful movement influenced Perth's civic agenda, so that the city's built fabric was complemented by an increased number of parks and gardens. By 1911 Perth was ringed with suburbs. Suburbia developed rapidly in the 1920s, fed by immigration from Britain and natural population growth and shaped by new tramways, bus routes, and the motorcar. The availability of transport, location of industry, topography, and the consequent price of land, meant that Perth's suburbs increasingly reflected a pattern of residential segregation by class.

Western Australia was saved from the worst of the Depression of the 1930s by gold. With the depreciation of world currencies, the price of gold began to increase after 1930 and investment funds were attracted into gold mining companies, leading to a mining resurgence in the Eastern Goldfields. The flow-on effect of the resulting boom meant that some Perth people were cushioned from the depressed economic conditions of much of the western world. Many new buildings were erected in the city, including numerous cinemas, and a suburban building boom began in well-to-do-areas.[4]

The year 1962 marked a coming of age for modern Perth, when astronaut John Glenn, the first American to orbit the Earth, dubbed it "City of Light," after Perth people left their lights on for him as he flew overhead. The city thus gained international exposure and Perth's lord mayor a place near the head of New York's tickertape parade for the homecoming astronaut. At the time Western Australia was on the cusp of another mining boom, this time based on nickel and iron ore and facilitated by government development policies. During the twentieth century the state's population rarely grew at more than 2 percent each year, but in 1961 it grew by 3 percent and in 1971 by nearly 6 percent. As the 1960s and 1970s unfolded, Perth's population grew by natural increase as well as by rising intrastate migration, with people from other parts of Australia largely attracted by opportunities in the mining sector. Investment in mining and industrial development and the prosperity generated by mining exports enabled major infrastructure to be developed for the city. At that time a metropolitan regional plan, which still provides the blueprint for urban development today, passed by state legislation. As a result, regional centers were planned and a major freeway system was constructed with new bridges spanning the river, facilitating the development of Perth's southern suburbs. The appearance of the city was once again transformed with the erection of high-rise office blocks.

Today's city skyline, however, is largely a creation of the past thirty years. The city center has always reflected the state's resource-based boom-bust economy, and its economic cycles can be discerned in the city's built fabric. In the mid-1980s the city was again remade, the population of Perth topped one million, glass towers dominated the city, inner-city suburbs had been gentrified, and suburbia sprawled farther out. The population again began to grow rapidly, this time fueled by the continued development of iron ore mining and the discovery of offshore oil and gas in the northwest of the state.[5]

Cast as a "lotus land" back in the 1970s, the sprawling city of Perth and its suburbs were described in 2001 as "a billboard for the Great Australian Dream" and, according the *Economist*'s 2002 survey, it had become the third most livable city in the world. At that time it was on the cusp of another boom, this time fueled by China's demand for mineral resources. In 2012 it was said that people were arriving in Perth at the rate of one thousand per week.[6] Five-year inter-censual figures reveal that the population of the metropolitan region grew by approximately 10 percent between 1991 and 1996, 7.2 percent between 1996 and 2001, and 8.4 percent between 2001 and 2006. In 2011 Perth had the fastest growth rate of all Australian capital cities. Growth rates of more than 25 percent were recorded in many newly developing fringe suburbs.

Although much more research is required to unpack these trends and to look more closely at changes in the city and its central business district (CBD), it is clear that the city has been heavily dependent on Western Australia's mineral wealth for its prosperity. It is difficult, however, to differentiate between the respective influences of its various mineral treasures—predominantly gold, coal, iron ore, and natural gas—and this requires a great deal more research and analysis. The history of the energy resources of the state, however, can be outlined.

Coal

Coal has been the major source of energy in Western Australia for more than a century and the city's power has been dependent on the burning of coal from the Collie coalfield, 204 kilometers (127 miles) south of Perth. First discovered in 1883, the field was declared in 1896, spawning a number of small mines. By 1920 they were rundown and in need of an injection of capital. Perth entrepreneurs Robert Lynn and Walter Johnson bought them, consolidating them as Amalgamated Colleries. The company was awarded the contract to meet the state government's coal requirements and monopolized the output of coal from Collie for a decade, making a fortune for Lynn and Johnson.[7] In 1927, Griffin Coal Mining Company, made up of local businessmen, miners, and prospectors, joined the company. A Royal Commission in 1931–33 investigating the production costs of coal reported that Amalgamated Collieries paid abnormally high wages to its miners and was severely critical of the firm, which it concluded had been making unfair profits at the expense of the government. A third company, Western Collieries, joined Amalgamated Collieries in 1949. Following the awarding of state government contracts to Western Collieries and Griffin in 1960, Amalgamated Collieries ceased operations. The last underground mine closed in 1994 and all operations are today open-pit. Today 80 percent of the coal produced at Collie is used for domestic power generation in Western Australia, and the remainder for industrial purposes including mineral processing.[8]

Strikes and union unrest in the years following World War II generated considerable concern at the vulnerability of a power system that relied on one source of fuel. Perth's electricity and gas supplies were dependent on coal, as was much public transport because the trains were fired by coal and the trams ran on electricity. The Collie coal companies could dictate prices, and the price of coal doubled between 1949 and 1953, continuing to rise throughout the 1950s. It was against this background that the search for other energy resources commenced.

The Transition to Oil

As a result of concern about the city's dependence on a single energy resource, the Kwinana Oil Refinery was established thirty-five kilometers south of Perth. Built by the Anglo-Iranian Oil Company (later British Petroleum) to process crude oil from the Middle East, it began operations in 1954. In the following year oil-firing capabilities were introduced to the major city power station at East Perth, and the South Fremantle Power Station was converted to use oil instead of coal.

In the mid-1960s at a price of about AU$2 a barrel, oil was cheap and the state government decided to build an oil-fired power station at Kwinana. It was completed just in time for the oil crisis of 1973, when the Organization of Petroleum Exporting Companies (OPEC) imposed an oil embargo and prices increased fivefold over a three-year period. As a result the Kwinana Power Station was converted to burn coal as well as oil and gas. The price of oil was so high after the second world oil price crisis in 1979, that the cost of the conversion was recouped within two years. The other major Perth power stations were converted back to coal.[9]

The oil crisis generated considerable interest in alternative sources of energy. Nuclear power was considered, but rejected, and, although on a small scale, wind and solar power trials began.[10] Far more significant, however, was natural gas.

Oil and Natural Gas

Oil and natural gas exports make a major contribution to Western Australia's export market. But although exploration was undertaken throughout the state for more than a century, no significant finds were made until the 1960s. It was not until the state's 1936 Petroleum Act was amended in 1951 to remove the onerous conditions that discouraged private investment in exploration, that a joint venture agreement between Caltex and Ampol was signed to form a company to undertake exploration in Western Australia. The new company West Australian Petroleum Pty Ltd (WAPET), headquartered in Perth, was registered in 1952.

It was WAPET (now Chevron-Texaco) that made Australia's first successful oil strike on the North West Cape, approximately 1,270 kilometers (794 miles) north of Perth. The first test hole, Rough Range 1, struck oil on September 5, 1953, to the astonishment of the drilling crew. At that time the nation imported AU£72 million (equivalent to approximately AU$2,234.5 million in 2011[11]) on imported petroleum products. Huge excitement followed the strike, with portly Sir Arthur Fadden, then federal treasurer, running through the corridors of Parliament House in Canberra, the nation's capital, waving a telegram and shouting "we've struck oil!" Following more drilling, however, WAPET was forced to conclude that the field was only a few acres in extent and that it had been a "hole in one." The hopes of the nation, to say nothing of shareholders who had seen their shares skyrocket in value (brokers in Perth had seen nothing like it since the gold boom of the 1930s), were dashed.[12]

WAPET had more success on other fronts. In 1964 WAPET discovered a small oil field on Barrow Island off the North West Shelf that went into production in 1967 and, although still producing oil today, peaked at 50,000 barrels per day in 1971. WAPET also discovered the Dongara gas field farther south, some 350 kilometers (215 miles) north of Perth in 1966. Although reserves were relatively small, in 1971 a pipeline was built from Dongara to Perth. From that date gas appliances were converted across Perth so that natural gas could be used widely by both domestic and industry consumers.

Major gas discoveries followed from the Barrow Island oil discovery, with the North West Shelf proving to have remarkable reserves of natural gas. Woodside Energy had been awarded exploration rights in 1963 and with Shell and Burmah Oil had formed the North West Shelf Venture (NWSV) to explore for offshore petroleum and gas. In the early 1970s the NWSV found significant quantities of natural gas and condensate at a depth of between 125 and 131 meters (410 to 430 feet) in the Carnarvon Basin off the Pilbara Coast approximately 125 kilometers (77 miles) northwest of Dampier in Australia's northwest. Further exploration indicated the massive scale of these discoveries.

Development slowed in the 1970s, as a result of the Federal Labor Government's policies that envisaged Federal purchase of all North West Shelf gas. With the return to power at the state level of Charles Court's development-minded Liberal National Party Government in 1974 and the fall of the Federal Labor Government the following year, an Action Working Group was formed at the state level, with representatives from the NWSV and the State Energy Commission to review future needs for natural gas. They concluded that needs would be less than originally thought and that the state government, rather than private enterprise, would have to build and operate a 1,500-kilometer (932-mile) pipeline from the northwest to Perth because it had access to funds at lower interest rates and could deliver gas at a cheaper cost-for-service basis. At this time there was considerable debate about the overseas sale of oil and gas reserves, but with the federal government's agreement that NWSV natural gas could be exported as liquefied natural gas (LNG) and gas liquids, the stage was set for development.

The infrastructure constructed to exploit these resources included the building of offshore production platforms, subsea infrastructure, and a network of pipelines and trunk lines. It culminated in the construction of the North West Shelf Gas Plant, loading facilities, jetties, and shipping and associated infrastructure, on the Burrup Peninsula near Dampier. The project, costing AU$12 billion, was then the largest private-sector construction project in Australia. It was commissioned in 1984 and has been progressively developed, so that by 2008 it had five LNG trains in operation, each providing between 2.5 and 4.4 million tons of gas per year.[13] The joint venturers entered into a twenty-year "take or pay" gas supply contract with the Western Australia Government under which the government also funded the building of the Dampier to Bunbury Natural Gas Pipeline, at 1,530 kilometers (950 miles) the longest natural gas pipeline in Australia. Gas turbine stations were then constructed at Mungarra, near Geraldton, and Pinjar, just north of Perth.

The NWSV project was Australia's first LNG effort; it grew the market and led the way for further exploration and development. It has now delivered natural gas to the domestic market for more than twenty-five years, has diversified electricity generation, and fueled the expansion of the mineral industry. The composition of the NWSV changed in 2008 when Woodside acquired Shell's interest in the project. It is thought that to date approximately a third of the NWSV reserves have been produced.[14] By 2008 it had generated investments totaling approximately AU$27 billion.[15]

In 2008 ACIL Tasman was commissioned to prepare a research paper to assess the economic impact of the NWSV. The resulting paper drew on the work of economists Ken Clements and R. A. Greig, who had used the NWSV as a case study in 1994 to model the economic impact of large resource development projects. Their research was followed by Access Economics in 1997 and 1998.

In each of these research projects, most of which were commissioned by the NWSV, the emphasis was on the national and state impact of the NWSV. The most recent research, ACIL Tasman's 2008 study, concluded that, during the period between 1989 and 2009, the NWSV project generated export revenues of AU$80 billion, increased gross domestic product (GDP) for Australia by more than AU$70 billion, and for Western Australia by approximately AU$90 billion. They also showed that in 2007 the total tax revenue, in royalties and excises, had been AU$887 million to the Commonwealth and was estimated to rise to AU$2,528 million in 2013, and to the state government, it had been AU$199 million in 2007 and was estimated to rise to AU$256 million in 2013. The impact on employment was estimated to have been more than ten thousand full-time equivalent jobs nationally in the peak construction period of 2000 to 2005, with the majority of these jobs located in Western Australia, and up to an estimated forty thousand jobs projected nationally in the years from 2011 to 2020, with about two-thirds occurring in Western Australia.[16]

More detail was provided on the contribution of the NWSV project domestic gas to Western Australia. Prior to the NWSV, natural gas consumption in Western Australia was only about 5 percent of 2008 levels and was limited to comparatively small amounts extracted from the Perth Basin. The development of the NWSV enabled the domestic market for gas to grow rapidly. The NWSV Plant is Western Australia's largest producer of domestic gas, providing 67 percent of the total. The second major source is the Carnarvon Basis, where Apache Energy's Varanus Island provides 30 percent of the total, with smaller amounts from Chevron at Thevenard Island and Origin Energy Resources at Onslow. The Perth Basin, which provides 2 percent, includes Australian Pipeline Trust's plant located at Dongara, AWE's plant at Woodada and Origin Energy Resources at Beharra Springs.[17]

The vast majority of natural gas in Western Australia is used for industrial purposes and power generation. Five customers—Alcoa (alumina refining),[18] BHP Billiton (resource processing), Burrup Fertilisers and Alinta and Verve Energy (power generation)—account for 90 percent of gas consumption. The residential and commercial sectors, reflecting that the state's population is only 2.26 million, use only 3 and 1 percent, respectively. Electricity generation accounts for 46 percent, nonferrous metals (alumina and nickel) 27 percent, basic chemicals 11 percent, other manufacturing 6 percent, gas pipeline transport 4 percent, and mining 2 percent.[19]

More recently the government has signed an agreement with Chevron for the development of the giant Gorgon gas field. Located on Barrow Island, the estimated AU$50 billion project, owned by Chevron (50 percent) and ExxonMobil and Shell (25 percent each), is the biggest LNG project being planned and indeed the largest development of any sort in Australia. It is expected to produce 15 million tons of LNG per year from 2014 and has already produced more than AU$6 billion in contracts and 3,000 jobs to industry. It is forecast to deliver an estimated AU$33 billion into the Australian economy during its lifetime. As a result, major companies, like BHP Billiton, have transferred their petroleum divisions to Perth, and there are now about thirty global oil and gas companies and more than forty service companies with offices in Perth.

The signing of the agreement to develop Gorgon came in August 2009, as well as other LNG projects coming online. These include the immense Browse Basin gas field off the Kimberley coast now in the early stages of development by Woodside Petroleum Ltd., led Colin Barnett, the state premier, to predict that Western Australia was "becoming the Saudi Arabia of natural gas" and, to quote an American petroleum executive who told him that, "'Western Australia' offshore oil and gas reserves were at the stage the Gulf of Mexico was 30 years ago."[20]

However, the location of Woodside's proposed gas hub for the Browse Basin gas field at James Price Point in the Kimberley has been mired in controversy. A vocal protest movement has delayed development, with opponents to the project citing likely environmental damage to the Kimberley, one of Australia's most pristine wilderness regions.[21] The state's Environmental Protection Authority (EPA) gave the green light to development in July 2012, but four members of the board bowed out of proceedings, citing conflicts of interest and leaving its chair to make a sole decision. State and commonwealth environment ministers, who have the responsibility for making a final decision on the precinct's environmental approval, have yet to consider the EPA report and recommendations.[22]

Minerals and Energy and Manufacturing

Looking back over the state's minerals and energy history, it becomes clear that attempts to establish resource manufacturing plants in Western Australia have not always met with success. This is of significance for the state's long-term economic future. The most successful attempts have been in gold refining, the manufacture of pig iron, alumina manufacturing, and nickel refining.

The Perth Mint was established in 1899 as a branch of the Royal Mint in London, to refine gold from the State's newly discovered Eastern Goldfields and strike gold sovereign coinage for use throughout the British Empire. In 1970, ownership of the Perth Mint transferred to the government of Western Australia, which created the parent company Gold Corporation to manage its operations and expand its activities interstate and internationally. The Perth Mint currently refines the total annual production of gold in Australia, as well as gold ore from New Zealand, Papua New Guinea, Fiji, Thailand, and Malaysia. For many years located in Perth's CBD, the Perth Mint moved its refinery to an industrial suburb in 1990.[23]

During World War II, in an attempt to develop an iron industry fueled by charcoal (from timber not suitable for logging) and locally mined iron ore, the State Labor Government established a pilot plant at Wundowie, 59 kilometers (37 miles) northeast of Perth in close proximity to rail transport and the Goldfields Water Supply pipeline, in the hope of supplying the state's needs for pig iron. Production began in 1948 but the quality of the local iron ore proved so poor that not until the discovery of high-quality iron ore at Koolyanobbing (more than 300 kilometers [186 miles] east) which could be railed to the site, could high-quality pig iron be manufactured. In the mid-1950s Wundowie supplied the state's needs for pig iron and had developed an export market. However, successive governments were unwilling to invest in modernizing the plant and a Liberal government sold it to a private company in 1976. A shortage of timber and increasingly obsolete plant marked the death knell for the manufacture of pig iron at Wundowie, which ceased operations in 1981.[24]

More successful have been alumina and nickel refining. Following the commencement of bauxite mining in the Darling Ranges just east of Perth by Western Mining Corporation (WMC) in 1960, the company developed a partnership with Alcoa to form Alcoa of Australia, then making an agreement with the state government to develop an alumina refinery at Kwinana and beginning production in 1963. A second refinery was constructed at nearby Pinjarra in 1969 and is supplied by bauxite from the world's largest bauxite mine at Huntly near Dwellingup south of the original mine. Nickel has also been a success story, with WMC's Kambalda nickel concentrates railed to the company's nickel refinery at Kwinana that commenced production in 1970 and is now owned by BHP Billiton.[25]

Less successful were attempts to establish a petrochemical plant and a hot briquette plant. During the 1980s the Labor government bailed out private entrepreneurs who were developing a petrochemical plant at Kwinana. The project collapsed and the government, discredited by this and a number of other failed business efforts, fell.[26] In 1996 BHP began construction of a hot briquette iron plant at Port Hedland, the major iron ore port. The plant was opened in 1999 and was hailed as the first of its type to undertake secondary processing of Western Australian iron ore. After a gas explosion at the plant in 2004 killed one worker and seriously injured three others, the plant was shut down, leaving BHP facing litigation and a major taxation dispute.[27]

Whether the development of energy resources will assist in the further development of manufacturing industries remains to be seen. Controversy surrounds the cost of power to the domestic market, with costs increasing notwithstanding the availability of immense supplies of natural gas. Although very considerable mineral refining takes place, mainly centered on or near Kwinana, at present Western Australia's manufacturing industry is predominantly light manufacturing for the domestic market.

Kwinana

In the 1950s, a single short-lived oil strike had raised and then dashed the hopes of a nation. Then the riches from natural gas exports were not even dreamed of. But the grave concern about the city's dependence on a single energy resource, coal, led to the establishment in 1954 of the Kwinana Oil Refinery on the shores of Cockburn Sound, 35 kilometers (22 miles) south of Perth's CBD. This is Australia's largest oil refinery.

This development was largely due to the drive and confidence of Russell Dumas, then the state government's coordinator of Works and Industrial Development and dubbed "the bulldozer" by colleagues. He struck the deal with the Anglo-Iranian Oil Company (which changed its name to British Petroleum in 1954), who in 1951 had approached all Australian state governments regarding the establishment of a refinery in Australia.[28] Generous concessions to Anglo-Iranian were the key to Western Australia "winning" the refinery: cheap land in large quantities, exemption from harbor dues, roads, a rail link, water (three million gallons a day for free), electricity (at standard rates), sanitation, a temporary construction camp, and a housing estate for workers, as well as assistance with materials and migrant labor. The cost to the state was estimated at A£12 million (today roughly equivalent to AU$370 million), a remarkable amount that meant that all other public works in the state had to be suspended while attention was directed to the developments at Kwinana. At the same time Dumas also negotiated first with BHP (Broken Hill Proprietary) for a steel rolling mill that was planned to provide the foundation for an iron and steel industry, and then with the British Rugby Portland Cement Company. Both companies set up factories in the same area and were granted generous concessions. "The great thing is that it will be a major step forward industrially for Western Australia," said Dumas.[29]

As historian Lenore Layman has argued, the development ethos that Dumas introduced to government came to full fruition in the late 1950s, and as a result of this the state government persuaded the federal government to lift its ban on iron ore exports in 1960, thus paving the way for successive iron ore booms. Charles Court, minister for Industrial Development and later premier (and the recipient of a knighthood), was the heir to this approach, believing that resource development would lead to more industrialization and would provide the state with a "great industrial future."

Ultimately this has been difficult to achieve, as noted earlier. A legacy of a "four on the floor" Labor government in the 1980s was a failed 1988 government plan to build a petrochemical plant at Kwinana and the failed BHP Billiton hot briquette iron plant in Port Hedland, discussed earlier. The latter, designed to create a market for the fine ore mined with lump ore by turning it into briquettes, was fueled by cheap gas, made possible by the state government's deregulation of the gas market.[30]

In the mid-1950s, however, hopes were high. Gordon Stephenson, the British town planner engaged in 1953 to prepare a plan for the whole metropolitan area, wrote, "The establishment of an oil refinery and steel rolling mill at Kwinana will result in a flow of new materials and by-products which have not previously been made in the State. The two large, new industries are symbolical of the growing interest shown by industrialists in the State . . . [and] are indications changes in the nature and scale of industry to be expected."[31]

At the time Kwinana lay outside the metropolitan area, but Stephenson planned to connect the port of Fremantle with Kwinana. The railway was to be extended south so that goods could be transported to and from the projected port development in Cockburn Sound at Kwinana.[32] This, plus the expansion of water and gas mains and the electricity grid, would also facilitate the expansion of residential areas to the south, "chiefly because of its very attractive building land and its pleasant situation adjacent to fine sheltered beaches."[33] He also forecast that the employment capacity of Kwinana would be fifty-four thousand, that the number of resident factory workers would be ten thousand and that the number of other factory workers would also be ten thousand.[34]

By the time Stephenson completed his plan, the newly created town of Kwinana, built by the state government to house about forty thousand workers and their families for Anglo-Iranian Oil Company (later British Petroleum) refinery, was well under way. The first two neighborhoods were almost complete. It was buffered from the oil refinery on the coast about two miles west by a mile-wide belt of land reserved as public open space that could be used for recreation.[35] Stephenson stressed the importance of the buffer area between industry and residential development and noted the potential problems of "smells carried by the prevailing South and South-West winds."[36]

The key decision makers in shaping the development of Kwinana were the Anglo-Iranian Oil Company, the State Housing Commission (SHC), and consultant planner Margaret Feilman (the state's first female town planner). In her study of the development of Kwinana, Sarah Brown noted that under the 1952 Oil Refinery Industry (Anglo-Iranian Oil Company Limited) Act (WA) a number of terms were agreed to that shaped the new town site. An area no more than two and a half miles from the refinery site was to be used "solely for the purposes of a residential area, social centre and recreation ground"; 333 houses were to be erected in each of the three years, 1953 to 1955, to the State Housing Commission's standard design; and "roads, septic tanks, sewerage, fencing and water and electricity services necessary for the reasonable occupancy of the houses" were to be provided.[37] Brown argued that, although Anglo-Iranian did not want an "oil town," its major considerations were economic: it required a stable and productive workforce for the refinery, and to gain this it needed to provide accommodation.[38] The State Housing Commission, which had only been brought into existence in 1947 and was responsible for the provision of housing for low-income earners across the state, was also governed by the developmental ethos of a state government concerned to provide a stable and accessible workforce for industry, as well as budgetary constraints.

It was Feilman, Brown argued, who had "the most pervasive vision for the development of Kwinana." She based its design on the British "new town" model and it was arguably the first comprehensive example in Australia. But she was also heavily influenced by the design criteria for neighborhood units as developed by American planner Clarence Perry. The town occupied 7,560 acres and she located it on two north–south ridges of land, rising to 200 feet, with two of the four neighborhood units, Medina and Calista, on the western ridge and the other two, Parmelia and Orelia, on the eastern ridge, with a parkway and open spaces located in the valley between. Each neighborhood was to have about 1,000 housing sites and a local shopping and amenities center. But she also planned for Kwinana to have "an individual identity," and to be "an experiment in methods of site selection, traffic segregation and house siting in relation to open spaces" that would provide a basis for social relationships. By 1960, the town boasted a cinema, a shopping center, a community hall (completed by 1955), a high school, and a maternity hospital as well as a new railway line to Fremantle. A large proportion of the immigrants attracted to Kwinana were British. In the 1971 census 5,210 of Kwinana's 12,224 residents were born in the United Kingdom.

As time and funding were limited, the project was fast-tracked. Expenditure and expediency became the key concerns. Feilman, nevertheless, gave full consideration to environmental concerns: undertaking landscape site analyses, considering the conservation of native flora and fauna, and taking into account refinery fumes and potential pollution.

A snapshot of Kwinana's community profile in recent years is revealed by census data. Kwinana had a population of 22,935 in 2006, the majority of whom were born in Australia. Thirty percent of the labor force was in trade or technical occupations, 14 percent were semi-skilled, and 13 percent were laborers; 5.8 percent of the labor force were unemployed in 2006, though this increased to 9.4 percent in 2009, the second highest level in the metropolitan area.

To the north of Kwinana, but within the subsequently created local government area of Kwinana, are two older suburbs, Hope Valley and Wattleup, which until recently were semi-rural areas, used as grazing and agricultural land, including market gardening. They lie directly in the path of air pollutants from industry on the coast at Kwinana. The metropolitan area of Perth is located on a coastal plain between the Indian Ocean and the 300-meter-high (984 feet) Darling Range rising to the east. On summer days, the winds are generally easterly. Hence air pollutants from industrial areas are carried out to sea in the morning. The pollutants return again in the early afternoon with the onset of a strong sea breeze and continue to move inland until, by midnight, the easterly winds overcome the dying sea breeze and the pollutants return to the coast by 6 a.m., not very far to the north of where they were at 7 a.m. and again at 2 p.m. the previous day.[39]

With the growing awareness of environmental issues, air pollutants from the Kwinana industrial area—most significantly sulphur dioxide from the processing of crude oil at the refinery and the combustion of heavy fuel oil at the Alcoa alumina refinery and the Kwinana Power Station—had become apparent by the 1970s.[40] It was gradually recognized that the Kwinana region, which by then included a raft of heavy industry—power generation, refineries (oil, alumina, and nickel), iron smelting, cement works, and titanium dioxide and fertilizer plants—was a major source of pollutants, both airborne and in the waters of Cockburn Sound, where water quality had deteriorated and there had been widespread loss of sea-grass beds. BP, the major polluter, developed a range of environmental improvements during the 1980s.[41]

An air-quality buffer for the area was established under the Environmental Protection (Kwinana) (Atmospheric Wastes) Policy (EPA, 1992). Ten years later, a review was undertaken by the State Government Department of Planning and Infrastructure, the Department of Environment, and the Department of Industry and Resources, which concluded that the buffer could be reduced in size. There was a public outcry, which raised a number of concerns: the review was limited in scope, did not address emissions beyond sulfur dioxide, and was not scientifically based. There was also great concern about the health of residents, particularly in Hope Valley and Wattleup. As a result further studies were undertaken on industrial emissions and air quality in the Kwinana area, and it was recognized that "industry and associated infrastructure generate a range of emissions including noise, air emissions (gases and odors), light-spill and public risk and therefore adequate separation distances (buffer) to sensitive land–uses are required."

The government then announced that it would purchase houses in the town sites of Hope Valley and Wattleup on a voluntary basis and provide a relocation allowance to residents.[42] Following the introduction of the Hope Valley Wattleup Redevelopment Act in 2000, LandCorp, the state government's land development arm, began purchasing properties in the area as part of its role in planning, developing, and promoting some 1,400 hectares of land to be redeveloped for mixed industrial and commercial uses. It was branded the "Latitude 32 Industry Zone" in 2006.

Latitude 32, according to the promotional literature, "presented a number of exciting opportunities," including: "Development of eco-industrial park concepts to promote more efficient use of infrastructure and encourage the reuse and recycling of industrial by-products as inputs for other activities. Relocation of established industries to purpose-built precincts where they can profit through colocation with other businesses, which have commercial synergy. Establishment of activities to complement and support current and proposed bulk cargo facilities. A logistics hub accessible by road, rail or sea it will service the rapidly growing southern suburbs of Perth and the state's South-West region."[43] It was also to cater to an estimated ten thousand jobs, an important factor in an area of Perth by then noted for its high unemployment. In 2006 tenants and homeowners were given notice that the demolition of rental properties, which made up 90 percent of the properties in Wattleup, would be undertaken in stages over a twelve-month period.[44]

The reality of the situation, however, and the long-term health issues for those who lived in Hope Valley and Wattleup are made clear by comments on an online forum:

> I am now a proud Aussie but before coming here to WA I was a fit policeman. I joined the WA Police and achieved Dux 1/97 recruits at the age of 32 one of the older recruits. Then I made the mistake of moving into a house where the prevailing winds plume fallout from Alcoa Kwinana on a regular basis. Within 18 months I had severe hay fever and terrible skin allergies and then Crohn's disease. I have now had it for 15 years and have had 7 re-sections, I believe the pollution in the air from Alcoa is responsible. As a school-based policeman I used to go to camps in the southwest and within 48 hours of arriving down south my symptoms would cease and my skin dermatitis problems would clear up. On return to here back they would come. It is so bad now I have been medically retired unfit to do any work at the age of 47.[45]

> Were you in Hope Valley? I understand school children there had the highest respiratory problems & Hope Valley has been effectively closed down.[46]

> My parents lived in Wattleup with Alcoa not more than a couple of kms from their home. They have since died of cancer. My mother's cancer was apparently from smok-

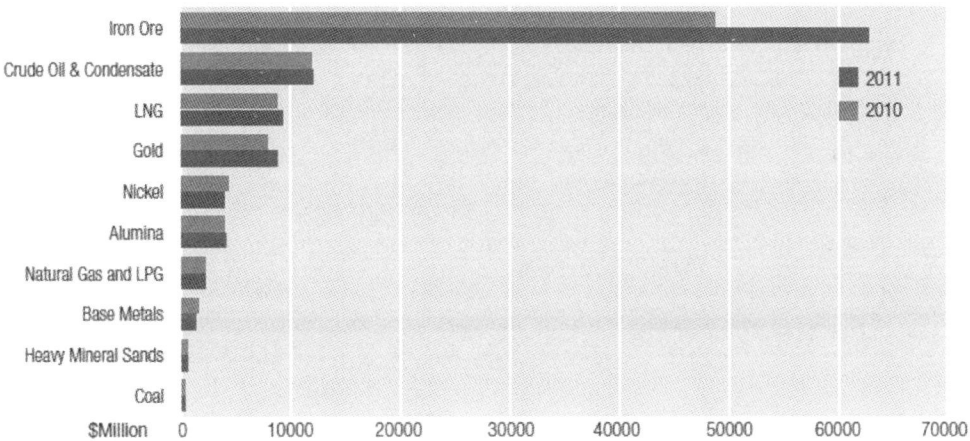

Figure 5.1. Major commodities in Western Australia by value, 2010–11.

Source: Perth: Department of Mines and Petroleum, Government of Western Australia, *Western Australia Minerals and Petroleum Statistics Digest 2011.*

ing but she never had a smoke in all her life. Our neighbour developed a serious skin allergy that was never diagnosed. Many other people in Wattleup have since passed away from cancer. I also wonder why it is that the Government are bulldozing the residential properties in Wattleup![47]

The local government area of Kwinana, which includes Wattleup and Hope Valley, is today home to approximately 30,000 people. According to the Town of Kwinana, its industrial area makes a major contribution to the wealth of the state, generating a combined annual output valued at $15.77 billion per annum, employing around 4,800 people, and providing indirect employment to another 26,000.[48] The Hope Valley story has been quietly forgotten, with most residents now living in other areas, and Latitude 32 Industry Zone now billed as "an economic powerhouse for the future."[49] The Kwinana Industries Council claims that its members have a high standard of environmental performance in the areas of cleaner production, waste minimization, energy efficiency, water conservation, and noise abatement, which provides significant benefits for the local community.[50]

A Research Agenda

This chapter has provided an overview of the history of resource booms in Western Australia and, in a broad sense, their impact on Perth; an outline of the history of energy resources in the state; and then has narrowed down to a case study of Kwinana, site of Australia's largest oil refinery and other industries, and its impact on the surrounding area.

What has not been discussed, except in passing, is the iron ore industry. In 2011, iron ore was Western Australia's major export, earning AU$62.8 million. It was followed by crude petroleum, earning AU$14.9 million, gold AU$11.5 million, and then natural gas AU$8.6 million.[51] Despite the hype that has accompanied recent natural gas projects, iron ore is still the largest resource industry in Western Australia, as is shown in figure 5.1.

What then is the research agenda for assessing the impact of energy resources on Perth, as one of the world's energy capitals? As has been indicated in this chapter, there has been virtually no research on the impact of energy resources on the city alone. The few research papers that exist focus on impact on the state and the nation. Furthermore, because of the dominance of iron ore, at this stage it is difficult to see how the impact on the city of energy resources can be separated from the impact of other mineral resources. The two are interlinked; further expansion fueled by the resource sector will increase the demand for energy, both now and into the future. This chapter represents an initial foray into the field. Clearly a major research project lies ahead—but whether it will focus on Perth as an energy capital or Perth as a resource capital is not yet clear.

Symbolic is the city skyline, thousands of kilometers south of the mineral and energy resources being wrested from the earth and sea and exported by ship from northwest ports to China and Japan. For the past decade the skyline has been crowded with cranes. The high level of building licenses granted in the city has continued to increase steadily since 2007 and 2008, contradicting the trends of the global financial crises.[52] The newly completed City Square has become Perth's second tallest building with forty-six stories, at 244 meters (801 feet). With its distinctive bulk dominating the city skyline, it has provided much-needed floor space for Perth's expanding commercial sector, but most significantly it houses BHP Billiton's Perth-based iron ore, alumina, and nickel headquarters.

At Arm's Length

Energy and the Construction of a
Peripheral Prairie Petrometropolis

Matthew N. Eisler

Rising from the southwestern foothills of the Canadian province of Alberta on the edge of the Western Canadian Sedimentary Basin (WCSB), the northern portion of a vast geological formation occupying the heart of North America from the Gulf of Mexico to the Mackenzie River delta, Calgary is the headquarters of the Canadian oil industry. Over the last century, efforts to exploit the energy resources of this vast area have shaped the economy, built space, mythos, and culture of this frontier city and modern Canada. But it has been only since the early 1970s that Calgary began a remarkable rise from provincial backwater to a center of power and influence rivaling Toronto, the nation's largest city and financial headquarters. Influenced by tropes of rural populism that originated in the American West and Southwest in the late nineteenth century and diffused into western Canadian political discourse by the early twentieth century, Calgary's boosters tend to interpret this history through the lens of the conventional regional narrative. Accordingly, they root the dichotomic tale of plucky frontier town versus imperialistic metropole in a larger conflict pitting "East" versus "West." Indeed, urban and regional frames of reference are elided in the city's trademarked motto: "Calgary, Heart of the New West."[1]

When the historical pattern of resource recovery in the WCSB is considered, however, an alternate geography can be discerned. This would situate the city at the epicenter of a gradually expanding regional energy-exploitation frontier moving along a south–north axis, one that confined the worst environmental collateral damage to the province's hinterland. And although the leaders of the Canadian oil industry have managed this frontier from Calgary, lending the intranational-regional frame some validity, they and the built space in which they worked comprised a relatively minor node in the global energy industry, a system that redistributed the bulk of the wealth outside the region. Regional discourse serves to obscure this reality and the collaboration between provincial, national, and foreign industrial elites and governments in shaping Calgary as an energy capital.

Defining the Energy Capital

Progressive scholars of the Canadian West have generally been more concerned with problematizing regional discourse than in exploring Calgary's place in it. They tended to interpret the Canadian energy industry as the product of intense conflict between local, national, and international elites.[2] Such work reflected the longstanding tension between critical theoretical approaches and urban studies.[3] However, some relatively recent literature has combined urban and environmental history, exploring natural and artificial spaces as they related to class, race, and gender. Often these studies focused on the socioenvironmental effects of heavy industrialization in urban spaces, as well as the ecology of urban space more generally.[4]

Such approaches can help elaborate the socioenvironmental contours of Calgary as an energy capital. The concept of the energy capital is a not-unproblematic unit of analysis and a critical reading of it helps clarify how changes in patterns of energy recovery and use have shaped cityscapes over time. An obvious question is how to distinguish energy capitals from other modern industrial cities. Virtually all use energy in vast quantities, of course, and have elaborate energy conversion and transmission systems serving local dwelling, manufacturing, service, and administration spaces and transportation systems. But energy capitals may be distinguished by hosting economies that are dependent in varying degrees on the extraction and export of energy. As such, they feature two parallel energy systems, generally controlled by different interest groups: one serving local markets and another serving external markets.

Over the last 150 or so years, the main kinds of primary energy traded in large quantities in national and international markets have generally been fossil-based owing to their abundance, versatility, and portability. As chemically fixed forms of solar energy in gas, liquid, and solid states, they are easily stockpiled and transported. Accordingly, their large-scale use both requires and reinforces an industrial and, hence, urban space. When did the first specialized energy capitals appear? The case of Britain helps contextualize this question owing to the country's vanguard role in industrialization and large-scale energy exploitation. Coal began to be used for domestic heating in Britain from the mid-sixteenth century but the fuel did not become a major industrial input until the early nineteenth century. Just as coal-fired steam power drove the heavy phase of Britain's Industrial Revolution from around 1815, the export of British heavy industrial inputs from around the 1840s helped establish an international coal-based energy market along with imperial hegemony.[5] In Britain, the United States, Germany, and other industrializing countries in this period and afterward, coal-producing areas were often subsumed within larger heavy industrial complexes or regions. Relatively isolated coal-mining towns, on the other hand, meet our baseline criteria of an energy capital, an urban area with a specialized economy based on energy exports.

However, another kind of specialized energy capital emerged from the expansion of the international energy market over the last century, one based on petroleum and natural gas. The economies of most of the energy capitals in this volume were or remain strongly linked with the extraction and export of these fuels. In most if not all cases, these cities were founded before such resources were discovered in their environs. Nevertheless, in different ways, the exploitation of fossil energy has had profound consequences on the pace and nature of their growth.

Energy capitals have varying characteristics. In constructing a typology, we might begin by identifying such cities in rich and poor countries. As energy and environmental scholar Vaclav Smil cautions, however, there is much intragroup variation within the rich/poor divide. Class variegations certainly exist within regions of relatively wealthy countries.[6] Accordingly, energy capitals may be further classified along lines of functionality and workforce composition. The energy sectors of some may be predominantly industrial, like Edmonton and Baton Rouge, or predominantly administrative, such as Calgary and Dallas. Houston is an exceptional case, having both attributes. Nevertheless, all of the latter three cities house large concentrations of head offices or major regional offices of oil companies and relatively large white-collar workforces that play an expansive role in the planning of energy markets. These three energy capitals also have cultural bonds, being dominated by conservative business communities that share a frontier agrarian heritage with which they strongly identify. To some degree, these communities constitute an international class of petroprofessionals that migrates between and works in these and other energy capitals.[7]

In notable ways, however, modern Calgary is distinct. Unlike Houston but like Dallas, it lacks significant petroleum processing, manufacturing, or transshipment facilities within its environs. Calgary was perhaps unique for the diversity of energy resources that once existed or remain in relatively close proximity including coal, oil, natural gas, falling water, biomass, and wind. However, most of the resources of the WCSB were located in central and northern Alberta, the eastern slope of the Rocky Mountains, and northeastern British Columbia. As elsewhere, the timing of their exploitation has been determined by their proximity to markets and ease of access, and the social relations of resource extraction. The discovery in 1947 of major oil reserves near the Alberta capital of Edmonton, three hundred kilometers north of Calgary, presented unique challenges to the southern city's petroleum class. Unlike Dallas or Houston, Calgary was and remains an isolated city with neither a large market of its own nor significant hinterland markets. If the city's petroleum entrepreneurs wished to flourish, they had to access the big metropolitan centers of the United States and eastern Canada. This meant transcontinental pipelines—projects dictated by the imperatives of the major integrated multinational oil companies. More generally, the Canadian oil industry followed in the shadow of its giant American cousin.

These factors combined to shape Calgary as a junior headquarters energy capital. Popular political discourse framed eastern Canada as an imperial antagonist of the West, yet Calgary oil elites and Alberta political elites were not aligned in a monolithic bloc.[8] Sharing conservative political views, they had significantly diverging material interests. Canadian petroentrepreneurs sought capital from a variety of sources and assistance from the provincial government, but also sometimes welcomed federal intervention if it worked in their favor. Despite promoting populist rhetoric, Alberta's conservative political dynasts of the 1950s and 1960s worried about the undesirable social effects of resource development, above all the possible formation of a traditional industrial base, a large working class, and, thence, working-class politics.[9] Reluctant to alter the status quo, they favored the international oil majors, as did the federal governments of the time.

As a result, these companies dominated the early Canadian oil industry, extracting the lion's share of the best and most easily recovered resources in the WCSB, including important deposits near the city of Calgary. Calgary-based independents did not come into their own until after most of the low-hanging fruit had been plucked by the mid-1970s. They required science-based techniques and technologies in order to extract the remaining marginal reserves and, accordingly, depended heavily on the state. In this period, a new, dirigiste provincial government bolstered the independents through a variety of means, supporting development of a petrochemical base and a large post-secondary research complex, specializing in geological (and health) sciences, and introducing a favorable royalty regime that incentivized marginal resource extraction. These sociohistorical and geophysical factors helped ensure that the most environmentally damaging energy recovery activities unfolded far from the city of Calgary itself.

Prairie Penury and Plenitude

Calgary only gradually acquired the traits of the energy capital. The exploitation of petroleum in its hinterland began as a by-product of the colonization of western Canada by eastern financial elites and was intertwined in what began as distinct interest-group politics coalescing around oil and gas respectively. As in other energy capitals, these groups have coexisted uneasily. As the Canadian Pacific Railroad pushed Canada's first transcontinental rail link through semiarid southern Alberta in the 1880s, its periodic, inadvertent discoveries of natural gas in its search for water set the stage for concerted gas exploration in the region in the first decade of the twentieth century. Gas strikes in 1908 near Calgary and then in 1909 and 1911 in the southeastern corner of the province led to the formation of a number of gas companies. For urbanites across southern Alberta, cheap and clean natural gas quickly became an important service, used in some cases for lighting before the advent of the electrical grid but above all for heating, crucial in a region subject to long and harsh winters. The disposal of this versatile, clean-burning fuel, popularly seen as the basis of future industrialization in the province, would become a politically charged issue.[10]

An even greater prize lay in the Turner Valley, some sixty kilometers southwest of Calgary. There, in 1914, William Stewart Herron discovered naphtha, a high-quality light oil that could be used in the automobiles of the day without refining. This led to a number of additional finds and booms into the early 1920s that spawned a multitude of small Calgary-based oil producers. It also attracted Imperial Oil. Originating in southwestern Ontario, the birthplace of Canada's oil industry, Imperial was formed in the city of London in 1880 by sixteen independent producers and refiners to protect themselves from John D. Rockefeller's Standard Oil trust. Nevertheless, lack of capital drove the conglomerate into the arms of the monopoly, which acquired a controlling interest in Imperial in 1898. In the aftermath of Standard's breakup in 1911, Imperial was awarded to the reformed Standard Oil of New Jersey. The company's arrival in Alberta swung the center of gravity of the Canadian oil industry to the western prairie frontier, a region it and other large integrated multinational companies would dominate for decades afterward.[11]

In important ways, these developments mirrored the contemporary political economy of oil in frontier areas of the United States. David Breen captured these dynamics in his monumental history of the Alberta oil industry. In an unregulated environment governed only by the law of capture, a plethora of independent producers desperately competed for a share of a market dominated by a few large integrated players, resulting in overproduction and periodic price swoons accompanied by massive wastage. Whenever an oil field was discovered, capital-poor independents scrambled to produce as much as possible to pay off their investments.

This quickly destroyed reservoir drive—the pressure of water or gas released by drilling that drove oil to the surface—rendering much of the deposit unrecoverable. In the late nineteenth and early twentieth centuries in both Canada and the United States, early prospectors mainly sought oil, first for illumination and lubrication applications and subsequently to power internal combustion engines. Unlike municipalities and the companies that supplied them, oil companies regarded natural gas as a nuisance and treated it accordingly.

Ironically, the field feeding Canada's first oil boom was comprised largely of wet gas, a hydrocarbon mixture with a significant proportion of condensate or compounds heavier than methane. A half century later, such a field would be viewed as a splendid prize for all of its various high-quality resources. In the 1920s, however, oil operators were prepared to waste massive volumes of gas, a product the supply of which then outstripped demand. This was especially true during the Depression era, a time when Calgary was still a city of only eighty-five thousand in a province of around seven hundred thousand. The result was prolonged flaring on such an immense scale that it scandalized even Americans accustomed to profligacy in the Texas and Oklahoma fields.[12]

Popular sentiment, however, opposed gas export. Moreover, provincial governments were, on the one hand, reluctant to impose production quotas on the independents, or, on the other, force the majors, price-setting Imperial above all, to relinquish a larger share of the market. Such was the pressure for a solution, however, that the provincial government established the Petroleum and Natural Gas Conservation Board (PNGCB) in 1938. Using legislation based on the Texas regulatory system and devoted essentially to the Turner Valley fields, then the focal point of the Canadian oil industry, the PNGCB eventually succeeded in curbing some of the worst excesses. But the government's preferred solution—voluntary conservation among the many producers—proved impossible as long as Imperial had a monopoly. And provincial bureaucrats were not prepared to challenge the company. Only in 1944 did they enforce conservation by allocating shares for all producers in a gas market hitherto dominated by Imperial. As elsewhere, considerable damage had already been done by the time authorities intervened to eliminate oilfield waste. Nearly 75 percent of the natural gas produced in the Turner Valley up to 1946 was destroyed.[13]

Nevertheless, this deposit fueled Canada's first modern petroleum boom. And its proximity to Calgary crystallized that city as the base from which the petroleum resources of Alberta, and, later, Canada, would be managed. In addition to hosting the offices of the PNGCB, Calgary became the headquarters of petroleum interests doing business in the WCSB. As Breen notes, the oil industry's proximity to Calgary shaped the city's early development a full generation before this sector marked Edmonton, the provincial capital.[14] As a result, the southern city's regulatory and business elite dominated political decisions relating to energy.

Urban Energyscape

Calgary's municipal energy infrastructure could be defined in terms of two chief characteristics. One is the enduring domination of the private sector in producing electricity, consonant with the city's pervasive free market ethos. The other is the general paucity of plants in the city's environs. There have been a few exceptions. In the early 1920s, Imperial built a refinery in Calgary connected by pipeline to the Turner Valley.[15] In the early twentieth century, a few power plants were also constructed within city limits. These were the product of fierce debate over the merits of city ownership of utilities, with proponents arguing that it would result in lower rates for users and provide an important revenue base. In 1904, a time when Calgary was served by the private Calgary Water Power Company, a citizens' petition bolstered advocates of public ownership on city council. The result was a decision to develop the City of Calgary Electric System. Over the next twenty years, this municipally owned utility constructed and operated several of its own small coal-fired steam plants in the city. Built in 1910, the Victoria Park plant, for example, was initially equipped with a 560-kilowatt unit. Demand grew so rapidly that two 5,000-kilowatt generators were added to this plant by 1914.[16]

Calgary electrical managers soon cast their eyes west to the falling waters of glacier-fed rivers flowing east of the Rocky Mountains. In 1911, the city began purchasing power from Calgary Power, a private company that had just completed Alberta's first hydroelectric plant at Horseshoe Falls on the Bow River. Founded in 1903 by influential easterners—the financier Max Aitken, later Lord Beaverbrook, and R. B. Bennett, a future prime minister—this company would grow to become Alberta's largest generator of electricity. In 1928, the city decided to end its experiment with public power generation. It leased its Victoria Park plant to Calgary Water Power, which in turn was acquired by Calgary Power, and the city assumed sole responsibility for distributing electricity in the city.[17]

In this manner, city leaders helped construct a private monopoly in municipal power generation. By the end of the 1920s, Calgary Power had built a number of plants on the Bow and hydropower remained the most important source of electricity for the company and the city until the 1950s. Alberta was then experiencing its second oil boom, centered not near Calgary but near Edmonton. Both cities grew rapidly, as did demand for electricity. But the remaining hydroelectric potential in semiarid southern Alberta was limited. An additional difficulty was balancing the load factor, complicated by changes in the seasonal flow of the Bow, a shallow river that ran heavily in the spring and low for the rest of the year; moreover, downstream water availability declined in winter months when the Bow froze over. This was partially ameliorated by the company's construction in 1941 of a new dam on Lake Minnewanka, a glacial lake in Banff National Park, creating a large reservoir that fed the Bow via the Cascade River, easing the task of water management throughout the year.[18]

By the 1950s, most of the best hydroelectric sites in southern Alberta had been developed. And although there were large coal reserves in Calgary's hinterland, little effort was made to exploit them on a large scale. Calgary Power looked north to the coal- and water-rich center of the province. In 1956, the company put its first thermal generation complex into operation on the shores of the eighty-two-square-kilometer Lake Wabamun, sixty-five kilometers west of Edmonton. Located near a deposit of tens of million tons of coal, the facility's two 66-megawatt units were initially configured to use then-cheap natural gas. Even after the construction of the Wabamun plant the vast majority of Calgary Power's output (240 of 407 mega-watts) was derived from Bow hydropower.[19] Changing energy economics accompanying the next oil boom would make coal increasingly attractive as an energy source for the base load.

Leduc and the Subterranean Frontier

Imperial Oil's 1947 strike near the small central Alberta town of Leduc and subsequent finds around Edmonton finally brought the oil bonanza Calgary oil companies had long sought. The problem, as it had long been with natural gas, was a surfeit of riches. Independents hastened after Imperial into the field and the subsequent orgy of overproduction tripled Alberta's output to nearly twenty million barrels in 1949, quickly overwhelming the small local market.

Imperial and the long-ruling populist-agrarian Social Credit government of Ernest Manning formulated a dual response with implications that bore heavily on Calgary's fledgling oil industry. Aware of complaints that the major integrated companies controlled the independents by purchasing their production, Manning instituted prorationing in 1949. In both Canada and the United States, production quotas eventually developed both objective good stewardship and subjective wealth-distribution criteria. Companies had an economic allowance that enabled operators to recover investments, and production beyond this was based on the science of preserving reservoir drive. With provincial and federal governments unprepared to challenge Imperial's near monopoly and to balance supply and demand, the alternative strategy, one favored by Imperial, was to export Alberta oil to the United States.[20] This, in turn, meant pipelines, a technology entailing complex financial and political relations between local, national, and international players.

The problem was that the U.S. market was already divided among American producers. In the wake of the Leduc strike, Imperial could find only niche markets in the Great Lakes and Puget Sound regions. The outbreak of the Korean War allowed the major oil companies to conflate national security with local and national interests in Canadian petroleum, facilitating the construction of transnational pipelines.[21] In 1951 and 1952, an Imperial-led consortium built the Trans Mountain Pipeline (TMPL) connecting the Edmonton fields to Vancouver and Seattle.[22] Completed in two years between 1952 and 1954, and crossing through Jasper National Park, its builders framed the TMPL as an "all-Canadian" engineering marvel. Populist politics complicated the export of Alberta natural gas, which the provincial government had prohibited. But here, too, national security served as a passepartout. Pressure from U.S. Defense Secretary Charles Wilson and Howe prompted Manning to authorize the release of gas from Pakowki Lake in the southeastern corner of the province to the Anaconda mining complex in Montana.[23] Shortly thereafter, the Alberta government approved construction of the Westcoast Transmission pipeline, a project of Calgary entrepreneur Frank McMahon, linking gas fields in Alberta's northwest with the Pacific Northwest, a decision also influenced by national security considerations.

National and international politics determined the construction of the Trans-Canada Pipeline, the next big gas project. It reconciled, on the one hand, Howe's political goal of building an all-Canada gas line to Ontario and Quebec as a grand national project in the vein of the Canadian Pacific Railway, with, on the other hand, the desire of Manning and Alberta gas producers to find markets as wastage reached "disturbing proportions" with the expansion of oil production in the province.[24] The resulting Trans-Canada Pipe Line (TCPL), built between 1956 and 1959, was a public-private hybrid, combining competing Canadian and American pipeline interests cemented by a temporary crown corporation created by Howe to subsidize building costs. Determined that export pipelines not be used as means of controlling Alberta gas, the provincial government established the public-private Alberta Gas Trunk Line Company (AGTL) in 1954, designed to control the collection of gas within provincial borders. As Richards and Pratt note, this company would form the basis of Alberta's first large petrochemical concern.[25] By the end of the 1950s, two oil pipelines and three gas pipelines connected the WCSB to markets in the United States and Canada, all but one of which crossed the international border.

Having accessed export markets, Calgary independents were now subject to the vicissitudes of American oil politics. Oil exports to the United States expanded every year from 1951 until 1957, peaking around the time of the Suez War, which cut Europe off from supplies of Middle Eastern oil. The resolution of the crisis hurt Canadian producers, for it freed up massive volumes of crude at a time when American consumption in the post–Korean War period was already in decline as the economy sank into recession toward the end of the decade. With the world awash in oil, American companies drastically curtailed their imports of Canadian crude. As oil markets declined, Canadian production briefly dipped but then rose again as independents pumped even more to stay afloat. This led to calls to open the Montreal market, a solution Imperial staunchly opposed.[26] The company had by then divided the Canadian market in two. It supplied parts west of the Ottawa River and the marginal American markets with Alberta oil, while Quebec and the Maritimes were reserved for oil produced by its Venezuelan affiliate, an arrangement profitable for Imperial. Using the threat of a Montreal pipeline, the Conservative government of Prime Minister John Diefenbaker managed to force open larger American markets while retaining the basic market structure.[27] The resulting National Oil Policy (NOP), instituted in 1961, was a tacit acceptance of the status quo. It came at the expense of the Calgary independents, for they remained locked in a subservient role to the majors, committed to exporting the WCSB's cheapest and best oil and gas resources to American markets during a time of low energy prices.

Pushing the Limits of Growth

The NOP had implications both for the way energy was produced for use in the city of Calgary and for the fortunes of city-based oil companies. Where the issue of municipal power was concerned, the question of public ownership as well as long-term supplies of primary energy continued to be raised. Several independent studies done in the late 1950s found that Calgary could considerably lower costs if it generated its own electricity, the practice of a number of Alberta municipalities including Edmonton, Lethbridge, and Medicine Hat. Some also noted that with natural gas prices tied to the export market and likely to rise over time as a result, coal-fired generation represented a means of allowing Calgary to break into the generation field.[28] City managers did not act on the advice. But as the 1960s progressed, Calgary Power planned to develop coal-fired electricity to carry the base load and reserve hydropower for peak load. In early 1964, the company converted one of Wabamun's two gas-fired sixty-six-megawatt steam units accordingly.[29] Following exploration of the southern area of the lake that established a deposit of more than one hundred million tons of coal, the company built a large generation plant near the site. Dubbed Sundance, it went online in 1970 and became Alberta's largest power plant through subsequent additions.[30] By the early 1980s, Calgary Power, renamed TransAlta, would produce over 80 percent of Alberta's electricity.

While Calgary Power expanded and prospered in the 1960s, independent petroleum producers did not. The story of Home Oil is instructive. One of the pioneering Turner Valley independents, Home was founded in the late 1930s by a group of Calgary entrepreneurs. In 1952, the company came under the control of Robert A. Brown Jr., another Turner Valley alumnus, who merged it with his own Federated Petroleums. By the early 1960s, Home had become one of the largest Canadian independents. Under Brown's direction, the company played a leading role in the campaign to lobby for the Montreal pipeline, joining the Independent Petroleum Association of Canada, an organization in which Brown served as provisional director. But exploration in north Alaska in the late 1960s overextended the company, leading Brown to consider ceding control to an American company in 1971. This was opposed by the Canadian federal government on grounds that Home was one of the country's largest petroleum concerns. Canadian-owned Consumer's Gas of Toronto ultimately agreed to acquire the company.[31]

The episode underscored the broader weakness of Calgary-based independents by the early 1970s. Richards and Pratt hold that this was a direct consequence of Manning's strategy of relying on outside capital and expertise in nurturing the Canadian oil industry in its formative years. The premier justified the policy by pointing to Alberta's relative poverty and lack of experience in this sector. But by the late 1960s, the burgeoning middle and professional classes wanted change. In 1971, they voted out the Social Credit dynasty and brought to power the Progressive Conservative government of Peter Lougheed. With the price of oil skyrocketing following the Arab oil embargo of 1973, the premier struck out to bolster the indigenous petroelite, setting aside some resource revenue in a Heritage Trust Fund but investing vastly more in industrial diversification. He reasoned that with the best and cheapest resources of the WCSB already consumed, Alberta's future lay in developing a petrochemical industry and exploiting the Athabasca tar sands, a vast deposit of bitumen mixed with sand, clay, and water covering over one hundred thousand square kilometers in the northeast of the province.[32]

Calgary's growth in this period was dramatic. Between the early 1970s and early 1980s, the city's population swelled from around 400,000 to around 620,000. A construction boom tripled office space to over thirty million square feet by the early 1980s.[33] This expansion took place as Pierre Trudeau's Liberal federal government was attempting to exert more control over the oil industry. Like Richard Nixon, Trudeau initially responded to the energy crisis with price controls, a move popular in the oil-importing east but not in oil-exporting Alberta. But he also introduced measures to Canadianize the oil industry, including the creation of a state-owned integrated energy corporation called Petro-Canada in 1975. Headquartered in Calgary, it grew mainly through acquisitions. In a business where physical symbolism was important, Petro-Canada was housed in a new 215-meter skyscraper, the tallest in Canada outside Toronto.

Ottawa's subsequent efforts to exert more control over the oil industry through the National Energy Program (NEP), introduced in 1980 as the nation's first energy policy, coincided with the end of the 1970s boom and the return of hard times. Aiming to restrict foreign ownership, claw back a larger slice of resource revenue for the federal government through various taxes, and control oil prices, the NEP was in fact relatively moderate, a far cry from the nationalizations sweeping developing oil nations in the 1970s. Nevertheless, both independent and major producers associated the NEP with the recession. The result was a political battle lasting into the mid-1980s. Local wags dubbed Petro-Canada's fifty-two-story red granite–faced headquarters "Red Square," a reference to the company's status as a crown corporation. But the appellation was equally fitting for AEC and Nova, companies nurtured with generous helpings of public cash. Conditions worsened with the 1986 oil price crash. A full 14 percent of Alberta's workforce was unemployed and 15 percent of Calgary's office space went vacant, underscoring the limitations of diversification when it was tied to an industry subject to global currents of supply and demand dictated in part by the largest producers.[34]

The accession to federal power of Brian Mulroney's Progressive Conservative party in 1984 marked the beginning of a new phase in the development of Calgary's independents. With the world glutted in oil, many majors abandoned the WCSB, throwing hundreds of the city's oilfield technicians and scientists out of work. Canadian oil firms remained as vulnerable to takeover as ever.[35] It was in this period that Dome Petroleum, then the largest Canadian independent and an iconic Calgary company, staggered under an unsustainable debt load and was acquired by Chicago-based Amoco in 1988.[36]

At the same time, the Mulroney government began a major restructuring of the Canadian oil industry. It dismantled the NEP, removed foreign ownership restrictions, decontrolled oil and gas, and gradually privatized Petro-Canada, paving the way for unrestricted energy export. Aided by generous subsidies from the Alberta government, a host of independent firms targeted marginal unconventional resources intended primarily for U.S. use.[37] Crucially, noted Pratt, this path was cleared by Conservative and Liberal federal governments. Each party promoted free trade in energy, enshrined in the North American Free Trade Agreement (NAFTA) in 1994. The deal removed the prerogative of provincial and federal governments to levy export taxes or otherwise restrict energy supplies for domestic use, and committed producers to "proportional access." This meant that if production was cut for any reason, the average proportion exported over the previous three years was fixed, so that shortages experienced by export customers would be felt by domestic consumers as well. The agreement paved the way for a continental market and the most intensive phase of petroleum exploitation in Alberta history.[38]

Arrhythmia in the Heartland

What were the effects of these policies on Calgary's business class, cityscape, and hinterland from the late 1980s to the 2000s? The results, in keeping with so much of the history of Calgary and the Canadian oil industry, were paradoxical. On the one hand, the city slowly began to recover in the era of energy free trade and deregulation. In the void left by the majors emerged a new generation of independents optimized for marginal resource development. Some, like Talisman Energy, Suncor, and Nexen, were gradually spun off from foreign parents (British Petroleum, Sunoco, and Occidental Petroleum, respectively). Companies founded with state assistance—the AEC (merged with PanCanadian Energy to form Encana) and Petro-Canada—became fully privatized. They circumvented the old U.S.-bound natural gas pipeline routes controlled by Nova and TransCanada with the new Alliance and Vector lines financed through the promise of markets guaranteed by free trade and completed in the late 1990s by Calgary-based Enbridge, the successor of the Inter-Provincial Pipeline company (IPL).[39]

Growing in the fits and starts typical of the resource boom city, Calgary at the turn of the millennium was a clean, attractive city with a high quality of life and relatively progressive transportation and energy-use policies. Located at around 1,050 meters above sea level, it has a dry climate with long, cold winters and warm, short summers. Subject to frequent westerly or northwesterly winds, Calgary has good air quality despite extensive use of automobile transport. Physically, the city boasted the second-largest volume of office space in Canada after Toronto, a skyline commensurate with grand metropolitan ambitions. With a population of around 1.2 million in 2011, Calgary had a well-developed mass transit system by North American standards.[40] The city promoted transit-oriented development, encouraging suburban employment near existing and planned light rail corridors and increasing job and residential density in the downtown by expanding transit services and creating use incentives through restrictions on automobile parking.[41]

With prosperity came a certain amount of diversification in a number of advanced technology sectors including telecommunications.[42] Widely considered a conservative bastion in other parts of Canada, Calgary has been described as "middle of the road" by American standards, a virtual "yuppie haven."[43] One manifestation of the city's commitment to liberal lifestyle values was its effort to develop a green energy brand. During the 2000s, Enmax, the city-owned gas and electrical utility, acquired some renewable generating capacity including 11 megawatts of hydroelectric power and direct or partial stakes in around 218 megawatts of wind power.[44] Calgary's transit system was the first in North America to register all operations to the ISO 14001 environmental standard. A related initiative was "Ride the Wind," launched in 2001 in partnership with Enmax to completely switch the light rail system from coal to locally produced wind-generated electricity, a goal accomplished by the end of the decade.[45] Enmax also developed the Downtown District Energy Centre, a plant that heated the downtown core through a network of underground pipes. This facility introduced important efficiencies by rendering unnecessary the construction of separate boiler systems in individual buildings.[46]

In stark contrast to these forward-looking urban policies was the approach to resource extraction of Calgary-based petroleum firms. Well before the renais-sance of the independents, the effects of industrial diversification and long-term exploitation of the WCSB were felt mainly in central and northern Alberta. By the late 2000s, Alberta hosted one of North America's largest petrochemical and petroleum-refining complexes in the Edmonton region, the fabrication and stag-ing point for the Canadian oil industry. Here were clustered three of the prov-ince's four refineries (Imperial having deactivated its Calgary refinery in 1976) and seventeen of its twenty-eight chemical plants, including one of the world's largest ethylene operations. Only one chemical plant was located near Calgary.[47] How-ever, the city did have a relatively small waste-oil recycling facility in its south-east that began operation around 1958. Owned by Hub Oil, this installation was destroyed in an explosion on August 9, 1999, that killed two people and caused considerable environmental damage in what was probably Calgary's worst petroleum-related industrial disaster.[48]

Electricity generation was also heavily concentrated around Edmonton. Of Al-berta's 13,898 megawatts of installed capacity in 2013 (of which about 5,690 mega-watts were coal-fired), about 5,167 megawatts were located near the provincial cap-ital, all but 535 megawatts of which were coal-fired. The Lake Wabamun area was the largest single such complex, with four plants in its environs. In contrast, the coal-fired plants nearest to Calgary were more than two hundred kilometers to the northeast near the towns of Hanna and Forestburg.[49] As elsewhere, concentrated coal-fired generation contaminated local lakes and rivers with high levels of heavy metals and polycyclic aromatic hydrocarbons.[50]

The worst environmental effects of the post-1986 recovery were felt even far-ther afield from Calgary. In the 1950s and 1960s, the Manning government had lim-ited industrial development of the tar sands so as not to compete with convention-al oil production. These deposits yielded a low-grade form of petroleum used as a feedstock for synthetic crude, a process that combined mining with petrochemical processing. With heavy capital, water, and energy requirements (between 250 to 2,000 cubic feet of natural gas and between two to five barrels of water are needed to produce one barrel of synthetic crude), this fuel was developed only on a limited scale from 1967 to the mid-1990s. In 1996, with oil prices languishing and Alberta's conventional crude production in decline, the federal and provincial governments drafted an incentive package to induce companies to invest in synthetic crude, charging only a 1 percent royalty until all capital costs were paid off and allowing 100 percent of those costs to be written off.[51]

A number of Calgary-based companies, both foreign and domestic, were attract-ed to the tar sands in this period. Suncor continued and expanded the operations of the Sun Oil Company, the oil sands pioneer, after it divested from the Ameri-can company. Syncrude, a consortium of Canadian and foreign majors that began processing tar sand in 1978, has been dominated by Imperial.[52] By 2010, a mix of foreign or foreign-controlled companies and domestic firms worked the bitumen deposits, including Devon, Husky, Imperial, Shell, Statoil, and Total in the former categories and Canadian Natural Resources, Suncor, and Nexen in the latter.

Producers prospered, but at an environmental cost dear even by the standards of the oil industry. Accounting for more than 60 percent of Canada's daily oil production of 3.35 million barrels by 2008, the tar sands complexes consumed as much water per day as a city of two million and enough natural gas to heat six million homes. Four major mining and processing complexes and numerous in situ operations pumped more than forty megatons of carbon dioxide into the atmosphere every year, constituting Canada's single largest source of emissions, one surpassing the carbon footprint of many countries. This volume is projected to triple by 2020.[53] Since operations began in the late 1960s, a vast expanse has been laid waste. Of some 530 square kilometers of forest and wetlands that have been stripped and mined, only 12 percent had been reclaimed by 2009, and hundreds of billions of liters of toxic wastewater tailings occupy another 130 square kilometers, held back by some of the largest dikes and dams on earth. Reclaiming the remainder, as required by law, poses major scientific and engineering challenges.[54]

Water and energy are the limiting factors in tar sands development. The power requirements for these operations have spawned plans for other energy megaprojects that threaten further environmental havoc, including nuclear reactors and a gas pipeline connecting the Mackenzie River delta to northern Alberta. In keeping with historical trends, these projects will likely proceed only with the aid of large government subsidies. Although the Conservative government of Prime Minister Stephen Harper has ruled out direct aid for the Mackenzie valley project, it supports the project in principle.[55]

Regional Energy Capital

In some twenty years, a handful of Calgary-based independent energy companies emerged as among Canada's largest enterprises, rivaling the dominant Toronto-based big five national banks. Thanks to Calgary oil entrepreneurs, supported by the provincial and federal governments, Canada became the main supplier of oil and gas to the United States, exporting more than two-thirds of the 1.3 billion barrels of oil and over half of the six trillion cubic feet of gas produced in 2008.[56] In the realm of regional energy politics, Calgary and the Canadian energy companies that called it home had become important players in the 2000s.

Yet the city and the industry remained dependent on external markets and sources of capital. By 2011, Calgary hosted the headquarters of most Canadian independents and branch offices of majors like Shell, Exxon (Imperial), BP, and Chevron, but only two Schedule 1 banks, both smallish, relatively recently formed, and not directly related to the petroleum sector.[57] Although the city possessed a diversified economy by some estimates, employment was strongly linked to the energy sector.[58] A century after the founding of the Canadian oil industry, the position of independents in the global energy economy remained tenuous. Despite further recent consolidation, notably Suncor's acquisition of Petro-Canada in 2009, the market capitalization and oil and gas reserves of the largest Canadian firms remained relatively small compared with the major national and international petroleum companies.[59] The possibility of foreign acquisition remained high.[60]

Conditions by the end of the decade posed many challenges for Calgary independents. Falling gas prices brought about by the recession and new U.S. supplies, produced from formerly marginal fields using hydraulic fracturing and horizontal drilling technology, made it difficult for them to compete with their relatively more expensive gas. As in past such situations, they struggled to raise capital.[61] The difference, of course, was that the WCSB's conventional resources were by then largely exhausted.[62] Only the economic downturn has slowed the extraction of the remainder. Calgary's physical separation from the worst environmental effects of these activities must be considered an important factor in the decision-making processes that informed them and has certainly played a role in shaping environmental politics in Alberta. Distance has hitherto enabled the city's municipal and business leaders to avoid seriously reflecting on the sustainability of current practices. This is unlikely to change unless the next boom brings the collateral effects of marginal resource development directly to bear on the lifestyle of Calgary's professional and political elites. Coal-bed methane activities present one such scenario if the accompanying groundwater contamination impinges on wealthy ranching communities, an important conservative constituency with traditional links to the city's petroleum entrepreneurs.

Oil Shocks in an Oil City

The View from Stavanger, Norway,
1973–2008

Gunnar Nerheim

During the twentieth century the oil industry became the world's biggest and most pervasive business. From World War I until today oil as a commodity has been intertwined with national strategies, global politics, and war. Cities close to the production and processing of oil have naturally profited from the oil business regarding population growth and wealth. Some of these cities grew to become national oil capitals. After World War II and until the early 1970s oil prices were stable at around $3 a barrel. During the last forty years, however, both producers and consumers have had to come to grips with volatile oil prices. This chapter focuses on the consequences of oil price shocks on the well-being and wealth of Norway's energy capital—Stavanger. Like other petroleum centers, such as Houston, Texas, and Calgary, Alberta, Stavanger found itself more intimately tied to global economic and political events as oil increasingly dominated its economy.

Stavanger and the North Sea

The town of Stavanger sits on the west coast of Norway facing the North Sea. Until 1875, the backbone of its economy consisted of the catching and salting of herring, and the transport and sale of herring barrels in small sailing ships to the towns along the North and Baltic Seas. Shipping became an important business in its own right, and by 1875 Stavanger was the second largest shipping town in Norway. The owners of these ships, however, delayed their conversion to steam, and when freight rates fell after 1880, they struggled to compete. Many ship owners and merchants in Stavanger went bankrupt.[1] A new industry grew out of the ashes of the old sailing ship business—the production and export of canned sardines. The first sardine factory opened in 1873, and fourteen canning factories were operating by the next century. Stavaner's population increased from 21,850 in 1865 to 36,202 in 1900.[2]

The sardines were caught in the fjords close to the city, processed in the factories, and exported across the world, although Great Britain, other countries within the British Empire, and the United States offered the primary markets. With the outbreak of World War I in 1914, an almost unlimited demand for canned sardines grew overnight. Canned sardines were in great demand in the trenches on both sides of the war. The sardine canning industry then fell on hard economic times between the two world wars. While it blossomed briefly during World War II and remained Stavanger's leading economic sector into the 1950s, the industry slowly faded away despite considerable investments in new technology and new factories. Changing consumption patterns in the western world, the growth of supermarkets, and the diffusion of refrigerators and frozen foods reduced the demand for canned sardines to such a degree that most Norwegian sardine factories had to be closed.

After World War II and until the early 1960s, the main job creator in the city of Stavanger was the Rosenberg shipyard, which experienced a strong revival as a builder of large oil tankers. The Rosenberg yard set several Norwegian records for the largest oil tankers built in the country in this period. When Bergesen d.y., the owner of the yard and one of the largest shipping companies in the world, began contracting the construction of even larger oil tankers at Japanese shipyards, it was a sign that the Rosenberg shipyard could soon be outdated.[3] Moreover general job growth remained very weak in Stavanger in the 1950s and 1960s compared with other major cities in Norway, causing concern among local politicians and businessmen.[4]

In 1965, an unexpected opportunity arose. Representatives from multinational oil companies arrived in Stavanger looking for a convenient location from which to explore offshore sites on the Norwegian Continental Shelf and to operate drilling rigs. Stavanger politicians and city boosters greeted them with open arms. Esso Exploration, the first oil company to locate its operations in Stavanger, bought a former herring oil factory as base camp for its activities. Other oil companies soon followed. In addition to its geographic proximity to the planned exploration sites in the North Sea, Stavanger offered amenities that Americans appreciated:

an excellent airport with direct connections to Amsterdam and London and an eighteen-hole golf course.

Many firms that pioneered the offshore oil industry in the Gulf of Mexico in the 1950s called New Orleans and Houston home. When offshore operations began to span the globe in the 1960s, drilling and service companies followed closely in the footsteps of the major oil companies.[5] Once the international oil community became aware of the significance of the discovery of the giant Groningen gas field in the Netherlands in 1959, oil geologists began searching for promising prospects beneath the North Sea. Oil companies contacted governments around the North Sea regarding rights to explore for and produce petroleum, launching within these countries the development of laws and regulatory systems for offshore resources. The first multinational oil companies approached the Norwegian government about oil exploration on the Norwegian Continental Shelf in 1962.[6] By 1964 Great Britain prepared to award blocks for oil exploration in the First Concession Round on the British Continental Shelf. One year later the Norwegian government awarded blocks in the First Concession Round on the Norwegian Continental Shelf. Esso Exploration Norway, Inc., drilled the first wildcat well off the Norwegian coast in July 1966 with the Odeco-owned, semi-submersible drilling rig "Ocean Traveler." As a nation, Norway had no experience in oil exploration, and in its first years in this new venture it imported the necessary skilled workers and technologies. The main agents in this technology transfer were the multinational oil companies holding the concessions, and the oil service companies under contract to them.

With no major discoveries on the Norwegian Continental Shelf, several companies planned to leave until Phillips Petroleum, the American oil company and the operator for a group of companies, discovered the giant Ekofisk oil field. Announcement of the find led to frenzied exploratory drilling in the area around Ekofisk and on geological structures farther north.[7] Weather conditions in the North Sea proved to be much harsher than those in the Gulf of Mexico. When drilling began east of Shetland Islands (UK), the offshore industry encountered greater difficulties there than anywhere in the world. Nonetheless, in 1970 British Petroleum identified the huge Forties oil field on the British Continental Shelf, followed in 1972 by the Frigg gas field which straddled the Norwegian-British demarcation line. Farther north on the British side the Brent field and the Piper field were located in 1971 and 1973, respectively.

After the discovery of Ekofisk in 1969, Norwegian politicians and business interests quickly recognized that Norway would become a net exporter of oil. Among them, a broad consensus emerged that Norway should not continue along the path it had followed since 1965 when foreign multinational oil companies had been given too much influence over the rules of the game. Norwegians desired fuller participation in their offshore industry. Government officials began to formulate new rules that favored national interests to a much larger degree. In June 1971 the Norwegian Parliament, Stortinget, unanimously approved ten political command-ments for offshore oil, which the Parliament later effectively translated into po-litical action. Twelve months later, Stortinget made one of the most far-reaching decisions in Norwegian oil history. It established the Norwegian Petroleum Direc-torate to control and supervise all petroleum activities on the Norwegian Conti-nental Shelf, dealing with subjects such as conservation and resource management on the one hand and safety on the other. Stortinget also voted for the creation of a 100-percent state-owned oil company named Statoil. This company became the nation's major political instrument in safeguarding a fair percentage of the eco-nomic rent. The voting majority in Stortinget, all the parties from Arbeiderpartiet (Labor) to the left and some of the parties in the middle, wanted Statoil to become a fully integrated national oil company as soon as possible.[8] From then on Statoil ob-tained at least 50 percent of all new oil concessions; Statoil also would be "carried" during the exploration phase.

Stortinget chose to locate both the Petroleum Directorate and Statoil in Stavan-ger, strengthening and securing the city's future as Norway's oil capital. While the economic impact of the exploration phase in the 1960s had been negligible in Sta-vanger, activities gained speed following the Ekofisk discovery. By late 1971, the offshore industry in Stavanger directly employed some 1,500 people, most of whom had no personal connection to the city. Six offshore drilling rigs employed another 700 workers; 80 percent of them carried foreign passports. Foreign oil companies and service companies occupied 43 offices, with a total staff of 500. Skilled person-nel were expatriates, while Norwegians performed office work and unskilled tasks offshore.[9]

The OPEC-Triggered Oil Boom

From the establishment of OPEC in 1960 through 1972, the price of oil oscil-lated around $3 a barrel. Taking into account inflation, its real value actually de-clined.[10] However, in 1973, OPEC experienced a sense of its strong and lasting pow-er to influence crude oil prices. On October 6, 1973, Syria and Egypt attacked Israel, launching the Yom Kippur War. The United States and many other western coun-tries supported Israel, prompting several Arab oil-exporting nations to impose an oil embargo on them and to curtail production by five million barrels per day. This curtailment reduced world oil production outside the Soviet Union by 7 percent. The ability of the United States to control crude oil prices declined, while OPEC's rose. The extreme sensitivity of oil prices to supply shortages became crystal clear. The price of crude oil quadrupled to more than $12 a barrel by the end of 1974.

Events in Iran and Iraq triggered another price shock in 1979 and 1980. The Iranian revolution resulted in the loss of 2 to 2.5 million barrels per day of oil production between November 1978 and June 1979. Iran, weakened by the revolution, was invaded by Iraq in September 1980. Within two months the combined production of both countries fell to only 1 million barrels per day—6.5 million barrels less per day than a year earlier. In response, crude oil prices more than doubled, rising from $14 a barrel in 1978 to $35 per barrel in 1981. The steep price increase had profound and negative consequences for the economies of Western Europe, the United States, and Japan. Increased inflation and unemployment, as well as severe balance of payment problems between nations, resulted in a worldwide economic slump in 1974 and 1975.

Oil Shocks and Oil Capitals

In the midst of this turmoil high prices benefited Stavanger and other oil cities. In Houston, for example, the rapid rise in energy prices contributed to phenomenal economic growth. A building boom changed the skyline while upstream employment nearly doubled from 1974 to 1981, adding some 167,000 jobs directly tied to oil production.[11] The energy-dependent portion of Houston's economic base grew to account for 84 percent of all economic base jobs divided between upstream and downstream sectors. Upstream activities—exploration and production, oilfield equipment manufacturing and wholesaling, as well as oil and gas transportation—benefited from rising prices. Downstream activities—petroleum refining and chemicals manufacturing, both of which were highly capital intensive—benefited from declining oil prices, but were also influenced by such factors as demand for plastics and the value of the U.S. dollar relative to foreign currencies. With the second oil shock in 1979–80, Houston-based oil company profits skyrocketed and budgets for exploration and production increased significantly. The rig count—the number of active domestic rotary rigs, a key economic indicator that had fluctuated around 1,000–1,200 before 1974—reached an all-time record of 4,520 at the end of 1981.

Elsewhere in North America, boosters had promoted Calgary, the financial center of Canada's oil industry, as a "haven of opportunity and optimism, particularly to the young, the well qualified, and the ambitious."[12] With rising oil prices, the city enjoyed its own building boom, with a record demand for office space. Employment in construction rose from fifteen thousand workers in 1971 to forty thousand a decade later. More than half of the population growth in Alberta occurred in Calgary.[13] Edmonton, Canada's production center, experienced strong, but less spectacular growth. As Americans increasingly worried about the security of its oil supplies, the first plans for investing in the oil sands north of Edmonton took shape in the 1970s. At the same time conventional oil exploration in Alberta also reached record levels, especially as oil companies aggressively explored the Mackenzie delta and Beaufort Sea.[14] Edmonton operated as the primary supply base for much of the exploration in the north. While economic downturns after 1973 led to de-industrialization and unemployment in eastern and central Canada, the oil boom in Calgary and Edmonton attracted thousands of migrants.

In the North Sea, high oil prices after 1974 provoked fierce competition among international oil companies trying to gain concessions on promising geological prospects. News of oil discoveries frequently surfaced in the first half of the 1970s. The development of these oil fields proved more complicated and costly than anticipated, but as long as oil prices remained high, the oil companies were willing to invest in the North Sea. The first permanent installations at Ekofisk came online in 1974, but the last installations were not in place until 1980. Ekofisk production increased from 10,000 barrels a day in 1971 to 427,000 barrels a day in 1980. The Frigg field began delivering gas to Britain in September 1977. The development of the largest oil field in the North Sea, the Statoil field, experienced a series of delays and cost overruns. Production did not begin until November 1979, but the new price hikes in 1979 and 1980 covered overruns on investments much faster than expected.

The value of Norwegian oil and gas relative to GDP grew from 0.5 percent in 1974 to 15 percent in 1981. During this same period, the export of oil and gas, as a share of total exports, grew from 0.8 percent to 31 percent. As early as 1974, some Norwegian politicians on the left began worrying about the moral consequences of Norway becoming so rich so fast on oil income. With such questions looming large in their minds, members of Stortinget decided to set an upper ceiling on the yearly production of oil and gas, fixed at ninety-million-ton oil equivalents per year, although the decision also caused uncertainty among politicians and oil companies about the appropriate pace of development of the oil and gas fields.

Even with the ceiling, the impact of oil revenues on the Norwegian economy was immediate and significant. Norway still struggled to develop the necessary knowledge, technology, and skills to allow it to gain independence in relation to the multinational oil corporations. In the early 1970s, for instance, Phillips Petroleum conducted the engineering and planning of permanent production structures for Ekofisk at its Bartlesville, Oklahoma, headquarters and its other offices in Houston and London. All large contracts for engineering and steel work between 1970 and 1972 went to companies outside Norway, mainly American engineering firms such as Brown & Root, Santa Fe, and McDermott.[15] Phillips showed little interest in constructing a substantial organization in Norway for the development and operation of the Ekofisk field. The French company Elf Aquitane, the operator for the French Petronord group on the Frigg gas field, similarly failed to demonstrate enthusiasm for a Norway-based organization. Elf conducted planning and engineering work in its Paris headquarters or at the offices of its subcontractors in London.

As work on the Ekofisk discoveries went into high gear after 1973, new jobs in the construction emerged, particularly in connection with a one-million-barrel concrete storage tank in a fjord outside Stavanger. The success of this tank served as inspiration for Norwegian engineers to invent, design, and build concrete gravity platforms as a technological and economical alternative to the steel structures traditionally used in the Gulf of Mexico. A new company, Norwegian Contractors, invented the Concrete Deepwater Structure (Condeep), and experienced its breakthrough in 1973 when Mobil North Sea ordered the first Condeep for the Beryl A platform on the British Shelf. A new order from Shell for the Brent B platform followed four weeks later.[16] Photographs of these massive Condeep structures towed by power tugs from the Norwegian coast to their final locations came to symbolize North Sea oil.

The Norwegian Ministry of Industry increased pressure on foreign operating companies, such as Phillips Petroleum, Elf Aquitaine, and Mobil, to "Norwegianize." Nationalization, in practice, meant increasing economic activities and employment. Oil-related jobs in Stavanger grew from 925 in 1973 to 12,223 in 1980. In the county of Rogaland, home to Stavanger, jobs increased from 2,111 to 19,298, and in Norway as a whole, expanded sixfold. For example, Phillips Petroleum's Stavanger operation numbered 315 people in 1973; one-third of the employees were expatriates, and Americans held all leading positions. By 1976 its number of employees exceeded one thousand, but expatriates constituted only 18 percent of the total. Elf Aquitaine and Mobil experienced similar trends.[17]

Norwegians expected Statoil to play a key role in nationalization. Between 1973 and 1980 it hired more than one thousand employees. Statoil owned 50 percent of the Statfjord field, the discovery that constituted the company's economic backbone. The petroleum licence for the Statfjord field contained several clauses whereby the foreign multinational companies agreed to help Norway and Statoil quickly close the technology gap.[18] A large majority of the Stortinget favored Statoil at every turn, whether in relation with competing Norwegian oil companies or with subsidiaries of multinational oil companies. After the second oil shock in 1979, which coincided with the Fourth Concession Round on the Norwegian Continental Shelf, foreign oil companies faced and accepted the strictest conditions yet.

Outside workers continued to play roles at Ekofisk, Frigg, and Statfjord. At the height of development on these fields in 1976 and 1977, and despite the decrease in expatriates at Phillips Petroleum, an American slept in almost every third hotel bed in Stavanger; others carried British and French passports. The presence of these workers and industry growth generally exploded the demand for housing and prompted an increase in rental costs. Prices matched Oslo levels and sometimes climbed higher. Stavanger and the surrounding communities were willing to accelerate the construction of new homes and offices, but national authorities decided the number of building permits and the amount of credit allowed in the region. Stortinget and the Labor government wanted to avoid the negative effects of the oil booms, and viewed rapid, uncontrolled urban growth as an evil.

Market forces shaped Stavanger, and national policies strengthened its position as Norway's oil capital.[19] Oil-related taxes from firms and expatriates on net salary increased Stavanger's tax revenues. By 1980, some 30 percent of city taxes derived from these sources. Stavanger bound itself to the global oil industry and any future shifts in oil prices. Well into the 1980s Stavanger enjoyed strong, continuous growth. With prices above $30 a barrel after the second oil shock in 1979, many Norwegians expected boom times to last forever.

By 1980 the offshore industry was Norway's leading economic sector and Stavanger its undisputed oil capital. The transfer of offshore technology from the Gulf of Mexico, and its adaptation to Norway's political, economic, and cultural context, was very successful. In those fields of the offshore business where maritime skills were required, Norwegian companies closed the technology gap. Their shipyards mastered construction of drilling vessels, permanent production platforms, and various modules for such installations.

The First Oil Bust

Oil capitals such as Stavanger benefited from surging prices; the oil shocks that caused the increases prompted industrialized countries to improve energy conservation and invest in alternative fuels. In the early 1980s a global recession also caused a reduction in demand. With these factors, crude oil prices fell. OPEC faced this lower demand for oil at the same time that recent explorations in Mexico, Alaska, and the North Sea resulted in a higher supply of oil from outside OPEC. From 1982 to 1985, OPEC attempted to set production quotas low enough to stabilize prices, but met failure as various members produced beyond their quotas. During this time Saudi Arabia played the role of swing producer, cutting its production below its quota in an effort to stem the free fall in prices. By August 1985, the Saudis were unwilling to continue this role. Saudi Arabia linked its prices to the spot market for crude oil, and in early 1986, increased daily production from two to five million barrels. By mid-1986, crude oil prices plummeted below $10 a barrel globally.

Declining prices undermined the economies of oil capitals around the world. For example, there was dramatic decrease in domestic exploration in the United States. Less exploration meant less demand for oilfield equipment. Oil service and supply companies in Houston had excess capacity and inventories. Houston's petrochemical industry also had to adapt to reduced demand for petrochemical products. Job losses followed. Many of these jobs were high-paying blue-collar positions. The first job losses were strongest in the upstream sector. Employment fell by more than 40 percent. Later nearly every sector of the Houston economy was hit by job losses. Like Houston, Calgary seemed recession proof for a long time. In the early 1980s, however, the instability of Calgary's continued reliance on the oil and gas industry was revealed.[20] Low crude oil prices, combined with sluggish demand for western Canadian oil and natural gas, created financial problems for large sectors of the oil industry.[21] Real estate sales in Calgary dropped 34 percent in 1981 compared with the year before.[22] City officials in Calgary were shocked when they learned that eleven thousand people had left Calgary in 1982; unemployment reached 15 percent by that summer. Office vacancy was estimated to be 23 percent.[23]

Norway and Stavanger, however, initially avoided the recessions that Houston and Calgary experienced in the early 1980s. Oil investments in Norway after 1980 increasingly influenced the nation's business cycle. Suppliers of capital goods to the oil industry saw their share of investments rise from 10 percent in 1973 to 60 percent in 1980. Ekofisk, Frigg, and Statfjord finally reached full production. The flow of tax revenues and royalties on oil and gas ran unabated into the coffers of the Norwegian state. Compared with other European countries, Norway had the capacity to fund social reforms. In addition to the large oil investments, the liberalization of the credit market contributed to a strong upturn in Norway's business cycle. After World War II and until the 1980s, Norwegian banks were reluctant to make loans. Oil changed this climate. Lending grew by 20 percent in 1984 and another 30 percent in 1985. The construction market for new homes and commercial properties boomed. Private consumption increased by 18 percent in three years, and the unemployment rate was down to almost 2 percent. On the other hand the inflation rate was 10 percent annually. The Norwegian economy was in the middle of an economic boom.[24] The construction and real estate markets were especially strong.[25] In Stavanger, between 1978 and 1984, builders set records for the completion of new homes several times, with a peak of 1,346 houses in 1983.

The Norwegian economic boom lasted from the first quarter of 1983 until the first half of 1987, but by the end of the year the international economic recession reached Norway. In 1988 the unemployment rate more than doubled. Oil investments had shown signs of weakening from 1985, but did not have any significant impact as long as the rest of the economy was steaming ahead at full speed. When oil prices fell to $10 per barrel in 1986, this reinforced the fall in oil investments. The recession that began in 1987 was the deepest in Norway since World War II. Investments in construction were almost halved in four years. The lending boom gave way to bank losses and a banking crisis in the early 1990s.[26] When the bottom was reached in 1992, Norway had an unemployment rate of 6.5 percent.

Falling prices undermined all oil industry activities, particularly exploration activities on the Norwegian Continental Shelf. Oil investments as a share of GDP fell from 8.5 percent in 1984 to 5 percent in 1988. Low oil prices also decreased Norwegian state income. The peak year for income from oil and gas was 1985 with 94 billion NOK (Norwegian Krone), 19 percent of Norwegian GDP. The following year, this income fell to 11 percent of GDP. By 1988, it was 8 percent of GDP, although it again grew to 13 percent in 1990.[27] The dramatic drop in state income had political, economic, and social implications. For example, Norwegians experienced their highest interest rates on loans. House prices fell by 40 percent between 1987 and 1992.

In spite of the fall in oil prices, employment in the Stavanger region and Rogaland held steady in 1986, and the next year there was even job growth of 4,000. The largest increases came in the secondary services and retail sectors, although construction and manufacturing struggled to find skilled workers. The number of oil-related jobs in Rogaland reached 30,400 in 1988, the highest number since the oil industry began. However, the national economic recession finally reached Stavanger in the fourth quarter in 1988. As is common in recessions, the construction industry first felt the pressure. Soon retail businesses, hotels and restaurants, and bank and insurance companies had to cope with layoffs. During 1989 the labor market worsened from one month to the next, and unemployment in Stavanger hit 4.6 percent.[28] Low oil prices after 1986 and their fiscal consequences proved a painful experience for Norway's government and its politicians. They became acutely aware of how dependent on oil revenues the Norwegian economy had become and how intertwined Norwegian oil had become with global politics.

At the same time, the recession in Norway strengthened Stavanger's position as the nation's undisputed oil capital. When prices had risen above $30 a barrel in 1981, national politicians believed that the best practice was to require oil companies to locate where the local economy would be most positively affected by oil money. The political consensus was that Stavanger had received a disproportionate share of oil-related growth and it was time to share the wealth with other coastal communities farther north by placing offices and facilities there to facilitate the development and operation of specific oil fields. In the early 1980s, for instance, the city of Bergen became the operational base for certain discoveries in the late 1970s. Foreign oil companies, as well as Statoil, wanted to expand their operations in Stavanger, but the Norwegian government forced them to locate new activities elsewhere. After taxes and royalties fell so dramatically in 1986, politicians focused more on state tax income than distributing oil-based operations. By 1990, it again became easier for the international oil industry to expand existing operations in Stavanger, reinforcing the city's status as Norway's energy capital.

Diversity in Recoveries

When the Norwegian recession was at its strongest, Houston was on the way to recovery. Houston regained jobs lost in the oil bust by May 1990. Houston's economy experienced new growth from the booming health care sector and expansion at Johnson Space Center with the construction of the space station, but also from the stabilization and local consolidation of the oil extraction industry and a construction boom in the city's giant petrochemical and refining complex.[29] Oil remained a dominant industry in Houston, and world oil consumption, most of it by Asia, increased by 6.2 million barrels per day between 1990 and 1997, fueled the city's economic growth.[30]

During the 1990s, Houston remained the center for oil and natural gas exploration, drilling, production, and marketing in the United States and emerged as the world hub for the international oil industry as well.[31] Driven by new technology, the American oil industry in the 1990s was focused on productivity and profits. New tools and systems, such as three-dimensional seismic, coiled tubing and measurement while drilling, lowered costs, reduced the risk of each prospect, and broadened the range of exploration opportunities. By 2000, companies at the forefront of the new technologies, as well as deep-sea exploration in the Gulf of Mexico, chose to locate in Houston.[32] During the 1990s Calgary also strengthened its position as the undisputed Canadian oil capital. An increasing number of businesses and Canadians moved to Calgary.[33] In 1997 Calgary was home to 103 major company headquarters, up from 61 in 1993 and second in number only to Toronto. When big companies moved to Calgary, many of their main suppliers tended to follow on their tail.[34] Personal income, real estate sales, and rental vacancy rates confirmed the vibrancy of Calgary's economy.[35] Between 1993 and 1998, the province of Alberta, due in large part to Calgary's oil success, outperformed all other regions in Canada with an annual growth rate of 3.8 percent. The province had the lowest unemployment rate in Canada. "Fuelled by another banner year in the oil patch, the Alberta economy kicked into high gear, growing by 6.5% in 1997, the best in the country."[36] Alberta consumer spending increased by 5.3 percent, and investment in residential construction grew 35 percent. The strength in the oil patch led to considerable spin-off activity in engineering services and manufacturing. Overall manufacturing output jumped 13 percent, twice the national rate.

In Norway, by comparison, the economy began climbing after bottoming out in 1993, and peaked in the third quarter of 1998. A significant decline in interest rates, strong growth in public consumption, and a marked rise in the international business cycle prompted the upturn. Households paid off much of their personal debt, and the demand for cars and durable household capital goods rose again. Due to its unique six-year-long economic slump, Norway experienced lower growth in prices and salaries compared with its most important trading partners, which, in turn, improved the nation's competitive position, contributed to greater profitability among private-sector firms, and prompted better capacity utilization and new investments. From 1991, except for a slump in 1994, petroleum investments increased. By 1998, petroleum investments reached a level that allowed them to again greatly influence the Norwegian economy. Norwegians found 230,000 new jobs between 1993 and 1998, and unemployment fell from more than 6 percent to a little more than 3 percent. These investments especially touched the Stavanger region. Since the mid-1970s, Norwegian Contractors manufactured the huge Condeep production platforms there. Around 1990 this company received large contracts from Shell for the Troll and Draugen fields and from Conoco for the Heidrun field. The building boom with Condeep platforms lasted little more than two years. After this era, subsea solutions and floating production ships appeared to replace concrete production platforms. Most observers initially viewed the closing of Condeep's manufacturing site in Stavanger in May 1995 as a significant blow to oil-related employment in Stavanger and nearby Rogaland. Norwegian Contractors and several of the largest oil companies, such as Phillips Petroleum, Statoil, and Elf Aquitane, reduced their payrolls by several hundred jobs during 1995. By year-end, there was new job growth in manufacturing, construction, retail and services, albeit not enough to cancel out losses in oil jobs. However, the next three years witnessed the development of new jobs across the entire economic spectrum. Some industries even struggled to find workers with the necessary skills.[37]

Were the Good Times Over for Oil Cities?

In 1998, a new global economic crisis presented more challenges to the world's energy capitals. The Asian financial crisis began in Thailand in May 1997, quickly spread to Malaysia, Indonesia, the Philippines, and South Korea, and eventually destabilized economies throughout Asia, Latin America, and Europe. The world faced the worst collapse of commodity prices since the Great Depression.[38] OPEC either ignored or severely underestimated this crisis. Ongoing humanitarian sales of Iraqi crude and OPEC's decision to increase quotas added crude oil to world markets. With oil prices around $12 a barrel by June 1998, companies reduced drilling in the United States and around the world. Prices did not improve until OPEC and several non-OPEC producers agreed in March 1999 to remove two million barrels of crude from world markets. Prices finally reached $35 a barrel in September 2000, but the impact of these crises lasted longer in some energy capitals than others. Diversification of local economies tempered some of the impacts. Houston lost almost twenty-five thousand energy jobs, but the strength of the national economy and the nonenergy sectors of Houston's economy allowed the city to continue to grow. For Calgary and Alberta, job growth in construction, manufacturing, retail, transport, and warehousing offset some losses in the oil patch. When the oil prices rebounded, Calgary experienced another surge. Construction cranes dominated its downtown skyline again, and urban sprawl continued in all directions.[39] More than sixty thousand new jobs were added to the workforce in Calgary by 2000.[40]

In spite of high oil investments in Norway in 1998, economic growth declined in the wake of the Asian financial crisis and the destabilization of international capital markets. The fall in oil prices hit Norway hard.[41] Oil investments declined by 13 percent in 1999, and another 30 percent in the course of the next three years. During the same years, investments were weak in the Norwegian economy more generally. While the Stavanger region enjoyed greater economic success than the rest of the country prior to 1998, the downturn prompted by the global crisis hit it harder as well.[42] The oil-related economic base, especially in manufacturing, had to adapt to a strong fall in oil investments. Unemployment in the region increased more than in the rest of the country. More than 3,500 people lost their oil jobs in Rogaland, and growth was weak in the rest of the regional economy as well.[43]

The New Oil Boom after 2000

The terrorist attacks in the United States on September 11, 2001, and the subsequent military actions in Iraq less than two years later once again revealed how global events and the oil price shocks that they caused dramatically altered economic and labor conditions in the world's oil capitals. Following the attacks, crude oil prices plummeted around the world in autumn 2011, although they rebounded to the $25-a-barrel range six months later. With this volatility, for example, America's energy capital, Houston, lost 13,300 jobs in 2002, a drop of 0.6 percent, although some losses stemmed from mergers among industry giants, such as Conoco and Phillips Petroleum. Weak growth continued in 2003 due to unexpected dips in oil and gas extraction, drilling support services, machinery manufacturing, computer manufacturing, and construction.[44]

When the United States and the other three nations in its coalition invaded Iraq in March 2003, low oil inventories in the United States and other nation members of the Organisation for Economic Co-operation and Development (OECD) contributed to surging crude oil prices that reached a record high, above $60, in June 2005 and gradually created flush times for many oil companies and oil regions around the world. Painful memories of earlier busts that had followed booms made businesses wary of basing investments on current prices. Many companies used the extra earnings to pay down debt and improve their balance sheets.[45]

Energy companies continued to migrate to Houston, but the office market remained slow.[46] The Houston economy picked up speed at the end of 2005, but job growth was still not strong enough initially to accommodate everyone who wanted to work, including some eleven thousand people from Louisiana who registered for work in Houston after Hurricane Katrina.[47] By 2006, however, job growth in Houston doubled the national average as the city enjoyed its strongest resurgence in more than twenty years.[48] Demand from oil and natural gas companies helped reduce office vacancies, sales prices for office buildings climbed 34 percent, and the housing market was one of the strongest in the United States.

Calgary followed a development pattern similar to Houston's. In 2002, the province of Alberta fell below the Canadian average growth of 3.4 percent for the first time in many years.[49] However, the Iraq War and other global events helped reverse these conditions. Oil investments in Alberta jumped 37 percent between 2003 and 2005. In particular, investments in oil sands north of Edmonton rose from CAN $400 million in 1994 to CAN $8.5 billion in 2004, while output from the tar sands grew to 41 percent of total Canadian oil production.[50]

By 2005, the Calgary economy experienced the greatest land and real estate rush since the late 1970s. The downtown vacancy rate was the lowest in twenty years and the lowest of any major Canadian city.[51] The residential housing market in Calgary was equally hot.[52] Calgary was a workforce magnet; since 2000 more than eighty-five thousand people migrated there. Calgary remained the epicenter of planning and finance for the oil sector, but Edmonton was the regional hub for production and distribution and matched its southern neighbour with the rate of job growth.[53]

Like its North American counterparts, Norway saw its business cycle reach bottom in early 2003; employment had hardly grown in four years. However, with changing global conditions, the business cycle picked up speed later that same year and the boom lasted through 2007. Oil investments grew by almost 60 percent between 2002 and 2006. The boom in construction and the market for homes matched the heyday of the middle 1980s. In Stavanger new building milestones were registered. House prices reached record levels. High oil and gas prices led to growth in oil revenues and taxes. Norwegian GDP grew at an average of 4.5 percent. Real wages grew by almost 15 percent. Norway enjoyed its most sustained period of economic growth since 1978.[54]

According to the Norwegian Bureau of Statistics, the economic boom reached its highest point at the end of 2007 and spring of 2008. There were warning signs looming on the horizon, however. Rising interest rates contributed to a decline in residential construction in the fall of 2007. In the first half of 2008 household consumption came to a near standstill. After strong investments in new technology and higher capacity during the boom, investments in manufacturing leveled off and began declining in the second half of 2008. The economic picture in Norway was dominated by an international financial crisis that absorbed Stavanger and the world's other energy capitals.

Toward the end of 2007 there had been many signs that a major economic recession would hit the world. In the United States, Houston seemed to be one of the last bastions for American optimism.[55] Demand for oil remained strong and the price reached a new record of $147 per barrel on July 11, 2008. In January 2008 Houston's 3.9 percent job growth rate was the highest among the twelve largest metropolitan areas in the United States. The national job growth rate in the same period was 0.7 percent. By the end of 2008, job growth leveled off and began to decline. The non-energy base and the secondary sectors experienced growth at their lowest level since the recession in 2004.[56] The unemployment rate rose to 5.6 percent in December. And the energy sector could no longer anchor the economy. By the end of the year oil was traded at $33 a barrel. Prices did not rebound to $70 a barrel until September 2009. The United States was in the worst recession in terms of job losses since the 1930s, and Houston felt its impact. The local economy lost jobs on a scale unseen since the mid-1980s.[57]

Like Houston and Calgary, Stavanger emerged as an energy capital not because oil and gas were discovered and produced at its doorstep, but because it developed the urban infrastructure and other amenities that oil industry executives valued highly when locating headquarters, oil services, and other facilities. Houston, Calgary, and Stavanger became oil network cities, intimately tied to developments in the global economy and international oil industry.[58] Like its counterparts, Stavanger experienced strong growth impulses from the oil shocks in 1973–74 and 1978–79. High oil prices hurt the national economies in the United States and Canada, but led to oil booms in their energy capitals, Houston and Calgary. In the case of Norway, a relatively new player in the world's oil game, high oil prices in the 1970s helped make Norway a very rich oil nation. Stavanger, as the oil capital, was situated perfectly to exploit the oil growth.

The timing of the oil bust in the 1980s was almost the same in Houston and Calgary and seemed to be felt more strongly before the large fall in oil prices in 1986 than after. While its impact was delayed until 1988, the global recession in Norway continued until 1993. In the 1980s, petroleum production in Norway grew from one year to the next, and even though oil prices fell slightly, neither the Norwegian state nor Stavanger felt this decline to any large degree. When low oil prices finally began to adversely affect taxes and the government's share of royalties, politicians became acutely aware of how closely Norway had become intertwined with international political events and the rise and fall in oil prices caused by political earthquakes. Stavanger felt the pain of low oil prices in the form of higher unemployment because international and national oil companies reduced their exploration budgets and shelved planned developments of petroleum fields. Nonetheless, compared with Houston and Calgary, Stavanger was the oil city least affected by the fall in oil prices.

In the end, Stavanger, Houston, and Calgary gained greater wealth and consolidated their relative position as oil capitals following each low ebb in oil prices by increasing their connections and building stronger networks within the global oil industry. At the same time, however, they have tied their fortunes more tightly to world events and oil prices than ever before. When something of world significance happens, oil prices are affected, and, each time, the resultant shocks produce consequences for the oil cities in both the short and long term. The last great energy boom from 2003 until 2008 led to very strong, almost unparalleled economic growth for these energy capitals. Despite earlier efforts at diversifying their economies, the economic forces during this last boom revealed that energy was the sector that drives their urban and regional economies. As long as oil continues to be the world's most important raw material, as long as the North Sea continues to produce oil, and as long as Stavanger can maintain its position within both Norway and the global industry, its business cycles, like those of other contemporary energy capitals, will be dominated by world oil prices. Despite periodic shocks and setbacks, there are no incentives for the residents of Stavanger to diversify their economy and abandon their most profitable source of income.

PART III   **Cursed by Oil?**

Tampico and Port-Gentil

The development of energy industries in any location carries with it benefits and costs. Whether a particular location can capture enough of the benefits—through the construction of infrastructure, the attraction of other industries, the recruitment and permanent residence of skilled workers, or the control of the generated wealth—to offset the costs depends, in part, on the economic and political maturity of the region when the key resource is discovered and developed, but also on the status of the larger industry at the time. Regions that possessed accommodating but ineffective or corrupt governments, lacked local financial resources to stake a role in the management of the industry, or quickly lost easy access to large pockets of oil, for example, found their natural wealth exploited by multinational corporations that invested little in their communities but left behind the detritus of the industry.

Myrna Santiago illuminates the complex factors that led to the diminishment of Tampico, which was once Mexico's most important oil port. Her essay reveals how earlier oil-led development tied the larger region to the fossil-fuel economy into the twenty-first century. A small port town handling the commerce of the southern half of the state of Tamaulipas, Tampico possessed a small multinational population when its oil resources were first developed on a small scale in the late nineteenth century. Developments in Texas and California soon prompted foreign investors to explore the rich oil resources underlying much of the Gulf of Mexico and the arc of land that borders the Gulf, including the hinterlands of Tampico. In anticipation of a sustained oil boom, foreign corporations rapidly altered the landscape, filling in wetlands, dredging the Pánuco River to allow for tankers, and working with local government to improve the municipal infrastructure. Construction drew laborers from all of Mexico, but companies recruited skilled workers and managers from the United States and England as the city's population grew fivefold in just twenty years. With World War I, the British- and U.S.-based companies drilling in Tampico drew the region more deeply into the international market to meet the demand. New refineries prepared oil for shipment and a small percentage for domestic consumption. The oil boom ended shortly after the war, however, due to declining demand and as a result of the industry's rapid depletion of accessible oil fields and problems of saltwater intrusion. The oil companies fired thousands of workers. Labor unrest led local officials to divide the city into two. Mexico's nationalization of its oil industry in 1938 left many multinational corporations uninterested in investment in the country. The industry focused on domestic needs. Organized in 1930 in the midst of this unrest, the new city of Cuidad Madera experienced a new oil boom as a production site for domestic consumption in the 1960s and later as a site for export in the 1980s as an alternative source of supply from OPEC. While Tampico's own fortunes as an energy capital declined, it played an important inaugural role in establishing the larger region as a persistent center of oil and gas production. As Santiago observes, "the case of Tampico therefore shows how petroleum production transformed the region in particular and lasting ways as it transformed the world economy in general."

As Tampico's fortunes waned, across the Atlantic Ocean, Port-Gentil emerged after World War II as the petroleum center of Gabon. Despite the mineral wealth that passes through the city, however, much of its population remains impoverished and angry over its economic marginalization. The source of this contemporary crisis, Douglas Yates advises, is found in the city's colonial past. The French government established the settlement that became Port-Gentil on Mandji Island in the 1880s, but it remained a small enclave until the 1910s when the French began to exploit *okoumé*, a soft form of mahogany that remained one of Gabon's primary exports into the twenty-first century. Geophysicists also identified rich onshore and offshore oil resources in Gabon as early as the 1920s, but the French state initially lacked the technology, capital, and interest to develop those resources. With access to abundant inexpensive oil from the Middle East, France had no immediate need for Gabonese oil. After World War II, however, the French financed a colonial oil company to exploit Gabon's resources. A nascent oil infrastructure of rigs and pipelines was in place when Gabon achieved political independence in 1960. As Yates observes, however, "Gabon was an African country, geographically shaped by the French, colonized by the French, forested by the French, led by a francophone assimilated elite who spoke French, read French newspapers, received French education, practiced French law, worked for French businesses, and who had adopted a French model of government which was additionally dominated through a system of French 'cooperation.'" In this neocolonial context, the Gabonese state acquired shares in the leading foreign oil companies and a portion of the revenues, but 85 percent of the production was shipped abroad. Moreover, these companies filled jobs with foreign workers, creating oil enclaves segregated by space and income from the residents of Port-Gentil. In the 1960s and 1970s, the Gabonese government worked with foreign companies to suppress local opposition and to ensure oil revenues went not to Port-Gentil, but to Libreville, the nation's capital and largest city. By the 1990s, the discovery and development of a new and larger oil field shifted the focus of the Gabonese industry away from Port-Gentil. The citizens of Port-Gentil experienced depleted forests, contaminated soils, and fouled drinking water, but enjoyed none of the economic benefits associated with the transformation of their land.

A number of scholars have argued that countries rich in natural resources generally, and in oil in particular, have been cursed, finding themselves unable to use that natural wealth to develop their economies and ensure the well-being of their citizens. Various explanations point to declining competitiveness in other economic sectors, the volatility of global energy markets, mismanagement of resources, exploitation by multinational corporations, and corrupt or inefficient domestic governments.[1] Have Tampico and Port-Gentil simply been cursed by oil? While some of these factors played out in Mexico and Gabon, comparisons with our other energy capitals suggest a more complicated picture. Unlike the U.S. energy capitals discussed herein, local interests in Tampico and Port-Gentil did not participate in all or even most phases of the energy industry, including ownership. These communities acted primarily as extraction and distribution locales with limited refining. Port-Gentil also entered the energy business at a much later date, when global commodity markets were well established. And unlike Calgary, Stavanger, and Perth, Tampico and Port-Gentil failed to emerge alternatively as the financial and administrative leaders of the oil and gas industry in their respective countries. In Mexico, after 1938, the central government dictated energy policy. In Gabon, foreign corporations with management headquartered far away had little long-term interest in developing the surrounding communities in these poorer nations. There was little integration of local laborers, and companies recruited foreign workers as temporary visitors rather than permanent community members. Lacking their own financial resources and an aggressive business elite to guide resource development, these regions became little more than resource colonies.

In the case of Mexico, nationalization policies led to an emphasis on domestic production, but at least Cuidad Madera and other Mexican oil centers were prepared to compete in the global marketplace when a series of international oil shocks and the ascendency of OPEC opened new doors, although Mexico's economy remains overly dependent on oil and thus is subjected to the volatile price shocks of that marketplace.[2] In the case of Gabon, however, a political culture of accommodation after independence facilitated the exploitation of natural resources by foreign corporations and filled the pockets of some government officials, but at the expense of the environment and the nation's own citizens. For Port-Gentil, there have been busts but without any intermittent booms. The city has borne the burdens of oil production but enjoyed little of its wealth. Perhaps indeed a curse.

Tampico, Mexico

The Rise and Decline of an Energy Metropolis

Myrna Santiago

There are cities in the world that have produced the energy that fuels the modern global industrial economies. Houston embodies the idea perfectly, as the premier energy capital in the United States today.[1] Other cities in similar positions in the oil sector include Calgary, Alberta; Lagos, Nigeria; Riyadh, Saudi Arabia; and Libreville, Gabon. Tampico, once Mexico's most important oil port is no longer on that list. Its moment as an energy capital was short-lived but intense. Between 1900 and 1924 the city was the magnet that attracted resources from a globalizing world. It consumed habitats and nature from its immediate surroundings as it expanded to accommodate the oil industry growing in its midst. From the hinterlands to the south (northern Veracruz) Tampico welcomed the crude itself. Labor came from the Mexican countryside in the east, Texas and Oklahoma in the north, and England and continental Europe across the Atlantic. As a result, cultural practices, architectural styles, ideologies, and the like mingled and competed in the port. Directly tied to an emerging world oil market, Tampico transformed human and natural energy into commodities: oil headed for refineries in Texas and Louisiana or English ports, and distillates for local consumption.

But the party ended suddenly. Rapid exploitation of the northern Veracruz oil fields led to saltwater intrusion by 1919. Companies fired workers by the thousands and fled to Venezuela, even though Mexican wells as a whole did not reach peak production until 1924. Labor unrest added to the metropolis's problems, leading municipal officials to split the city into two in 1924. Thereafter the troublesome eastern neighborhoods and their rowdy unions were on their own. With them went the refineries. Villa Cecilia, which became Ciudad Madero in 1930, became the oil town. Tampico reverted to its old identity as a commercial hub, a cramped port city battered annually by hurricanes and floods, trying to survive the ups and downs of the global economy, but with "New Orleans style" buildings to remind its chroniclers of its "cosmopolitan" era and good times.[2]

Still oil persisted in the city's future. In the 1960s, Ciudad Madero became an important site for petrochemical production for national consumption, which in the 1980s expanded for export to Central America and the Caribbean, shipped through Tampico.[3] The contiguous city on the north, Altamira, an industrial and petrochemical corridor with its own port, began competing with Tampico for cargo in the 1990s and gained status with the inauguration of a liquefied natural gas (LNG) plant in 2006.[4] Thus even though Tampico had lost its place as an energy capital early on, its legacy was to tie the locality to hydrocarbons for the rest of the twentieth century and beyond. The case of Tampico therefore shows how petroleum production transformed the region in particular and lasting ways as it transformed the world economy in general. The city was a node where oil extraction and production generated immediate environmental transformations at the local level, without actors ever imagining the ecological changes on a global scale (i.e., climate change) ushered in by burning this particular fossil fuel. Tampico does not figure among Mexico's most important six ports today, but it has not disappeared into oblivion.[5] Today the port touts itself as a tourist destination, marketing its rivers and lakes as natural wonders—although they once shined with rainbow-hued oil slicks. This oil metropolis did not persist as such for long, but it is not an exaggeration to say it transformed the world while it was on top.

Location, Location, Location

Location was all-important for the development of Tampico. The port is located at the midpoint of the circle that makes up the Gulf of Mexico from Florida to Yucatán. The arc that the Gulf creates and the Gulf itself sit atop underground rivers of oil exploited from Louisiana to Tabasco at different times during the twentieth century. It was the discovery of oil at Spindletop in Texas in January 1901, in fact, that alerted foreigners to Mexico's potential as an oil producer.[6] Above ground, Tampico shared the same hot and humid tropical climate that afflicts or blesses the geographical half-moon from Houston to Yucatán. Tampico is located at the juncture of one of Mexico's largest river systems, where the Tamesí and the Pánuco Rivers meet, about eleven miles upstream from the Gulf of Mexico. The port was surrounded by lakes and lagoons, which created marshes, bogs, and swamps all around and along the rivers, fed by a long rainy season that included hurricanes and a dry season that often witnessed drizzle or slight rains.[7] The site was a transition zone between scrublands to the north of the swamps and a tropical rainforest to the south, with mangrove forests flanking the lagoons that formed between the mainland and the Gulf. Before the oil prospectors arrived, commerce was the lifeblood of Tampico and a small multinational population reflected such economic reality. The port itself was not terribly exciting. Fanny Calderón de la Barca, the Scottish wife of the Spanish ambassador to Mexico, sailed up the Pánuco from the Gulf in 1848 and recorded her first impressions:

> [T]he first houses that meet the eye have the effect of a number of coloured band-boxes—some blue, some white—which a party of tired milliners have laid down among the rushes. On leaving the boat and walking through the town, though there are some solid stone dwellings, I could have fancied myself in a New England village: neat shingle palaces, with piazzas and pillars—nothing Spanish—and upon the whole, an air of cleanness and cheerfulness . . . There are some good-looking stores . . . not much to see. There are many comfortable-looking large houses, generally built according to the customs of the country whereof the proprietor is a native.[8]

In 1893, the American Thomas L. Rogers visited Tampico and found that commerce still dominated the economy and colorful houses were the norm, but there still wasn't much to see. The buildings around the plaza were few and made of wood, the houses were "pink or green or cream or other color . . . it is said to be against the law to paint any wall white."[9] The population hovered around six thousand souls. The black, sticky, and odorous *chapopote* dotted the rainforest to the south and southeast, but its uses were few and artisan: caulking canoes, decorating pottery in black, or burning as incense for fragrance.[10] All that was about to change and rapidly.

Local and Foreign Markets, 1900–13

The first man to spy an opportunity for oil in Tampico was Henry Clay Pierce, an American. As minority partner with Standard Oil, he opened the Waters, Pierce Oil refinery in Tampico in 1887 to provide fuel oil to the railroad companies just then laying tracks in the region and to satisfy a small local demand for kerosene as illumination. He did not drill for oil in Mexico, however, importing his crude from the United States instead. According to observers, Mexico used "no more than 700 barrels of refined oil a day" at the time.[11]

Another early twentieth-century capitalist and true oil baron, the American Edward L. Doheny, entered the oil business in Mexico in 1900. He had made millions in California oil already, but knew that Mexico had no market for the black ooze. He decided to create one as he acquired much land in the Tampico hinterlands and built the infrastructure for export: he contracted with his friends at the U.S.-owned Mexican Central Railroad to convert their locomotives from coal to oil fuel for the Tampico-Aguascalientes five-hundred-mile route. The contract would last until 1920. In 1905 he also convinced Mexican officials in various cities, including Tampico itself, to pave their streets with asphalt from his first refinery at Ebano, in San Luis Potosí. The rest of his production, the overwhelming majority, went abroad.[12] Production increased steadily, from ten thousand barrels in 1901 to one hundred twenty-five thousand in 1904, to five hundred thousand in 1906, and jumping to one million in 1907. By 1911, production had hit twelve million barrels.[13]

The stage of extracting oil for a limited internal market and a growing foreign market lasted about a decade. It began to change Tampico's infrastructure quite rapidly, however. The port housed the first refinery and its first five hundred workers. It also boasted of paved streets and a plaza illuminated by kerosene lamps. New companies scouting for office space encouraged the rise of new multistory buildings around the downtown plaza. Infrastructure projects transformed the city's environs further. In partnership with local authorities, the companies undertook a project to gouge a canal ten feet deep, thirty feet wide, and eight miles long to connect the Pánuco River to the Tamiahua lagoon for the anticipated "fast gasoline launches" that would ply what became known as the Chijol Canal between Tampico and the oil fields of the Huasteca peoples in northern Veracruz.[14] In anticipation of the oil bonanza to come, the companies filled in a marsh and built a large oil storage area thereafter known as "Tankville" right alongside the Pueblo Viejo lagoon, across the Pánuco River from Tampico, while they planned the dredging of the river itself to welcome oil tankers. Other profound changes were social: the city was growing in leaps and bounds, as men began to hear news of the construction boom.

The trickle of migrants became a deluge after 1908, when the Englishman Weetman Pearson struck oil at San Diego de la Mar, some seventy-five miles south of Tampico. Companies and labor recruiters proliferated. Company agents contracted men from Monterrey to Jalisco by the thousands. They also imported drillers and craftsmen from Texas, Oklahoma, and California and a few newly minted geologists from American universities eager to test the new discipline in Mexico.[15] From 17,569 inhabitants counted in 1900, Tampico grew to 23,452 in 1910. By 1921, the population had jumped to 94,736 people, making the port the fifth largest city in Mexico. Between 1910 and the late 1920s, the physical space the city occupied grew in tandem, from 432 acres to 3,461, filling in marshes and bogs and every inch of the riverbank.[16] Those very local ecological changes were tied to changes in global oil markets and international politics.

The Oil Boom, 1913–21

World War I gave enormous impetus to Mexican oil production. In 1913, England began seeking sources of fuel oil for its navy to make the shift from coal to hydrocarbons. Weetman Pearson, a Lord in Parliament since 1910, was happy to oblige and place his company, El Aguila, at his majesty's service.[17] In the United States, New Mexico's senator Albert B. Fall, had a similar idea in 1913, wanting to use Mexican oil for the American merchant marine. Doheny, his close friend, offered the production of his Mexican enterprises.[18] The war, then, turned petroleum into a strategic commodity with guaranteed markets for the foreseeable future. World War I was the first major armed conflict to run its death machines wholly on oil. Petroleum fueled the trucks, tanks, ambulances, buses, battleships, motorboats, and planes that transported men across the battlefields.[19] The demand was immense, prompting a rapid increase in Mexican oil extraction. Production rose from twenty-five million barrels in 1913 to fifty-five million in 1917, the year Mexico became the second oil producer in the world behind the United States, despite the ups and downs in prices created by gluts in global production for the duration of the war.[20] The acceleration in extraction meant a construction bonanza not only in downtown office buildings in the New Orleans style of wrought-iron balconies and windows attached to colorful wooden structures, but also in industrial infrastructure. What would distinguish the oil boom from those in other industries would be the heavy environmental footprint and long ecological shadow petroleum would cast upon the city and beyond.

In the 1910s Tampico began to consume the landscape in earnest. The oil companies developed the city into a world-class oil port, with hundreds of miles of pipeline connecting it to the Huasteca oil wells. The pipelines emptied the crude oil onto hundreds of storage tanks and dozens of terminals in Tampico, while fleets of tankers lined the Pánuco to transport the "black gold" to the United States and England. The changes in the land were massive. The mouth of the Pánuco changed. The river was dredged deep to accommodate the tankers, and the beach was broken up by long jetties that guided ships into the proper deepwater channel. The course of the Tamesí River also changed, as construction companies filled in sections to reclaim land for the refineries and to expand the city's limits. South of Tampico, the industry's environmental shadow loomed large. The forest was consumed and replaced by wells, workshops, and workers' camps. Mangroves were uprooted for wharves, piers, and docks, while marshes were filled for housing, recreational spaces, storage tanks, and wharves.[21]

While the armies of the Allied nations were consuming most Mexican oil, small local markets also created a demand; one was the industry itself. The oil companies used petroleum to pave roads from Tampico to the camps, to fuel boats, ships, and tankers, and to cook the oil in refineries (see below). They also fueled all of Mexico's railroads, which from 1914 to 1920 were commandeered by the two sides fighting in the Mexican Revolution (1910–20), the rebels, and the federal army. Thus ironically enough the oil industry fueled Mexico's own most important military conflict quite literally.[22]

Amid the Mexican Revolution (1910–20) and World War I, the oil companies built five refineries within and next to Tampico. Their principal task was to separate the most volatile compounds from the petroleum to make shipping it abroad less hazardous.[23] However, the refineries also satisfied the Mexican market for industrial lubricants (oil and greases), kerosene, gasoline, gas oil, solvents, and waxes.[24] Until the Great Depression, the refineries only used about 10 percent of the total oil extracted locally. The bulk went both to destroy and to develop the Western economies.[25]

Even so, the refineries had a deep environmental impact on Tampico and its inhabitants. Their demands for resources were heavy. They required a great deal of energy to function themselves. Although oil fueled much of the distillation and refining process, the refineries demanded their share of the electricity produced by the British-owned electricity company. They also sucked up millions of gallons of water from the Pánuco River.[26] Their most profound effect, however, arose from what they produced in tandem with oil: pollution. The refineries dumped their toxic waste and by-products into the air and water. The Pánuco River was so loaded with hydrocarbons that it combusted, twice reaching tankers loading fuel at midstream and exploding them.[27] Local beaches were ruined. As the Tampico Chamber of Commerce complained in 1920, the sand and water were "totally covered with *chapopote* and one can neither walk one step without one's shoes getting soaked by such bothersome and sticky object nor take a bath without one's body being totally tarred."[28]

The entire oil-producing complex centered in Tampico also cast a deleterious ecological shadow on the local marine environment. Petroleum wastes ended up in the Mexican Gulf. The ballast from oil tankers polluted rivers and shores. Spills from the Huasteca wells to the south were channeled toward rivers and streams and into the open ocean. Undoubtedly marine creatures, from plankton and crabs to marine mammals and birds, suffered the consequences, but no one documented the facts.[29]

The oil production boom created by the demands of American and European war economies ended soon after the drums of war fell silent. The reasons were complex and multiple, but one of them was ecological recklessness. The companies exploited the oil fields south of Tampico at such speed and with such carelessness that they exhausted them by 1921. Saltwater intruded the wells and although the geologists knew there was oil deeper underground, the technologies available did not allow them to reach those deposits. The oilmen moved to blacker pastures, east of Tampico, and to Venezuela. The refineries continued processing crude from the remaining wells and from new ones in the eastern fields of Ebano, but the port felt the shock. The companies fired half of their forty-thousand-strong workforce in mid-1921. Tampico stopped growing at an accelerated pace, but the political and social upheaval that ensued resulted in the division of the city into two municipalities: Tampico and Villa Cecilia. Tampico would never regain a top spot in energy production, losing its identity as a "ciudad petrolera" to its neighbor.

Two Municipalities, One Declining Energy Metropolis, 1921–38

Tampico's oil industry remained tied to international petroleum markets through the 1920s and 1930s. As such, its fortunes sagged, but the social changes that the oil industry also produced guaranteed the city a privileged position in the politics of oil. Thus, Tampico played a significant role in the mitosis of the city into two municipalities in 1924 and the nationalization of the oil industry in 1938.

Despite the 1921 bust, Tampico's oil still had buyers. Growth in the American automobile industry that the internal combustion engine made possible opened up a new market for Mexican petroleum. The oil companies not only fueled American cars, they brought the first passenger vehicles and trucks to Tampico and the oil fields, creating new demands for gasoline, and paved roads and the first city highways. With the automobile, Tampico grew even more. The city's wealthy moved to outlying neighborhoods (away from the stench of oil and its products) or small towns on the outskirts. Cars and taxis appeared to provide transportation for those who could afford faster service than trolleys and tramways. Commerce flourished as fleets of cargo trucks moved goods throughout the region.[30]

Tampico was also the first city in Mexico to enjoy air travel, thanks to the petroleum industry. That innovation came about as a local American banker came into possession of two decommissioned planes from the U.S. Army and set up an "air taxi" service. George L. Rihl installed a runway on an island in the middle of a lake, Moralillo. Having secured contracts with the oil companies to transport payrolls and officials, Rihl inaugurated Mexico's first airline and airport, the fourth in the world, on July 12, 1921. The planes were fueled with oil from the same fields it served. By 1924, the Mexican Transportation Company was flying between Tampico, Mexico City, and Matamoros delivering mail, money, and oil moguls.[31] The company moved the airport to a much larger location on Tampico's north side in 1929, where it remains to date.[32]

As the nucleus of energy production, Tampico endured its share of social and political conflict. The port hosted a politically active working class, first focused on the longshoremen and the railroad workers. Both had marked anarchist roots that were strengthened by European migrants who joined the oil industry over the course of World War I. The oil workers aggressively entered into the political fray, as seasoned craftsmen rushed to organize the thousands hired for the new industry. Strikes, demonstrations, and street battles over wages and working conditions became commonplace. The oil industry generated changes so quickly that the city was not equipped to absorb them.

The inadequacies were glaring. The housing stock could not handle the sudden swelling of the population, so improvised hotels and flophouses lined the floodplain to fill in the gap. Forced to provide some remedy, the companies housed craftsmen (mechanics, carpenters, welders, electricians, etc.), leaving masses of construction workers to their own fate. Workers squatted in every empty piece of land they found, creating working-class shantytowns around the refineries and the lagoons. Similar inadequacies plagued other areas: running water, drainage, sewage, electricity, garbage collection. Pavement, potable water, and street lighting did not extend beyond the fashionable downtown or the fancy neighborhoods.[33] Poverty, thus, increased faster than wealth and it was more widespread. The banks overlooking the main plaza did not invest their money in the city at all, repatriating it to the United States and Europe instead.[34]

At the same time, the Mexican Revolution that started in 1910 had economic repercussions for the growing energy metropolis. Inflation, the typical result of rapid and uncontrolled growth, was made worse by the revolution. Food shortages plagued Tampico through the 1910s, fueling social unrest.[35] While the oil industry consumed nature and human energy, Mexican workers ingested less than their bodies needed to be healthy. Pushed by militant men to address poor conditions, the companies opened commissaries and sold American foodstuffs and goods on credit. Instead of alleviating the problem, however, the company stores made it worse. Workers' wages were never enough to cover their debts at the end of the six-day workweek.[36] The result was, in the end, more social conflict.

Tired of a decade of protests, strikes, and marches, Tampico "resolved" the issue of class struggle by splitting the city into two in 1924. The initiative came from the workers themselves. In the aftermath of massive unemployment created by the 1921 bust, workers' organizations came up with the idea of lopping off the eastern end of Tampico for a new city. That side included three refineries and, thus, the overwhelming majority of working-class neighborhoods. The idea was that a separate municipality would grant land for farming to the unemployed and for home building to the crowded. Nothing came of the proposal until an unexpected turn of events in late 1923 and early 1924 resuscitated it. First, Edward L. Doheny found himself embroiled in a scandal over drilling concessions obtained illegally by his buddy the congressman from New Mexico, Albert B. Fall, at the navy's oil reserve at Teapot Dome. The news sent jitters through Tampico's commercial elite, dependent on the stability of the oil industry for their income. Then, the electric company workers went on strike in November, bringing the port to a standstill until January 1924. The conflict escalated when the refinery workers at El Aguila, which Weetman Pearson had sold to Royal Dutch Shell in 1919, protested the docking of pay for the time the refinery remained closed during the blackout the electric company strike provoked. In March 1924, El Aguila workers walked out, while some twenty other Tampico unions threatened to stage a general strike. To prevent the total paralysis of the port, the city council decreed that all the neighborhoods east of Tampico constituted a separate municipality. Villa Cecilia was thus officially born on May 1, 1924. The new municipality covered forty-four square miles of marshland and mangrove in various states of degradation and urbanization and included nearly fifty thousand workers. It also inherited a myriad of problems in addition to militant labor unions: infrastructure, public health, pollution, flooding, unemployment, and an oil industry in frank decline.[37]

The period between 1926 and 1931 is described as a time "from depression to Depression" in the Mexican oil industry.[38] Overproduction in global markets led to company mergers, plant closures, layoffs, and labor-saving policies and technologies that devastated Tampico and Villa Cecilia. The stock market crash of 1929 put the nail in the coffin of oil production in both locales. By 1930, the twin cities showed the devastation that the disruption in global markets caused in the region: "miserable hovels made out of rotten wood, semi-nude children with starved bodies, meat stores with a couple of slabs covered by flies . . . immobile men sitting by the side of the road; misery, filth, indolence. On the other side of the river, houses that used to be two or three stories high, abandoned and crumbling, boilers turned upside down, walls falling apart, broken storage tanks."[39] The population of both cities combined was 70,000. The port was crumbling. The rain and the periodic floods eroded the jetties, clogged the Chijol Canal, and deposited tons of silt at the bottom of the Pánuco River. Measured in 1930, the sandbar at the mouth of the river had risen to levels not seen since before 1910. The oil tankers that just ten years previous clogged its sinuous curves were few and far between, "lone ghosts" in a "graveyard" of a city.[40] The final blow came in September 1933, when two hurricanes struck the cities on the fifteenth and twenty-fourth. The twin storms left the twin cities "in ruins," according to the *New York Times,* with some 67,156 people adversely affected.[41]

It was not that oil production had ceased to be in Mexico. Not at all. The companies continued extracting black gold, albeit at a much slower pace, but the oil fields were not in Tampico's hinterlands anymore. They had moved south to central and southern Veracruz and Tabasco. The companies also had a new market: the growing fleet of trucks, cars, and buses of Mexico City, not yet choked on soot and smog. Until nationalization in 1938, Mexico's oil consumption steadily rose from 30 to 39 percent of national production.[42] But that the national market was not to be supplied from Tampico.

Despite the generalized economic depression and because of it, labor conflict remained constant in Tampico-Ciudad Madero. Workers who survived plant closures organized unions; layoffs led to impromptu strikes or work slowdowns in protest. And even though the workers reaped more defeats than victories, they did not give up. In 1936, oil workers from all over Mexico unified their organizations into one single industrial union, the Sindicato de Trabajadores del Petróleo de la República Mexicana. The half dozen Tampico-Ciudad Madero locals were no longer the most numerous, but they commanded political power due to their historical importance and their unwavering militancy. The battle over the first contract was fierce and violent, not only in the twin cities but at new refineries and fields elsewhere. It lasted two years, until circumstances pushed President Lázaro Cárdenas to resolve it by nationalizing the industry on March 19, 1938. The era of wholesale foreign ownership of the oil industry had come to an end a decade after Tampico had lost its position as Mexico's energy metropolis, although hydrocarbons (and pollution) would persist as a key sector in the regional economy and enjoy a revival in the early 2000s.

Tampico's Legacy

In the post-nationalization period, Ciudad Madero enjoyed more political clout than Tampico in oil matters, but neither enjoyed an economic renaissance. Production flattened as the foreign companies boycotted Mexican petroleum and the new owners, Petróleos Mexicanos (PEMEX), reorganized the industry. It would not be until the 1960s that the industry would enjoy a local revival with the installation of a petrochemical plant in Ciudad Madero. In the 1970s offshore drilling brought high hopes but few results. It was Altamira which became important, hosting petrochemicals and the first Mexican LNG-receiving plant in 2006. Thus the entire region continued to be tied to hydrocarbons, however haltingly, in the new century.

It took time for PEMEX to stabilize the industry after nationalization. Loss of foreign markets, strained relations with Britain and the United States, and the repatriation of foreign experts took their toll. Labor relations oscillated between distrust and hostility, as well, as PEMEX expected obedience from a workforce with a history of defiance of authority.[43] From millions of barrels extracted during the boom of the late 1910s and early 1920s—190,000,000 at the peak in 1921—Mexican production limped between 200,000 and 350,000 barrels per day through 1964.[44] By 1976, production had climbed to 890,000 barrels per day, still less than in 1911.[45] What those numbers meant for Tampico was the final closure of oil installations. Only one plant remained, the former El Aguila (Royal Dutch Shell) refinery, and it

Table 8.1. Population of Ciudad Madero and Tampico, 1990–2005

	1990	1995	2000	2005
Ciudad Madero	160,331	171,091	182,325	193,045
Tampico	272,690	278,933	295,442	303,635

Source: Instituto Nacional de Estadística, Geografía e Informática, "Mexico: Tamaulipas," http://www.citypopulation.de/Mexico-Tamaulipas.html.

was in Ciudad Madero.[46] Thereafter, Ciudad Madero would be the focus of hydrocarbon production until Altamira surpassed it in the early 2000s.

Tampico meanwhile, developed as any ordinary third-world city. It had spent a decade as an energy capital without a single secondary school.[47] Not until 1950 did Tampico inaugurate its first high school, only to see it severely damaged in a destructive 1955 hurricane.[48] Through the 1950s, the port's main economic activity, beside naval construction and collection of customs duties, focused on light industry, such as ice making, bottling carbonated beverages, and making rattan furniture. Commercial fishing was also important and sustained a number of canneries.[49] Tampico got a shot in the arm in 1970 with the discovery of offshore oil just eighteen miles into the open waters of the Gulf. The find spurred economic activity, but production was a disappointment: sixty thousand barrels per day in 1970 dropped to forty thousand by 1979. By the 1980s, all new investment in the oil industry had moved south, although Tampico served as the locus of hydrocarbon transport to northeast Mexican markets.[50] Thereafter Tampico and Ciudad Madero grew slowly, as table 8.1 shows.

Altamira, fifteen miles north of Tampico and nudged between the Champayán lagoon and the Gulf, took the lead in the local hydrocarbon industry. Between 2005 and 2009, twenty-seven foreign and domestic petrochemical factories and assembly plants operated in the city, catapulting it to one of the top four ports in the country. In August 2006, in fact, Shell, Total, and Mitsubishi inaugurated the first LNG plant in Mexico in Altamira. The LNG plant processes gas not from Mexico's own natural gas fields (which remain "underdeveloped" due to lack of capital and technology), but from Shell's oil fields in Nigeria. The market for the product is the Mexican electrical company and the United States.[51]

The cycle of growth, inadequate services, and environmental destruction started anew.[52] In 2001 Greenpeace reported that one of the petrochemical plants at Altamira was "emitting significant quantities of vinyl chloride to the environment," at concentrations that are "over 15 times the concentrations that would be acceptable for a European factory" and "between 58 and 91 times the daily maxima set by the US government for plastics manufacturers." In addition, the scientists testing the water reported that volatile organochlorine chloroform, hydrocarbons, phthalate esters, and zinc, all toxic and some known carcinogens, were being released into water and air.[53] In 2009, a local news magazine reported similar problems. "Reddish water emanating from industrial wastes turn black, frothy, or bright green, staining the group of interconnected lagoons in Altamira," read the article, which pointed out that there is no local water treatment plant. The news alleged that up to three hundred seventy thousand cubic meters of waste was dumped into the water on a daily basis, as recorded in official documents. Among the contaminants were heavy metals such as arsenic, cadmium, copper, chrome, mercury, nickel, lead, and zinc.[54] In the days following, the public echoed the news reports, arguing that they had seen massive fish die-offs, in addition to contaminated fish, crustaceans, and reptiles. Municipal authorities and the Association of Industrialists of Southern Tamaulipas vigorously rejected the allegations.[55] Thus, 150 years into oil extraction and refining, sophisticated advances in technology and manufacturing notwithstanding, oil development in Mexico followed the same pattern of consuming nature and producing uncontrolled pollution.

As other ports took the lion's share of oil exports, moreover, the erstwhile energy metropolis of Tampico realized that it faced great odds to compete. The port was "corralled" by the city itself and could not expand without destroying neighborhoods and substantial investment. Ships, likewise, kept growing in size and volume, which demanded dredging and new terminals beyond the city's budget.[56] Searching for alternatives, Tampico bet on tourism. City officials and entrepreneurs today highlight the beauty of the city's lakes and lagoons and the open waters of the Mexican Gulf. Oblivious to the history of punishment of the port's water bodies and apparently unconcerned with restoration or the persistence of toxins, boosters invite visitors to hunt, fish, race boats, swim, and practice water sports.[57] They ignore the smell of petrochemicals and forgo mentioning that the shells the waves toss onto the beaches are not collectable. Instead of housing living beings, the shells are plugged with hardened asphalt.

Energy metropolises are contradictory places. They produce just as much as they endanger and destroy. They created and to this day sustain the oil-driven world economy with all its conveniences, rapid transit, and innumerable consumer products, yet just as definitively they changed landscapes beyond recognition as they fouled water and air. Most significantly, they have done their part in generating global climate change. Tampico, the port at the edge of the rainforest and the mouth of the Gulf, was no different—creating and destroying simultaneously like the gods of ancient times. Its legacy of job creation and economic dynamism lives in Altamira, while the dark side, ecological degradation and destruction, is altogether hidden: the Huasteca rainforest is no more. Whether being an energy capital for one decade was worth the cost is for Tampiqueños and the world to decide.

Port-Gentil

From Forestry Capital
to Energy Capital

Douglas A. Yates

On September 3, 2009, violent riots broke out in Port-Gentil. Suddenly the world focused its attention on this undersized Atlantic seaport with its diminutive hovels and dumpy squats. Touted as the "oil capital" of Gabon, global television cameras instead revealed dirty markets, streets paved with garbage, pigmy doghouses, ramshackle sheds, and puny porticos from which swarmed town dwellers dressed in rags, marching in anger against their corrupt and patrimonial regime. "GABON RAMPAGE AFTER POLL RESULTS" headlined the BBC, giving these riots a political interpretation: "Opposition activists clashed with security forces, after election results confirmed Ali Ben Bongo with 42% of the vote. Critics say the poll last weekend was fixed to ensure Ali Ben Bongo succeeded his father, Omar Bongo, who ruled the oil-producing nation for 41 years."[1] But these riots were more than an isolated incident provoked by fraudulent elections. There were deeper causes of discontent. Angry mobs stormed the municipal jail and freed its inmates. They rampaged through the streets and set fire to the French consulate. They also attacked installations belonging to the French company Total. "GABON LOCKS DOWN CITY AMID RIOTS" announced the news later that evening after the government declared a nighttime curfew: "France, Gabon's former ruler, has told its 10,000 citizens to stay inside."[2]

The next day French businessmen read in their newspapers that protesters had ransacked their supermarkets in the city, set fire to the French consulate in Port-Gentil, and damaged part of the ground floor of the headquarters of Paris-based Total Gabon. Total's director of human resources said, "There was an upsurge of violence last night, and vehicles were burned in the parking lot, but the damage was limited by the curfew established by the police and Gabonese army."[3]

On the third day of rioting Total pulled out its expatriate personnel because it was clear that some rioters were targeting company buildings in the city, while others fought with security forces in the streets.[4] Protestors burned down a company sports center. "Some Total employees and their families have been transferred from Port-Gentil to Libreville in a temporary move for their safety," said a Total spokeswoman in Paris. French state television showed images of French families with children disembarking from boats in Libreville. One mother, who did not give her name, said she had wanted to leave Port-Gentil because the situation there was frightening. A French resident of the oil city, who gave his name only as Pascal, told state television that he had seen the Total leisure center go up in flames from his residence. "We could see the flames very well and we could hear gunshots and fighting. The building was completely burnt down," he said.[5] Prisoners who had been set free or escaped from a local jail were roaming the streets and set fire to a Total gasoline station, a resident told French state radio, France Info: "There are thousands of people. There is the opposition, but it is not just the opposition," he said. "Everyone is taking advantage of this, not just the prisoners."[6]

The government reacted with predictable violence, calling out its infamous "red berets." Hundreds of protestors were arrested, and at least fifteen people were killed during the violence, according to Pierre-André Kombila, an opposition deputy, suggesting the final tally could be dozens of deaths. Information from the morgue at the Ntchengué hospital, south of the oil city, reported fifteen corpses. "But we still have to visit Hospital Paul Igamba," he said. "There were several dozen deaths in Port-Gentil, according to different eyewitness reports I received. There was one rumor that the army was taking bodies and throwing them in the sea by helicopter. An international investigation needs to be put into place, interrogating honest members of the security forces, in order to arrive at the truth."[7] Ali Bongo did not remove the curfew until December 30, 2009. Over sixty protesters were tried for vandalism and other crimes against public order. No investigation was launched into the deaths of the protesters.

The Colonial Legacy

From its colonial origins as a French forestry enclave to its neocolonial present as an oil enclave, this place has long been the object of exploitation and of economic marginalization. Pent-up anger at two centuries of injustice is now boiling over in this equatorial African Atlantic seaport. Why is Gabon's second-largest city, the center of its petroleum and forestry, and the major port for some 80 percent of its total exports, a hotbed of political opposition? Why is this city, so rich in natural resources, still so very poor? Why does it suffer from the "paradox of plenty?"[8]

It is not because its abundant natural resources must be shared with a very large population (as is the case, for example, in the nearby Niger Delta). The population of Port-Gentil has never been very large: only 4,500 inhabitants in 1947, growing to 10,000 (in 1955), 21,000 (in 1960), 31,000 (in 1970), 60,000 (in 1985), 90,000 (in 1993), and to an estimated 150,000 today.[9] By comparison, over half of the population of 1.3 million lives in the capital city of Libreville. Nor is it because Port-Gentil is located in a central geographic area where its neighbors can exploit its natural resources. This oil capital is located on an island. No bridge connects it to the mainland. Port-Gentil is located at the deepwater mouth of the Ogooué River, on a large peninsular outcropping twelve kilometers long and six kilometers wide, where an immense low-lying bank of sandy clay opens to the Atlantic on one side, and on the other, along the bay, receives the waters of the Ogooué, which is the largest navigable river in Gabon. To find it on a map, just follow the line of the Equator to the coast of West Africa. There you will find Port-Gentil on the Cap Lopez peninsula, a little south of the equator, at an extreme western part of the country. The nearby mainland is a remote forested area unconnected by road to the rest of the country. A small airport serves as its main channel of communication. The solution to the paradox of Port-Gentil's poverty is to be found in its colonial history.

The original colonial French settlement of Port-Gentil was established on "Mandji Island." But the city takes its name from Émile Gentil (1866–1914), a French naval officer who had conducted the first hydrographic soundings along the coast between the Gabon Estuary and the mouths of the Ogooué River (1890–2), before being named lieutenant governor (1902–4), then commissioner general (1904–8) of the entire "French Congo."[10] Retaining the name of this French colonial officer is Port-Gentil's most visible legacy of French neocolonialism. This colonial history, and subsequent establishment of a neocolonial oil enclave economy, has offered scholars the best explanation for why it is still so underdeveloped and poor. Emile Gentil, who had conducted many investigations into abuses by French "concessionary companies" in 1905, also organized the very first public education in Gabon in 1907.[11] But he also remains an important figure in the dark history of French colonialism in the region, involved as he was in bloody military expeditions that established French control over Chad and Ubangi-Shari (today Central African Republic), two other equatorial territories that were later merged in 1910, with Gabon and Congo, into the vast colonial federation of AEF, Afrique Equatoriale Française (French Equatorial Africa).[12]

The French had first established a settlement at Port-Gentil in 1880 after signing two treaties with the indigenous Orungu who granted France the right to set up a trade outpost at Cap Lopez peninsula, being strategically separated from Gabon's mainland by numerous rivers that are at the mouth of the Ogooué. Who were these original indigenous inhabitants of Mandji Isle? The Orungu of the Ogooué delta were a Myène-speaking people who lived along the coastal creeks of the river and traded with numerous other Myène speakers living on the Atlantic coastline.

As the slave trade came to an end they signed a treaty in 1873, allowing the French to establish a post near the delta, its harbor protected from western ocean swells, which offered a vast basin for storing lumber floated downriver, an ideal location for a port.[13] The Orungu called the beach on Mandji Isle where Port-Gentil is now located, "Abendja." The name "Mandji" itself was taken from the name that local fishermen had given to a giant tree beneath which they built their shacks and which had also served as a landmark for vessels navigating around the island. This was how the island came to be known as "Mandji Isle," designating, at first, only the islet itself, then the whole maritime region of the Ogooué delta.[14] By the eighteenth century, mapmakers adopted the Portuguese appellation "Cap Lopez." By the twentieth century maps tended to locate a dot with the name of "Port-Gentil."

Although this place has over three centuries of recorded history, it really only developed into a city under the French, becoming an important center of exports during the twentieth century thanks to the establishment of operations by multinational enterprises, which for the most part were linked to natural resources, principally wood and oil. Of these two resources, it has been the exploitation of oil which authors tend to link to the organization and management of the modern urban space. But forestry came before oil. And forestry shaped an enclave economy.

From this colonial outpost there emerged a city whose subsequent history never deviated from the original pattern of settlement: a small port serving the needs of foreign businesses; urbanization reflecting the racial/cultural separation of European from African populations. Implanted on the island was a small European community, exploiting its natural resources and serving the interests of imperial France.

Forestry and the Rise of a Colonial Enclave Economy

At its origins, this port was a landing wharf, a simple beach where dugout canoes landed with a few French naval vessels anchored offshore. The east coast of the island was a vast plateau with satisfactory ventilation, and not very swampy, unlike the rest of the region. From there on its location on the delta of the Ogooué (called "Bay of Cap Lopez"), the town transformed into a center of trade, an important unloading point for cargoes from the interior of the country. In 1910 the French built a cement pier, two hundred meters long and twenty meters wide, fitted with iron pillars, around which they built three paired hangars for merchandise, a small workshop, and a coaster for loading cargo. The town only covered a dozen hectares, with an indigenous population of some two hundred inhabitants who settled along a plain living in shacks made of tree bark, and a small community of European merchants living on a grid of streets traced near the wharf.[15] Then in 1912 a Catholic priest named Théophile Klaine (1840–1911), an amateur botanist, became attracted by the apparent virtues of an abundant species of a tree called "okoumé," which had qualities that made it a useful construction material. Okoumé—a soft mahogany used in the production of plywood—was later named *Aucoumea klaineana* in honor of father Klaine.[16] The Orungu had been using this tree to make their dugout canoes, and they burned its resin on torches. In 1912 the French Société Commerciale, Industrielle, et Agricole du Haut-Ogooué (SHO) sent seven tons of this wood to France, where it sold well. They shipped it to Hamburg to replace Cuban cedar in the making of cigar boxes, thereafter to German and Dutch firms to make plywood. By 1930 okoumé production was 381,000 tons. Prosperity due to this one species profited the nascent colonial economy, as the immense basin of the Ogooué contained considerable quantities. Navigable parts of the river were used to float thousands of tons of okoumé logs toward Port-Gentil.[17]

In 1940 the harbor was fitted out by the French lumber company, which installed itself there. The local population had grown to three thousand inhabitants, half of whom were engaged in activities related to wood. By 1942, the town's commercial prosperity necessitated some kind of sea facilities that could enable naval vessels to berth. The French constructed a breakwater, 216 meters long, which remains today, permitting the berthing of barges with a maximum capacity of 200 tons, pulled by tugboats. Its quay possessed some 6,250 square meters of open-air storage space and 4,473 cubic meters of hangar space.[18]

Société Commerciale, Industrielle, et Agricole du Haut-Ogooué (SHO) was the most important concessionary company in Gabon. It held a commercial monopoly over 104,000 square kilometers during the high colonial period. Founded at the urging of Savorganan de Brazza, its presence was based on the 1899 French policy of developing AEF through some forty such concessionary companies. Because it was better funded and directed, it possessed more extensive power, was burdened with fewer responsibilities, and lasted twenty years longer than most of the other concessionaires. SHO had its own private police force and administered justice in its territories. It was one of the infamous agents of the "red rubber" trade excoriated by the British from which it took exorbitant profits.[19]

In the 1940s the company was presided over by Luc Durand-Reville (1904–98) during whose era the SHO became one of the three largest commercial groups in the French African colonies. Durand-Reville had been born in Cairo, where his father was an accountant for a trading company. He studied business in Paris at the Hautes Études Commerciales (HEC) and then studied law at the Sorbonne, in preparation for a career in business, before voyaging to Gabon in 1942 and thereafter becoming the president of SHO. When the French Fourth Republic created a few representative seats for its overseas possessions, Durand-Reville had become so influential that he abandoned his functions at the SHO to represent the colony of Gabon in the French Senate (1947–58) or rather, represent French lumber interests. He intervened on many important issues, like financial assistance to French businesses in overseas territories. He used his business connections and political offices primarily for the benefit of French foresters.[20] For example, he opposed the penetration by non-French foreign firms and blocked African family forestry operations, or *coupes familiales,* from exploiting their own forests.[21]

Even today the wharfs of downtown Port-Gentil are heavily occupied by the forestry firms. Okoumé became the single most important commodity of Gabon, and by 1957 the shipments of the tree were furnishing over 87 percent of Gabon's total exports. By independence in 1960 production of okoumé had risen to 737,000 tons. By the 1970s its production generally exceeded 1 million tons per year. This one single species of mahogany still accounts for around two-thirds of all of Gabon's wood exports today.[22]

French colonialism had a profound social and cultural impact on the people of Gabon. Political and power structures were altered as France began to establish a centralized state. Rural peoples were encouraged to alter their livelihoods from those based on local, small-scale agriculture to plantation agriculture and extraction of natural resources. The colonial government relocated many rural villages, moving them closer to transportation routes in an effort to increase the profitability of large-scale capital-intensive projects and to encourage employment in mining and timber camps. Many rural Gabonese still resent these forced relocations and blame French policies for the hardships they endured in the process. Migrants to Port-Gentil outnumbered the original Orungu inhabitants. All adopted French language and customs. Colonial administrators encouraged their elites to learn French and undertake French-style education. Fluency in French became an important element of assimilation and advancement. A network of elite Gabonese developed with strong cultural and political ties to France and French interests. Conversion to Christianity became common. Mission-educated Gabonese found well-paying work with the French commercial firms. Employment opportunities were also available in the French colonial administration. Positions as clerks and administrators helped educated Gabonese earn money and prestige. This type of employment also provided an avenue toward becoming politically active and helped create the elite Gabonese political class.[23]

When political independence came in 1960 the foresters needed a willing pro-French collaborator who would protect their business interests. The man they chose was an assimilated Fang politician, Léon Mba (1902–67) who was the first black mayor of the colonial capital city. Mba had been born in Libreville, the son of a Fang chief. His brother had been the first Fang to be ordained a Catholic priest. The numerous Fang immigrants were held in low esteem and feared by the indigenous Mpongwe who had been privileged by colonial administrators during this period. But they were being actively evangelized by the missions and educated by the priests, giving them a path to advancement. Léon Mba was educated in Catholic schools and became a Fang interpreter in the colonial administration. Later he worked for the John Holt Company and made contacts with the influential French foresters. Two key elements thus explain how he rose to power in Gabon: The Fang were the largest ethnic group, an arithmetical advantage in mass elections, and Mba worked for the foresters, a source of support. With their backing, his party won the first legislative elections (1957) and he became the first prime minister (1958), then the first president of Gabon (1960).

De Gaulle's adviser on African affairs was Jacques Foccart (1913–97), a white from the French Antilles who had joined the Resistance under de Gaulle and, after being on the winning side of the war, ran an import-export business in the empire. By the 1950s he was an important financial contributor to the RPR, the Gaullist political party, and organized the SAC, an informal secret intelligence network of spies and commandos to defend French power and interests throughout its African empire. Foccart had been involved in efforts to bring de Gaulle back to power in the murky events of May 1958, after which de Gaulle named him as the presidential adviser on "African affairs."[24] It was Foccart who had pushed de Gaulle to break up the two large colonial federations of French West Africa (AOF) and AEF into a dozen smaller countries before granting self-rule. He argued that it would be easier for France to control and dominate a dozen smaller African states than two large federations. He thought that Gabon, where petroleum resources were being developed at this time, would be easy to control because it was small and because it lacked roads. From his office in the Elysée he carried out clandestine operations through a shadowy "network" known as the réseau Foccart. With agents at every port watching and listening as his eyes and ears, Foccart masterminded numerous coups against independent-minded nationalists while installing in power a new generation of African rulers loyal to de Gaulle and himself. In Gabon, he orchestrated the victory of Mba with the collaboration of the French foresters under the leadership of Roland Bru (1916–97). Bru was another decorated war hero in the Resistance who had risen to prominence in Gabon during a struggle over okoumé prices that arose between the colonial wood office and the timber lobby Syndicat Forestier du Gabon (SFG). The colonial wood office had been created after the war to keep wood prices cheap in accordance with the needs of France. In response the foresters organized themselves behind Bru, who butressed okoumé prices and even acquired aid from the state to construct sawmills along the Ogooué. A side effect of large-scale mechanization was to drive out small concessionaires and to concentrate even more power in the hands of the French timber syndicate.[25]

During his first years in power President Mba tried, at first, to promote real autonomy for his country, but after a failed coup d'état in 1964, when French paratroopers restored him to power, he was reduced to a puppet who took his orders from Foccart.[26] Here is the neocolonial context of the early period of independence. Gabon was an African country, geographically shaped by the French, colonized by the French, forested by the French, led by a francophone assimilated elite who spoke French, read French newspapers, received French education, practiced French law, worked for French businesses, and who had adopted a French model of government which was additionally dominated through a system of French "co-operation." Yet the foresters never modernized their sector, which essentially exported logs. Large trees were felled near the rivers, and then simply floated downstream to the ports, to be shipped off, mostly raw and unprocessed, as timber. This meant that Port-Gentil never really enjoyed the integrated activities that might have evolved out of a rich timber industry, but remained an entrepôt, where logs were stored and loaded onto cargo vessels to be processed beyond the seven seas. Were this to have remained the only economic activity of Port-Gentil, it would not have become an important city.

Two elements characterized colonial urbanization in Port-Gentil. First there was the racial separation between the European part of the city and the African part. Streets, boulevards, and neighborhoods were nonexistent during the late colonial period, with the exception of a few sandy paths that were unusable during the rainy seasons. Colonial authorities only managed to meet the needs of the European quarters by installing some primitive roads. A second element was the development of markets along the coast, the oldest being the "Marché de la Ville," followed by "Marché de la Balise" and "Marché du Grand Village."[27] These markets developed local petty commerce where growing immigrant populations, not only immigrants from the rural areas but also from other French African colonies, established themselves. Locals tended, on the contrary, to concentrate their activity on fishing, gathering bananas, and producing taro or other produce consumed locally at that time. In the boutiques that multiplied along the roads, Senegalese, Malians, and other West Africans came to control the bazaars, while European and Lebanese businessmen took advantage of their access to capital and foreign imports, gradually building cement markets and selling to white expatriates who worked in the forestry sector. This was the situation until the development of petroleum in the late 1950s.

The Rise of the Oil Industry

The existence of petroleum in Gabon, both onshore and offshore, had long been known to the French. In 1929 geophysical crews from the French Compagnie Générale Géophysique (today Schlumberger) had demonstrated its presence with electrical prospecting techniques. But the metropolitan French did not invest capital in developing these finds, in part because their privately owned national oil champion, Compagnie Française des Pétroles (today Total) was purchasing abundant cheap oil from its partners in the Middle East, and in part because the French state lacked the capital and technology in the 1930s to exploit colonial oil by itself.[28] The first prospecting mission sent into the region in 1932 found superficial indices and led to the creation of a Syndicat d'études et de recherché pétrolières, but never however managed to attract investment in the finds.[29] Only after World War II did the French state develop that capacity and capital, under the aegis of the Bureau de Recherche du Pétrole (BRP), which financed a colonial oil company, Société des Pétroles d'Afrique Equatoriale Française (SPAEF), to conduct drilling operations in Gabon and the Middle Congo.

SPAEF purchased a heavy rig from the United States and shipped it across the Atlantic to Port-Gentil on Mandji Island. The first wildcat wells were drilled in 1947 around Port-Gentil, and by 1951 a small oil reservoir was found. Two more rigs were brought in to determine if the wildcat was an isolated find or a trend. For the next five years SPAEF mapped out a trend of one hundred shallow salt-dome structures that ran up the spine of Mandji Island. While individually each inverted dome contained relatively small amounts of oil, collectively they represented a viable reserve. By 1957 eight SPAEF rigs were pumping enough crude to justify construction of a nine-mile-long oil pipeline to Cap Lopez, where the first cargo of "Mandji" crude was loaded for shipment to Le Havre. This was the beginning of the Gabonese oil industry.[30]

During the late 1960s petroleum became Gabon's leading export and source of government revenue. By 1967 spectacular growth had occurred with the coming into production of the Gamba-Ivenga onshore south of the Ndogo Lagoon and the Setté-Cama, and of Anguille offshore just south of Port-Gentil. Port-Gentil became the site of the refinery of the Société Gabonaise de Raffinage (SOGARA), which was Gabonese in name only. A second refinery was opened nearby at Point-Clairette. Over the decade crude oil production increased eightfold from 1.4 million tons in 1966 to 11.3 million tons in 1976, with offshore accounting for 80 percent of this total. Eighty-five percent of the production was exported, the remainder being used by Gabon and its francophone neighbors.[31] The transformation of the Gabonese economy from a forest enclave to an oil enclave resulted in greater Gabonese interest in petroleum, although true "Gabonization" of the sector was never achieved. During the 1970s the state did acquire a portion of the shares in the leading oil companies and a larger percentage of the revenues from production. In the expectation that these reserves would be exhausted by the end of the 1980s, the government gave companies incentives to undertake new prospecting for additional deposits. But a downturn in world oil demand began in 1977 that resulted in a decade of financial crisis for Gabon. The oil industry remained largely unrelated to the people who lived around it. Oil enclaves were gleaming sites of modernity surrounded by undeveloped wilderness and unemployed locals. Unlike the forestry sector, which at least needed some local labor, the oil industry was a capital-intensive sector that tended to hire skilled foreign workers to run its technical operations.

The rise of the petroleum industry in Gabon coincided with a political transition of power from Léon Mba, a Fang (the largest ethnic group) to Omar Bongo, a Téké (the smallest) and a clan of Franco-Gabonais conspirators who ruled the country by coalition of ethnic minorities. Omar Bongo (1935–2009) was born in Lewai—now Bongoville—a tiny village located in what was then the French Moyen Congo. After studies in Brazzaville, he joined the French colonial army, serving in Chad, Central African Republic, and the Congo.[32] In 1960, blacks like him who had served in the colonial army were sent back to their nominal countries of origin. Bongo found himself sent to a Libreville he had never really known. Fortunately he had army contacts in the Foccart network that got him a posting in the capital's new foreign ministry. Demonstrating a Machiavellian prowess, he was soon promoted to the presidential staff. In 1966 President Mba became ill with cancer and flew to Paris for treatment. Foccart paid frequent visits to his deathbed to pressure the ailing old man to change his constitution, introduce a post of vice president, and appoint Bongo as his running mate in upcoming 1967 elections. Bongo was a complete unknown to Gabonese voters. So Foccart paid for radio and press slots, flying the young Bongo around the country (which still lacked basic roads) to give speeches legitimating himself as Mba's running mate. When President Mba died in 1967, Vice President Bongo was installed in office. Bongo remained loyal to the men who brought him to power. He personally admired Foccart and de Gaulle, and he regularly declared his gratitude to the French for their aid in the "development" of his country. For the next four decades he cultivated strong personal ties with all subsequent French presidents and worked closely with the French businessmen who exploited Gabon's natural resources.

By the 1980s oil production was concentrated in four areas. The first was Port-Gentil, controlled by Elf-Gabon, and accounting for 76 percent of the national production. The second was Gamba, controlled by Shell-Gabon (also partly state owned), accounting for 8 percent of national production. The third was Maymba, also owned by Shell-Gabon, accounting for 7 percent. The fourth was Oguendjo, controlled by Amoco-Gabon, then owned by Standard Oil, providing the remaining 9 percent. In other words, after a quarter century of political independence, Gabon's oil industry was still dominated by the French company Elf, which controlled three-quarters of national production.[33] The longevity of French domination of the oil sector can be explained in part by abusive concessionary terms signed during the colonial era. SPAEF had been given a seventy-five-year concession to exploit Mandji Isle. Another concession at Point-Clairette, signed in 1957, gave exclusive rights to SPAEF for thirty years. By 1987 this field had been exhausted, the drilling equipment was cleared away, and the concession was left empty and in a state of pollution. Nothing had been "developed."

Port-Gentil did urbanize during the oil boom years. It was during the 1980s that Elf Gabon helped finance the construction of paved roads connecting the growing city and the port. These roads were named after powerful members of the ruling family, French statesmen, and French corporations, for example, Boulevard du Président Bongo, Avenue Charles de Gaulle, and Boulevard Elf-Gabon. It was Albin Chalandon, the president of Elf (1977–83), who had funded the construction of Boulevard Elf-Gabon to connect the forestry offices located on one side of town to the new Elf headquarters located on the other side. No street signs indicated the names of these paved roads. Only if one had a map, could one even know that no streets were named after local heroes, or that no street names made reference to the people who had once inhabited the region. The major urban projects of Port-Gentil were infrastructure related to the French businesses who were exploiting the country. Everywhere Port-Gentil had buried pipelines, oil wells, and petroleum terminals and—dominating the landscape—a refinery. As new oil was brought online, some 2,218 kilometers of pipelines arrived at the Cap Lopez terminal, dominating the north part of the island.[34]

In order to house the thousands of foreign workers, mostly French, who occupied the well-paying jobs in the oil sector, the company built large housing projects called "concessions." The largest of these was called La Grande Concession, a white community with twelve ivory towers, a modern housing project for expatriates, surrounded—in descending order—by Cité du Général-Leclerc (a French general from World War II), Cité Victor-Houroq (a French oil geologist), followed by Cité Namina, Cité Akosso, and Cité Matanda (denoted for black staff members and their families who lived in them). At the bottom of the economic scale lived unfortunate families of nonstaff contract workers in Camp Roger-Buttin, some three hundred wood and corrugated steel shacks, a diarrheal shantytown constructed along unpaved roads that turned to mud during the two rainy seasons, which is most of the year at the equator, "built to lodge the *caca Elfique*."[35] The Boulevard Elf-Gabon separated the two communities, running parallel to an old colonial canal that had been built to separate the white city (strictly forbidden to blacks) from the African villages. On one side of the road/canal was Cité Namina, for the poor blacks, and on the other side was the Grande Concession, for the rich whites. The architecture reflected the contractual realities, pay scales, and social status of the two communities. Failure to Gabonize the oil industry had resulted in neocolonial urban development in Port-Gentil, which created deep resentment against the affluent, privileged, largely French expatriate residents who were posted there for a limited duration of years.

Architectural apartheid was symbolized by facilities like "Club Sogara," a luxury straw-thatched recreation center outfitted with tennis courts, volleyball pits, modern playgrounds, and gardens for the children of Elf-Gabon staff members. No such facilities were built for the black staff or for the nonstaff workers of Camp Roger-Buttin, where most homes lacked running water. (The company installed some outhouses and an open-air water pump, to meet the needs of these people.) A few supermarkets were constructed by the French, filled with French products, and provided the expatriates with all their food from home. The city itself, sprouted up along the roads, had no municipal facilities to speak of. For entertainment, locals went to bars, or sat by the road watching the rich foreigners and their black collaborators (who usually originated from somewhere else) drive in company cars between their guarded places of work and their walled communities of residence. There seemed to be an inability for oil to generate meaningful development outside these enclaves.[36]

Of course the expatriates could not spend all of their time in the walled concessions. Boredom created by such sterile suburban architecture quickly overcame foreign residents, who were forced to go out and to frequent nightclubs, restaurants, and bars that serviced the cash economy generated by wood and oil. Boulevard Elf-Gabon sprouted public houses, described with ingenuous charm in the handful of tourist guides that have attempted to eulogize the city, where foreign oilmen and foresters, with more cash than places to spend it, would spend their days off getting drunk and having sex with poor young women who had no other kind of employment in the extractive enclave economy. One of the landmarks of this kind of urban phenomenon was the hotel and bar Printemps, which became a meeting place for oilmen, foresters (who hated the oilmen and considered them parvenus who had destroyed their sector), and locals. It was a meeting place of cultures and an urban frontier. The longer one lived in Port-Gentil, the more time one spent frequenting such places. Eventually whites would return to France and brag about how they understood the locals. Every night the blacks would go home, knowing they would never be accepted as equals. Unwanted mulatto children were sent to African shantytowns to live in relative deprivation with their black families, while temporary white residents watched their children play in the company facilities. White children went to usually all-white private schools funded by the companies and then finished their higher education back in France.[37]

African children, when they had schools, knew a second-class education, with no real job prospects after graduation. A handful of blacks received scholarships to study in France (there being no higher education in Port-Gentil itself) but upon graduation would discover that they were unwelcome in France (because they were Africans), yet had no opportunities back in Port-Gentil (except if they were lucky in getting a job with a French firm). Elf-Gabon built a few prestige projects, like a movie theater and a national museum (housed in the Elf-Gabon headquarters), but these lacked any integration with the educational system, and rather existed as token symbols of the purported civilizing mission of the whites. Since French language and European culture were the only subjects taught in the primary and secondary schools in the city, and since local customs and traditions were largely denigrated, a few African masks hanging on the walls of the oil company headquarters did little to provide the people with an authentic African identity. Of course, efforts were made to provide dance troops to entertain visitors, but local traditions and culture were sterilized by a commercial urbanism without culture or tradition. Instead, cultural domination of foreign business was perpetuated, and the Africans internalized a sense of inferiority, which eventually erupted in rage and riots; or they masked their African identity and launched wholeheartedly into the assimilationist projects of the foreigners.[38]

The dominant money culture of the international business environment provided a limited metropolitanism. There was the oil business, the forestry business, the hotel and restaurant business. But there were no cultural spaces for the flourishing of civic culture. There were no bookstores, or community centers, or sports arenas, or public parks. Nothing was created to make the agglomeration of dwellings cosmopolitan. Municipal services were underdeveloped, with the kids hanging out in the streets. People tended to cluster in ethnic communities, less out of a desire to conserve their authentic traditional cultures and more out of a need to find others who spoke the same language or shared enough of a common identity to provide them with rudimentary mutual support. Many had migrated to the city only looking for work, and lacking the ability to commute because of the lack of roads, did not invest themselves culturally in the town's neglected commons, preferring to bring their money back home to their villages where the real culture of Gabon was found. Drinking alcohol, saving a little money, living a kind of transitory existence, civil society barely emerged in Port-Gentil among the blacks. Thus, despite its isolation from the outside world, Port-Gentil was not a very interesting place to visit for anyone interested in learning about African civilization or history. It was more like a beach resort that reproduced kitsch culture for white tourists, with black locals cleaning the rooms and serving the drinks. Beyond the pale of walled communities were poor, angry, deprived locals who had only whites to fill their imagination of what might constitute a better life than this. Thus the blacks who rose to elite status in Port-Gentil came in two kinds: those who put on suits and ties (sweating profusely in the sweltering equatorial heat) and worked for the foreign companies, and those who rejected this foreign culture and opposed the regime in Libreville that reduced them to a status of servants to their former colonial masters. In other words, the African urban elites of Port-Gentil became either collaborators or dissidents.[39]

To maintain order, the French built a small military base, where the 4th marine infantry battalion was stationed. Whenever the dissidents became too strong, the collaborators called out the French troops to put down the rebellion. This is precisely what happened in 1991, when Port-Gentil first appeared in the headlines of the international press. The event that triggered these riots was political: an assassination of a local dissident, Joseph Rendjambe (1936–90), who was found dead in his Libreville hotel room a few years before the first multiparty elections. The government claimed that the needle marks found on his belly were caused by his diabetes, but the people believed that he was injected with poison because of his political opposition activities in this strategic oil capital. Rendjambe had been born in Port-Gentil, to a family of Nkomi, a Myene-speaking people who had migrated to Port-Gentil during the colonial period. He earned a scholarship to study in France, where he headed an expatriate student group, Association Générale des Étudiants du Gabon, which conducted dissident rallies against the Bongo regime. Returning to Gabon after completing university, Rendjambe found work in Libreville as a professor of economics in the newly created Omar Bongo University. There he became implicated in an event known as the "professor's plot" (1972). The government claimed that Marxist-Leninists were distributing treasonous literature on the two-year-old campus. Rendjambe and three other professors were arrested, tried, and sentenced to eight years in prison.[40]

In 1976 President Bongo released them, and Rendjambe went back to his teaching post for a few years, but, frustrated by the realities of the system, returned to Port-Gentil where he helped to found a new opposition party in 1990, the Parti Gabonais du Progrès (PGP). The president of the new party was his university colleague and cousin, Pierre-Louis Agondjo-Okawe (1936–2005). It was expected that Rendjambe would become the PGP candidate directly opposing President Bongo in the 1993 presidential elections. When his body was discovered on May 23, 1990, in a Libreville hotel room, riots broke out in Port-Gentil the next day. For the next five days militants in Port-Gentil took to the streets, spontaneous protests spread to other cities, including the capital Libreville, and led to a strong French military intervention in Port-Gentil on May 29, 1990. French troops protected the French oil facilities, evacuated French expatriates, and crushed the spontaneous protests. The death of Rendjambe was the proximate cause of the riots, but it should be clear that there were deeper, structural grievances among the peoples of Port-Gentil.

After the death of Rendjambe, his cousin, Agondjo-Okawe became one of the two most important leaders of the PGP. Elected head of the party, he led the PGP to victory in the coastal areas during the country's first multiparty legislative elections in 1990. As deputy representing Port-Gentil in the National Assembly, he quickly became the voice of political opposition to Omar Bongo's ruling Parti Démocratique Gabonais (PDG), which monopolized power under a one-party system. Agondjo-Okawe organized a successful general strike of opposition groups in June 1991 in order to put pressure on the government to accelerate implementation of a new constitution enacted that year. He also urged Bongo to resign as a necessary step toward solving Gabon's problems. In 1993 he ran unsuccessfully against Bongo in the country's first multiparty presidential elections. Deeply fraud-ridden, these elections provided a common base for mobilization of opposition parties that rallied around Agondjo-Okawe (who had received fewer votes in his own bastion Port-Gentil than Bongo!). Leading a united coalition called the Haut Conseil de la Résistance in nonviolent protests he created a political impasse that only ended with his signing of the Paris Accords of 1994, but refused to accept the post of justice minister, and remained a strong opponent of the regime. Reelected as deputy for Port-Gentil in 1996 and 2001, Agondji-Okawe was elected mayor of Port-Gentil in 2000, but when the law was changed to prohibit holding office at the local and national levels, he retired from political life.[41]

If the PGP developed into a genuine opposition party in Gabon, its ethnoregional base was limited to the south around Port-Gentil. So PGP has been unable to create a sizeable presence in the National Assembly. Although Rendjambe and Agondjo-Okawe were both Nkomi, the plurality of the population in the southern region is Punu-Eshira, and therefore as the party struggled to increase its membership it had to cope with the ethnopolitical realities of the south. The PGP won municipal elections in Port-Gentil and managed to build a party machine in the city government that allowed it to select the major. Its secretary general, Marie-Augustine Houangni-Ambouroué, an Orungu, was elected mayor and mounted a challenge for the leadership of the party, leading to her ousting by the Nkomi faction. (She went on to form an Orungu party.) For his part, Agondjo-Okawe decided to throw his support behind the Punu-Eshira politician Pierre Mamboundou in the 1998 presidential elections, and in so doing, passed the mantle to this larger regional ethnic group. Because of the split between the Orungu and Nkomi factions, the PGP saw its representation reduced in the National Assembly from eight to three seats in the legislative elections of 2001. Because of the rise of Pierre Mamboundou, the symbolic opposition in the south became identified with the political fortunes of Mamboundou, who has unsuccessfully run three times for president, coming in second place in 1998, 2005, and 2009. It was his supporters who took to the streets in 2009 when he, once more, was defeated by the Bongo family.

One consequence of the opposition PGP controlling the municipal government of Port-Gentil was that the city was not favored by the regime's largesse. Gabon is a system where the local economy is largely a question of national government spending. Since the PGP held Port-Gentil, it became a political strategy for the Bongo regime to starve the city of resources. President Bongo would only provide resources in exchange for declarations of loyalty, and he used his control of the country's oil and timber rents to manipulate, reward, and punish all political forces in the country. Unable to take control of the National Assembly or win the presidency, the PGP found that its control of local municipal government did not translate into sufficient funds to provide for necessary urban services.

Gabon's oil capital was an opposition city. Despite its importance to the export-oriented economy, it suffered the consequences of political dissent. Oil installations may be located in Port-Gentil, but the oil revenues go to the government in Libreville.

The Decline of Port-Gentil

A symbolic turning point in French domination came with the publication of Pierre Péan's *Affaires Africaines* (1983) a book by a French investigative journalist that revealed, in great detail, the workings of the Foccart network and the corruption of a business and politics that it had cultivated. From that point onward a humiliated Bongo started to look beyond France for new partners in development. But the oil sector itself began to shift its focus away from the city. The expansion of the oil industry from its original base on Mandji Island to shallow water, then deepwater offshore facilities, had preserved the basic role of Port-Gentil as the country's oil capital. For decades any new oil field discovered would be connected by pipeline to Cap Lopez, becoming "Mandji crude." The terminal at Cap Lopez, which has dominated the north part of the island since 1957 (the site where Duarte Lopez landed in 1471), is still the terminus of 218 kilometers of pipelines loading oil on giant ships. After offshore derrick crews finish their tours, they are shuttled back to live in Port-Gentil.

But in the late 1980s, a major oil field was discovered onshore by Shell Gabon: Rabi-Kounga. This oil field was an elephant of major importance, and it soon was providing half of Gabon's total output by 1991, with Shell briefly replacing Elf as the leading oil producer in Gabon. To maintain its dominant position, Elf bought half of the equity in Rabi-Kounga field, but Shell remained its operator. This huge discovery brought in European, American, Japanese, and even Korean investments to explore the area onshore and offshore. It also shifted the center of oil production south of the traditional fields around Mandji and the terminal and refinery on Cap

Lopez. Finally, the discovery of Rabi-Kounga also meant that the most important oil facility in the country was no longer connected to Port-Gentil.

SOGARA had been constructed by Elf in 1968 on Pointe Clairette, near the 1957 discovery made at the Clairette field. Clairette field is exhausted, but the refinery remains, transforming "Mandji crude" into gasoline for local use. This local refinement of petroleum products at SOGARA had never been large. Mismanagement and the corruption of the regime left the refinery in desperate need of refurbishing, which remains unlikely, as Gabon's oil is largely exported. Like forestry, oil is an export commodity. In many ways, it is probably cheaper to import petroleum products than to produce them in this antique facility, especially now that most oil production is no longer adjacent to the site.

In the end, the demise of Port-Gentil is linked to a larger decline of Gabon's oil industry. Gabonese oil production peaked in 1997 at 364,000 barrels a day, and lacking any new discoveries of the magnitude of Rabi-Kounga, thereafter entered into a long, slow, and ineluctable fall in output: from 337,000 barrels per day (1998) to 340,000 (1999), 327,000 (2000), 301,000 (2001), 295,000 (2002), 240,000 (2003), 235,000 (2004), 234,000 (2005), 235,000 (2006), 230,000 (2007), and 235,000 (2008).[42] Oil prices have been high in recent years, but were extremely low during most of this period, only averaging around twenty dollars per barrel. The traditional actors, that is, the French and Anglo-American oil companies, were hesitant to invest in Gabon, where geological factors tended to make oil fields small and unprofitable. So they invested instead in horizontal drilling and subsurface injection to extract more oil from the existing fields, thus allowing the petroleum sector to gradually decline toward exhaustion. No new big find has been made since Rabi-Kounga.

Another turning point in the history of Port-Gentil was Omar Bongo's construction of the Transgabonais railroad. In order to connect the interior of the country with the Atlantic coast, Bongo invested massively in the construction of this railroad, which diverted the mineral and forestry resources from the interior to the northern port of Owendo, resulting in a permanent and ineluctable decline in the Port-Gentil forestry sector, which had relied on wood being floated downstream. Now logs from the deep interior are being transported to another port. The decline in the forestry sector in Port-Gentil directly affected the local population, as the forestry sector was one of the few direct sources of employment for locals (the oil industry mostly hiring highly skilled foreign expatriates). Combined with the decline of the oil sector, these changes in the strategic economic location of Port-Gentil have structured a long decline of the economic importance of both the city and its French elites. The economic decline of Port-Gentil gave rise to a multitude of grievances and sociopolitical unrest, which have translated into a culture of opposition to the Bongo regime.[43]

It was during this phase, the late 1990s, that another scandal broke out in the global media called the "Elf Affair." While it was a public enterprise, Elf Aquitaine was the victim of considerable diversion of funds, carried out by the company's director of exploration and production, André Tarallo, who later confessed when the matter was brought to trial that retro-commissions from oil had been diverted to secret bank accounts in Liechtenstein, Luxembourg, and Switzerland, and to several dozen offshore bank accounts around the world. A few pennies on every barrel were put into these *caisses noires* (literally "black safes") systematically over several decades. Part of this embezzled money had been distributed to political actors in France, as well as to Omar Bongo and his collaborators, causing an enormous political scandal in Europe. The worst abuses were committed during the presidency of Loik Le Floch-Prigent (1989–93), when several billion francs were drained from the company treasury. This resulted in a financial weakening of Elf that would result in its acquisition by Total in 1999; but it also broke the neocolonial system by which Elf had dominated the oil industry in Gabon.[44]

The aftermath of this political and economic scandal was the context into which China has entered Gabon. Diplomatic relations between the two countries had been established in 1974, but it took thirty years for those relations to translate into tangible economic results. The truly emblematic event was the first official visit of a Chinese head of state to Libreville in 2004. During that historic trip, Hu Jintao and Omar Bongo signed an oil-exploration-and-production agreement, and the chairman of China Petroleum and Chemical Corporation (Sinopec) led a business delegation to negotiate a "Memorandum of Understanding" for an oil-sales contract between Total Gabon (formerly Elf) and subsidiaries of Sinopec and China International United Petroleum and Chemicals Company (Unipec) involving a million tons of crude oil per year. China has become the third largest importer of Gabonese oil, after the United States and France.[45] President Hu also gave Bongo a cash grant of two million dollars and an interest-free loan of six million dollars, with "no strings attached." In addition to the crude oil purchases, Sinopec signed a technical agreement with the oil minister Richard Onouviet for three onshore oil fields around Port-Gentil, who explained that the agreement was designed to facilitate the investments Sinopec would have to make under a production-sharing contract because, "prospecting for oil on land in Gabon is very difficult and costly because we are a heavily forested country." He implied that environmental regulations, which protected these forests as nature reserves, would be relaxed: "We hope that Sinopec will discover a large deposit that is lying dormant under Gabonese soil."[46]

Despite its longstanding domination, France is slowly losing control of the oil sector, and the Chinese have penetrated the economy of Gabon. The question that is now on everybody's mind is whether China will replace France. The answer depends on what kinds of relations one believes really matter. If only trade statistics matter, then it is entirely plausible that Gabonese exports could be sucked into the vortex of Chinese consumption. But how many Gabonese will ever learn Chinese, go to China, study there, write books in Mandarin, adopt Chinese lifestyles, settle there, and so on? The sublime role of French cultural imperialism should become clear. Patterns that have taken centuries to form will not disappear overnight. In the larger scheme of things, Gabon is located in the Gulf of Guinea, geographically closer to Europe and America than to China. It plays a small but important role in Western security, as the site of a regional French military base, and possibly a future American one. Only the conscious abandonment of the country by the French or the Americans would open the door to greater Chinese strategic influence. That being unlikely, one should expect to find Chinese activities developing under the aegis of enduring Western influence, or at best, under depoliticized globalization. But there is no reason to predict that China will establish a relationship anything like the neocolonial system created by France, at least not in Gabon. For the present, Total is still the dominant firm in the oil sector.

Environmental pollution has long been experienced by the indigenous peoples of Mandji Isle, from the earliest days of the foresters, when the Africans watched in horror as their "sacred trees" were chopped down, to the recent invasion of land and sea by petroleum operators. When the first seismic tests were conducted in the 1930s, forests were cleared and locals pushed off the concessions. Geologists exploded dynamite in the bush. The first well drilled in 1934 missed the oil reserve and instead provoked an eruption of sulfurous waters. After the hole was sealed and measured, the site was abandoned. Cleanup was never part of the business plan in those days. As the operations moved offshore, such spectacles became less visible, and the pollution onshore shifted to abandoned equipment, dirty beaches stained by offshore oil spills, and contamination of the soil and freshwater by the SOGARA refinery. The villagers who lived on fishing found themselves and their catch swimming in oil that would not come off with soap. Drinking water was often contaminated.

This kind of dirty business was the norm in the first half of the twentieth century, but now that the regulatory environment is much stricter, have practices changed? Consider the case of Addax Petroleum and its subcontractors, which have been exploiting a concession near Mandji Isle in the department of Ndolou. It has recently been revealed that the company has been discharging its toxic chemical wastes into an artificial lake that feeds directly into the river Obangué. This river is the principal source of freshwater for the local inhabitants. "We are highly exposed to numerous sicknesses," testified a local official at Fouanou, an oil town situated a hundred kilometers from Mandji, "every day we use the water from the Obangué. People are regularly getting sick when they drink this water, or when they bathe in the river."[47] When they complained, Addax installed some public fountains, but these were usually broken, and the villagers have been reduced to drinking the polluted river water. According to Brainforest Coalition, an environmental non-governmental organization (NGO) headed by Marc Ona, this situation has been going on for almost a decade. Besides the oil, other sources of pollution from the Addax facilities like sewerage from the camp kitchen and toilets have been dumped into the river. During a recent drilling operation at well T48 water mixed with drilling agents flowed into the river. Then on December 7, 2009, a company pipeline started to leak, releasing a mixture of oil and water into the river. Local fish have been contaminated by these activities, which is disturbing to villagers who eat the fish. Besides the pollution of the river, foreign operators flare gas into the air, which burns surrounding trees. Animals consuming these trees have been found dead, which is especially troubling for locals who live on game.[48]

Like other oil capitals, the city of Port-Gentil has had an effect on the surrounding region, and a half century of development of its petroleum industry has extended throughout Mandji Isle. It would be imprecise to call them "suburbs," because of the low level of infrastructure and development, but there are many settlements attached economically, politically, and socially to the port city. Here live the poorest of the poor, those unable to find employment in the enclave economy, mostly locals driven out by foreigners and migrants from other parts of Gabon, who continue to try to eke out a living by fishing in industrial wastelands. On the southern part of the island, down the peninsula from the port, is Ntchengué, a vast oil concession that has become a de facto garbage dump for locals. Miserable people installed along a fifteen-kilometer road, which is "fifteen kilometers of chaos, past an abandoned gasoline station, going down to Ozouri, where the first oil wells were discovered."[49] The road, more potholes than pavement, passes the Société Gabonaise d'Assainissement, a water purification facility that has become a place where people abandon cars, charred bus remains, and old tires. This is not a public road, but a private road belonging to the oil concessionaires, who hold a seventy-five-year lease, a symbol of colonialism's long legacy.

Riots in Port-Gentil in 2009 reflected a deep anger on the part of local residents about the exploitation of their natural resources by French businessmen and Gabonese collaborators in the capital Libreville. For forty-two years Omar Bongo ran Gabon as a neocolonial enclave of enduring French interest. When he died in 2009, and his son Ali succeeded him to power, there was a feeling of desperation on the part of the people of Port-Gentil. Seventy-five years is a long time for foreigners to exploit your nonrenewable resources without bringing growth. Forty-four years is a long time for one family to get rich on that neocolonial exploitation. How long will Port-Gentil's oil revenues be used to enrich a tiny clan of collaborators, while the people are left to live in squalor? If Ali Bongo and France have their way, the answer is, for a very long time to come.

To keep the people tranquil while he steals the wealth from beneath their feet, Ali Bongo has announced a huge development package for Port-Gentil of 43.5 billion CFA ($87 million), consecrated to the construction of a large drainage canal, in order to clean up the city and make it more attractive. He promised to create jobs and eradicate endemic illness like malaria. Port-Gentil airport will also have its landing strips enlarged, thanks to a $10.5 million grant from his government, allowing it to meet international norms. He also promised construction of a floating maritime station, the creation of an advanced professional school of oil and gas, a new business school, and a mental health center; an antimalaria program would be included in this ambitious program, which additionally hopes to recruit 350 youths into the national security forces. His government promised to implement all these measures rapidly and claimed that it had already identified all the sources of financing.[50]

Ali Bongo is pursuing a development policy he calls "Gabon Emergent," a phrase that had first served as his campaign leitmotif in the summer of 2009 before evolving into his new government's official program in 2010. "Emergence" denotes a new approach to economic development that, he promises, will elevate Gabon to the rank of an emerging country by the year 2025. The International Monetary Fund (IMF) defines an "emerging country" as one with low-to-middle per capita income that has undertaken economic development and reform programs and has begun to "emerge" as a significant player in the global economy. To accomplish this ambitious goal, Bongo is pursuing administrative reforms, as well as legal reforms, notably in fiscal policy, designed to attract foreign investors and permit the economy to increase its competitiveness. Notable measures include valorizing the potentials of the mining sector with the construction of a world-class "metallurgic pole" able to locally transform iron and manganese ore into metal, a national forestry sector capable of locally transforming raw tree trunks into lumber, and development of new motors of growth such as the digital economy and tourism, which it is hoped will take over for oil, as reserves become exhausted. Modernization of infrastructure, within an environmentally sustainable framework, is also promised.

Not only has the slogan "Gabon Emergent" become the moniker for Ali Bongo's new political era, but the neologism "Emergents" has come to denote his new government elite. First among these "emerging elites" was Prime Minister Paul Biyoghé Mba who has presented a strategic five-year plan (2011–16) that includes construction of two thousand kilometers of roads. But most of his announced projects were still in their formulation stage by the year's end, including plans to refit Maymba airport in the southwest, and the construction of a new international airport at Andeme, outside Libreville. One of his beacon projects had been the Grand Poubara hydroelectric dam, estimated to cost around €300 million and heralded to become the largest power generator in the country, if it ever gets built.

Obviously this is all good news for those places, at least if the regime keeps its promises. But the ability of the government to compensate for the declining forestry and petroleum industries in Port-Gentil by spending oil and wood revenues is questionable. There has been a long-term failure by the regime to channel extractive industry revenues into its long-term, sustainable development. It is frequently announced that tourism will replace oil and timber as the principal sources of revenues in Port-Gentil, but with no international airport, little cultural interest, and a hot and humid equatorial climate, mass tourism will probably not come to this Atlantic enclave of enduring French interests. And as France withdraws from its old colonial empire, the postpetroleum future is not bright.

CONCLUSION

Comparative Perspectives
on Energy Capitals

The phrase "energy capitals" seems to strike a chord with scholars. In a world of oil shortages and debates over alternatives to oil, it has the ring of importance, the promise of relevance. But does it have analytical power? Can it help explain why some regions have benefited from energy-led development and others have not? In the recent past, a growing literature on the "oil curse" has focused on nations that have not benefited from the discovery of oil, but have instead paid significant social costs.[1] The majority of the cases in this volume examine regions in which the production, manufacture, and use of fossil fuels have been mixed blessings, not curses. These successful energy capitals have reaped significant economic benefits while gradually learning to manage the often severe environmental, social, and political costs that have accompanied their growth.

To acknowledge the obvious: different sets of cases yield different comparative perspectives. Four of our nine cases (Pittsburgh, Houston, Los Angeles, and the region outside Baton Rouge) are in the United States, where the development of coal and oil went forward under a well-established constitution, a representative democratic political system, and a capitalist economy with considerable freedom of action for individuals and businesses. Somewhat similar conditions prevail in three other regions included in this volume: Calgary, Perth, and Stavanger. Only in the cases of Tampico and Port-Gentil did oil bring serious long-term problems while yielding few economic benefits. These two regions are quite different from the other seven cases and from each other.

Our case studies suggest several key factors in the growth of successful energy capitals. Most significant is the region's preparedness to participate in the development of resources on the eve of their discovery. Of particular importance is the maturity of political and economic institutions. Is there a government with the legitimacy and the potential authority to assert a measure of control over the promotion and regulation of the resource and to prevent its domination by outside interests? Are existing economic conditions and ideals likely to encourage broad participation by local interests? Does the legal framework concentrate control over resources in the hands of big businesses or government, or does it disperse such control more broadly through concepts such as individual ownership of subsoil rights to those who own the land above oil and gas deposits? Preparedness also includes the capacity of regional political and economic institutions to adapt to rapid growth by defining new regulatory powers when needed, while encouraging participation in the energy sector by local interests.

Geology, geography, and timing also have significant impacts on potential energy capitals. The richness of a region's resource endowment helps determine how long it continues to produce substantial amounts of coal, oil, or natural gas. The location of these resources is also important, since good access to infrastructure, markets, and an adequate workforce shape the pace of regional development. As shown by the extreme case of Houston, timing can be critically important; at the turn of the twentieth century, giant new oil discoveries in the region marked both the birth of modern Houston and the birth of the modern petroleum industry as an important new source of energy. The two then grew together, making Houston one of the most fully developed oil capitals in the world. The timing of the emergence of an energy capital helps determine its place and importance within the global energy industries, thus influencing its prospect for sustained expansion.

Energy Capitals: Local Impacts, Global Influence

Oil capitals prospered in the long term because they came to play important roles within one of the largest, most dynamic, and most important global industries: petroleum. The expansion of the international oil industry became a potent engine of growth in those regions that were favored by geology, geography, and timing and were prepared by their history to participate in its operations. Both industry and government invested heavily in the global search for oil, and investments in exploration and production created jobs in oil and gas fields and in oil-related manufacturing in most energy capitals. Some of these regions also attracted substantial investments to build and operate refineries, research facilities, and office buildings, and, in the process, created general construction jobs and manufacturing jobs that attracted new workers. Building the vast transportation infrastructure for oil and natural gas also required large, sustained expenditures on pipelines and ports. Infrastructure and manufacturing facilities could not be easily moved, and once they had been built companies often concentrated future investments near them. Tankers and long-distance pipelines tied these specialized regional transportation systems into the global petroleum economy, as did the regional headquarters and operations of international corporations active in energy.[2]

The evolution of the international petroleum industry gradually altered patterns in investment—and the impacts of the industry on energy capitals around the world. The most substantial change in the global industry came after the 1960s, when major producer nations individually and through the collective actions of the Organization of Petroleum Exporting Countries (OPEC) asserted greater control over the price and the production of oil.[3] As these nations took control over their own oil, new oil capitals emerged in major producing areas from the Middle East to the former Soviet Union and China. This shifted more investment decisions to the producing nations, which seized the opportunities of the new era to create government policies and national oil companies that opened new space in their borders for the emergence of new energy capitals.

In old and new energy capitals alike, economic benefits from oil-led development came with often high environmental, economic, and political costs. These costs were most visible in the case of environmental problems. Refined oil products flowed out of major refining centers, for example, while pollution from their manufacture remained concentrated in these regions. In practice, energy capitals absorbed much of the pollution generated by the production of energy for national and international markets. Less obvious in the short term were the potential regional costs of imbalanced economies so dominated by energy that the decline of a key source of energy, or sharp swings in its price, could be devastating. In seeking solutions to these problems, citizens of almost every energy capital confronted still another problem: the dominance of regional politics by those who favored promotion of energy over its regulation. From Houston to Port-Gentil, regions dependent on oil for growth had to struggle to assert local political control over a large and powerful industry whose interests were increasingly global, not local.

Tampico and Port-Gentil: The Cursed?

The struggle for political control was particularly difficult for regions lacking well-established political institutions that were considered legitimate and representative by their citizens. One of the most useful concepts put forward in the oil curse literature is that of the "successful failed state," that is, a government that succeeds in keeping power through its control of oil revenues but fails at its fundamental tasks of providing basic public services and ruling in the broad interests of its citizens. Petroleum history is filled with successful failed states in which public officials placed their personal self-interests over national interests after the discovery of oil.[4] This sort of corruption seems almost inevitable in developing nations with immature political institutions and government control of subsoil rights and, thus, with petroleum revenues passing through the hands of public officials.

The oil curse literature often seeks lessons from the struggles in Venezuela in South America and Nigeria in Africa to explain why similar nations have not prospered from the development of large sources of oil and natural gas.[5] A tone of despair and a sense of profound pessimism mark much of this literature, particularly that which deals with African nations.[6] Other scholars have responded with suggestions of ways that emerging petroleum-producing nations might avoid the imbalanced growth, political corruption, and foreign domination that often accompany the growth of oil production in their nations.[7] Few of these studies include discussion of more successful energy capitals, leaving the reader with limited comparative perspective on different regions and on individual regions over time. In addition, few studies discuss fully how the interest of a region and a nation might differ, even though the impact of tensions among regions and ethnic and religious groups artificially brought together into a nation under colonial rule does stand out in the history of Nigeria—the most studied example of the oil curse.[8]

The two regions discussed in this volume that most clearly exemplify the oil curse are Tampico and Port-Gentil, which unsuccessfully encountered great challenges in managing oil-led development and asserting effective governmental authority over international oil companies. Several key historical factors distinguish these two regions from the cases of successful absorption noted above. Both Tampico and Port-Gentil were poorly prepared to manage the process of change introduced by the discovery of oil. The foreign oil companies in control of the process had overwhelming economic, technical, and political advantages that gave them great power to pursue their own self-interests with little concern for the interests of the citizens of the host nations. The national governments of Mexico and Gabon were in flux at the time of the arrival of foreign companies, and for decades they struggled mightily to assert even a measure of control.

Comparisons of the Texas–Louisiana Gulf Coast with the Mexican Gulf Coast are particularly telling, since oil-led development in these two regions began at roughly the same time and was dominated by many of the same oil companies.[9] Affiliates of the major oil companies created at Spindletop (Texaco and Gulf Oil) joined those of Pan American (a part of Edward Doheny's large Los Angeles–area company), Standard Oil, and Royal Dutch Shell in a rush to Mexico in the early twentieth century. These companies moved across the Rio Grande into Mexico with little consideration of politics, much as they had moved across the Sabine River from Texas into Louisiana. Porfirio Díaz, the president of Mexico from 1886–1911, encouraged these and other foreign companies to enter his country, hoping that foreign direct investment in vital industrial sectors would spur sustained economic growth. Instead, such investments heightened the economic tensions between haves and have-nots that hastened the coming of the Mexican Revolution in 1910.[10]

Over the next quarter of a century, the revolution brought a succession of governments, all of which faced determined opposition from the foreign oil companies on issues ranging from the treatment of Mexican workers to tax policies to the basic question of government ownership of the nation's subsoil rights. By the time the government of Lazaro Cardenas nationalized the properties of the foreign oil companies in March 1938, they had moved much of their investment to Venezuela, where the government at least temporarily allowed the companies great leeway in their operations and effective control of oil-led development.[11]

One fundamental difference in Houston and Tampico was that Houstonians took ownership positions in many of the oil-related companies in their region; Mexicans in the Tampico region did not. Segments of the population in the Houston region also owned oil reserves through their legal control of subsoil rights and participated in the oil boom through their receipt of royalties. In contrast, the Mexican State, not individual citizens of Mexico, owned the nation's oil reserves. Workers in the Houston region often found long-term jobs with room for advancement in the Texas petroleum industry, but the early Mexican oil industry remained a symbol of foreign domination with quite limited room for advancement for Mexican workers in the Tampico region.

This distinction provides a useful reminder of the flexibility of the phrase "energy capital." Foreign capitalists controlled the evolution of early Mexican oil. The booming demand for refined products made Mexico the largest source of U.S. imports of crude oil in the 1920s, and companies based in the United States and Europe determined the pattern of investment in production and refining. Several of these companies owned established coastal refineries in Texas and Louisiana that were short tanker trips from Tampico. In the 1920s, they built technologically advanced units capable of producing more and better gasoline in their U.S. plants instead of building modern refineries in Mexico. There they generally constructed rudimentary "topping plants" that limited their financial risks in a politically volatile region while also limiting the long-term economic impacts of their operations in Mexico. The expansion of U.S. refineries to process Mexican crude solidified the role of the region from Houston to New Orleans as national and even international refining centers.[12]

Capital, not technology, was the key source of the power the international companies exercised over Mexican oil. In the early decades of the twentieth century, these companies found the vast reserves of Tampico's "Golden Lane" primarily by drilling shallow wells at locations with clear surface indications of oil, not by applying any sort of technically advanced seismic processes beyond the capacities of the Mexican workforce.

In some cases the critical knowledge for finding oil simply was understanding Spanish; in the Tampico region, place names that suggested the presence of oil— "hill of tar," for example—often pointed the way toward potential oil fields. Drilling required no major technical advances. The foreign oil companies had access to growing markets for oil products in the United States and Europe. With capital to build rudimentary infrastructure to find and produce Mexican oil and then transport it to the Mexican coast for export, the foreign companies were well positioned to practice unabashed exploitation. For decades they resisted the Mexican government's regulatory efforts, polluted with impunity, and discriminated against their Mexican workers. In 1938, only after more than thirty years of conflict, did the Mexican government finally assert control through the extreme response of nationalization.[13]

Myrna Santiago's chapter offers glimpses into the operations of the international oil companies in Mexico, where their investments had few permanent economic impacts. One exception to this pattern is telling. Foreign oil companies in Mexico invested heavily in "infrastructure for transporting crude from the camps in Tampico and then shipping them out to the United States or England." One result was the construction of a "world-class port" in Tampico, but it came with a high environmental price from "massive changes in the land" such as filling marshes and dredging rivers and harbors to make them more useful for the oil industry. Santiago's extraordinary book, *The Ecology of Oil,* includes a much fuller description of the environmental damage caused by rapid oil-led development in the Tampico region.[14]

One short section in Santiago's chapter seems particularly important in understanding the limited participation of Mexicans in the early years of oil industry in the region. Santiago reports that Tampico "developed as any ordinary Third World city. It has spent years as an energy capital without a single secondary school. Not until 1950 did Tampico inaugurate its first high school."[15] Long before that, Houston had well-developed public and private schools, the beginnings of a community college system, and several rapidly improving universities. Migrants from Houston's hinterlands often came to the region in search of better education for their children, but even small farming areas far removed from the city had public schools that provided the opportunity for students to attend at least eleven grades free of charge. Surely, such differences had long-term implications for the "human capital" of the Tampico region and for the capacity of its citizens to gain the education needed to share in the economic benefits of oil-led development.

The differences between Tampico and Port-Gentil seem even greater than those between Houston and Tampico. Douglas Yates's chapter on Port-Gentil is a bleak look at the realities of oil-led development in a small, impoverished African nation whose citizens as yet have received few lasting benefits from the development of oil. The end of the long era of French colonial rule in Gabon in 1960 left the nation poor and ill-prepared to manage its own oil. The initial development of oil in Gabon began in the 1950s and accelerated sharply in the 1960s and 1970s under the direction of foreign oil companies. By then the international petroleum companies had the capital, highly sophisticated technology, and advanced management systems needed to boost Gabon's production of oil and move it to market in the highly competitive global oil industry. But the Port-Gentil region and the nation as a whole lacked the economic and political institutions and leadership to assert control over the foreign oil companies or manage the efficient use of the tax revenues generated by oil. Even after the rise of OPEC in the 1970s and the nationalization of reserves in many major producing nations, Gabon still struggled to gain control over its own oil.

The overview of the perils of oil-led development in Port-Gentil offered by Yates adds still another chapter to the oil curse literature in Gabon, the rest of Africa, and other less developed nations in the late twentieth century.[16] In this volume, he discusses a range of factors that limited the benefits to Gabon from its oil reserves and that also raised their societal costs, making them more a curse than a blessing to the nation. Colonial rule left Gabon very poorly prepared for self-governance when it gained independence from France. Governments in Pittsburgh, Houston, Calgary, Perth, and Mexico also faced this difficult set of problems, but Port-Gentil had much weaker political institutions or governmental authority to manage its oil than any of the other regions discussed in this volume.

Norway: The Capacity to Govern

Gunnar Nerheim's chapter on Stavanger, Norway, suggests the stark contrast between Port-Gentil and other regions cursed by oil on the one hand and, on the other, a region much better prepared to manage its newly discovered oil wealth. Foreign petroleum companies discovered oil in the Norwegian sector of the North Sea even later than they found oil in Gabon, yet the outcome of oil-led development could not have been more different. Gabon and Norway are both relatively small nations, but there the similarities end. As oil development began, Norway had a balanced and prosperous economy and a well-established social democracy with a broad consensus on the need to manage its newfound oil wealth in the long-term interest of the nation. Early on, the Norwegian government took charge, defining tax policies, labor policies, and local content requirements while also creating a "rainy day fund" from oil revenues that made clear its intent to ensure that Norwegians benefited from oil development. The timing was favorable to Norway, since the nationalization of reserves in many of the OPEC nations in the 1970s sent international oil companies scrambling to find new non-OPEC sources of oil. In the North Sea, they found oil and gas—as well as a Norwegian government with the legitimacy, the authority, and the will to manage the benefits and costs of oil production.

In the long term, the most significant decision of the Norwegian government was the creation of a strong national oil company, Statoil. As the North Sea entered the boom years of the 1970s, the national government granted Statoil a majority interest in all newly discovered Norwegian oil fields. The new company used joint ventures with international oil firms to learn the oil business. The nation's strong educational system trained individuals to take on the technical challenges of offshore oil; advanced degrees and work at Statoil or with its joint-venture partners added special technical and managerial knowledge of the petroleum industry. In effect, Norway left enough potential profits and space for participation to keep the international oil companies involved while also encouraging Statoil to expand steadily. When the national oil company matured, the government gave it more leeway to grow through partial privatization. With its oil revenues and the continuing impact of industries other than oil, Norway could afford to be patient as Statoil emerged as an international oil company capable of competing globally.[17]

Norway's approach to oil-led development presents an interesting model for nations with balanced economies and governments that were well prepared before the discovery of oil to set and enforce basic ground rules that mandate local participation in many phases of the oil-related economy. The nation's approach also shows the benefits of a well-established educational system ready to produce technical specialists capable of building and operating the advanced technical systems used in the modern petroleum industry. Unfortunately, in the last fifty years few emerging oil-producing nations have possessed the strengths enjoyed by Norway as it took charge of oil development.

Calgary and Perth: Centers of Finance and Administration

Unlike Stavanger, Calgary and Perth were not near the major sources of oil in their region. This fact of geography had far-reaching ramifications for both economic and environmental impacts in the two regions. Major centers of refining, oil-related manufacturing, and chemical production, which served markets outside the region, did not grow near Calgary or Perth. The two regions thus did not reap the economic benefits of oil-related manufacturing jobs—nor did they pay the high environmental costs associated with such jobs. Instead both regions evolved as centers of finance and administration for broader areas that stretched far beyond their borders.

In the case of Calgary, the production of both oil and oil sands moved to the north in the late twentieth and early twenty-first centuries.[18] In many ways, the large triangle encompassing Calgary and Edmonton and Fort McMurray (the town nearest the major oil sands projects) can be analyzed as an Alberta oil region. Over time these three centers of the province's oil and oil sands industries came together, with each performing important functions. Calgary provided important specialized financial, administrative, and regulatory functions; it also served as a primary point of contact with the foreign companies that dominated important aspects and eras of development. Edmonton grew as an oil production and refining center, as well as the seat of the provincial government of Alberta and thus the regulatory heart of the region's petroleum industry. Late in the twentieth century, Fort McMurray boomed as the heart of the oil sands industry that altered both Alberta's economy and the energy mix of Canada.[19]

Perth's history as an energy capital poses even more complicated questions about the city's historical roles in energy production and consumption. As the only major city in the giant state of Western Australia, Perth has been the center of all sorts of economic and political activities for the much smaller towns that dot the state's vast landscape. Coal played a central part in the Perth region's growth in the twentieth century, with oil making a much more recent impact. As oil production began to grow in the 1950s, business and government leaders encouraged the creation of a complex twenty miles south of the city that became the region's refining center. This complex served primarily the needs of the region around Perth and parts of Western Australia; the fact that it was not a major exporter of refined products to outside markets limited its expansion and thus its impacts on the region. Perth proper thus escaped the worst of the environmental impacts of refining and oil production. With its relatively small population and isolation from other urban centers, Perth became the financial, administrative, and regulatory center of a very broad section of Western Australia; the Perth region has been well-positioned to manage the costs and benefits of the production of both coal and oil.[20]

In the twenty-first century, its capacity to continue to do so will be tested by its involvement in the development of giant reserves of natural gas in northwestern Australia, more than a thousand miles from the city. Its success will be determined in part by the capacity of all levels of government in Australia to maintain a cooperative and profitable relationship with the international oil companies developing these natural gas reserves. In this endeavor, the companies, not the governments, are the key actors. The state and national governments will oversee the construction and operation of the LNG projects in far northwestern Australia, serving as both their promoter and regulator.[21]

Matthew Eisler paints a portrait of Calgary as an "ancillary energy capital" and its business leaders as sort of junior partners with international oil companies with offices in Calgary. He notes that "Canadian [independent] firms remained relatively small compared to the major national and international petroleum companies." His description of Calgary perhaps also applies to the evolving situation in Western Australia, where international oil companies are guiding a new wave of development by supplying the capital, technology, and access to markets needed to develop giant offshore natural gas fields in Western Australia, far from an established city in northwestern Australia. A delicate and durable balance of promotion and regulation must be struck between business and government to secure the interest of the nation while also giving the companies sufficient incentives to continue to grow.

Managing Costs: Politics and Pollution

The difficulties in finding a balance between promotion and regulation are evident in the case studies of the U.S. energy capitals of Pittsburgh, Houston, Los Angeles, and Baton Rouge. Each of these regions expanded for decades before the creation of strong environmental regulations in the United States, and each initially struggled to define regional remedies for new types of pollution produced by coal and oil. The general lack of technical knowledge about pollution and its control lagged behind new technology of energy production and use. No regulatory institution at any level of government had a clear mandate and sufficient authority and resources to address this relatively new set of environmental problems.[22] Decades of neglect of energy-related pollution resulted in all of these regions. When they finally sought to respond, all faced similar difficulties in finding effective ways to regulate ongoing sources of energy-related pollution while also seeking to address the accumulated impact of years of neglect from past pollution.

The resulting political dramas played out differently in each region, depending on the nature of local politics and the attitudes of local citizens toward the balance of economic growth and environmental quality. Pittsburgh's coal-led development had a fifty-year head start on oil-led development in Houston; for half a century, Pittsburgh grappled with its new source of pollution in the heyday of capitalist industrialization that was largely unfettered by government regulation.[23] As Joel Tarr and Karen Clay show in their chapter, for a brief time in the late nineteenth century, natural gas became both available and economic, providing a brief respite from the worst of the coal-related air pollution. Around the turn of the twentieth century, municipal regulations made some regulatory headway, but the breakthrough in air-pollution control did not come for another fifty years, when the coming of southwestern natural gas via newly constructed cross-country pipelines combined with strong public sentiment to begin to clear the skies in the Pittsburgh region. Houston's experience with oil-related pollution was somewhat similar. It took almost a century for the upper Texas Gulf Coast to grapple seriously with its pollution problems, and then only after strong federal regulations in the 1970s forced action.[24]

Los Angeles, where oil development began at roughly the same time as it did in Houston, took a much different path than Pittsburgh or Houston. Sarah Elkind's chapter shows that in the Los Angeles region, oil enjoyed the sort of largely unregulated growth that characterized its history for much of the twentieth century in Houston. But by the 1910s and 1920s, representatives of other industries and citizens' groups pushed for government restraints on the oil industry's actions. Increased drilling on Southern California's beautiful beaches brought protests that jelled into political action. Drilling in residential areas that spread rapidly to keep pace with the region's growth also produced political protests. Long before strong federal laws regulated oil pollution, new local and state laws in California created controls that proved increasingly successful in prohibiting oil production in residential areas and on the beaches.

On this issue, Los Angeles differed fundamentally from Houston and Baton Rouge. Much earlier in the process of oil-led development, the political system in Los Angeles and California as a whole questioned the path of political accommodation of oil industry interests and moved toward stricter regulation, especially on environmental issues. Texas had no equivalent regional regulations for decades, perhaps because it was a poorer region more strongly dependent on oil for economic growth. In contrast to the one-party, pro-segregation, states' rights politics of Texas, California's two-party system probably also encouraged the rise of oil pollution as a cutting-edge political issue. On an even more basic level, the physical beauty of the California coast encouraged citizens to act more quickly to curb pollution than did the flat coastline south of Houston, where brown sand met brown water.[25]

Craig Colten's chapter on the Baton Rouge area examines a region in a traditionally one-party state with pro-business politics similar to Texas. Oil-led development began in Louisiana just after the Spindletop discovery, with major refinery construction beginning after that in the Beaumont–Port Arthur area, but before the opening of the Houston Ship Channel. As in the Houston region, chemical production followed in the years before World War II and then boomed during and after the war. A giant chemical complex grew in what had been a rural area in the hinterlands of Baton Rouge. Government promotion in the form of massive public expenditures on flood protection made this possible, continuing a pattern evident on the Texas–Louisiana Gulf Coast, where promotion of profits and jobs traditionally outweighed government's commitment to the regulation of environmental damage. Through the middle of the twentieth century, the management of the "mixed blessings" from oil refining and petrochemical production remained much the same in coastal Texas and Louisiana.

Events in the chemical corridor in Louisiana took a distinctive turn when citizen groups gradually asserted a stronger voice in the pursuit of environmental justice after the 1980s.[26] This change reflected in part the extreme environmental damage absorbed by the once rural communities that housed the major chemical plants in the region. But it also illustrated differences in the politics of oil and pollution at the local and state levels in Louisiana and Texas.[27] The comparison of the Houston and the Baton Rouge regions serves as a useful reminder: although, for purposes of economic analysis, the concept of a refining region from Corpus Christi to New Orleans is useful, political distinctions at both the local and state levels in Texas and Louisiana shaped difference in both the timing and content of regulatory policy.

The pace of adaptation, the level of independence, and the amount of corruption differed in developed regions around the world. A Texan might point to Louisiana as the region that lagged behind in its development of a neutral government with an acceptable level of corruption. Californians might lump Texas and Louisiana together as examples of how the attitude of growth at any cost slowed the emergence of an independent regulatory state; Canadians or Australians or Norwegians might argue that the U.S. cases as a whole reveal a pattern in which promotion far outweighed effective regulation for most of the twentieth century. Observers in Tampico might regret that their earliest efforts in the 1920s to control the behavior of the international oil companies did not succeed as well as the contemporaneous efforts of Californians. Those in Port-Gentil might look at all of these regions and envy their demonstrated capacity to adapt—even if effective environmental regulations were a long time in coming.

Managing Costs: Oil Price Swings and Imbalanced Economies

Another serious problem facing energy capitals appeared as the global oil system has become more tightly interconnected and drastic swings in oil prices periodically have swept through world markets. Because of the scale and importance of the global oil industry, these unpredictable price swings threaten the health of entire national economies, but the world's energy capitals have been at ground zero in absorbing their impacts. Gunnar Nerheim's chapter compares the impact of the boom and bust in oil prices in the 1970s and 1980s on Stavanger, Houston, and Calgary. He shows that oil prices rose by a factor of ten in the decade after the Arab Organization of Petroleum Exporting Countries (AOPEC) embargo of 1973 before plunging below $10 a barrel in 1985 and 1986. The resulting oil depression of the 1980s devastated the oil industry while also pulling down prices of most other sources of energy. Not surprisingly, Nerheim shows that the region most dependent on oil, Houston, took the hardest economic hit. After decades of steady growth in oil capitals with strong links into the global oil industry, these regions were caught off guard when the same links transmitted to them a devastating and unpredicted economic downturn.

Their economies gradually recovered as the petroleum industry adjusted to what turned out to be an era of much lower energy prices. The great uncertainties brought by the oil bust gave a new sense of urgency to ongoing efforts to diversify the economies of oil capitals. In the 1990s the growing possibility that public policies to curb climate change might mandate reductions in the use of fossil fuels heightened the incentive for fossil-fuel capitals and companies alike to explore diversification into other energy sources and industries outside of energy. Economic diversification became the order of the day as fossil-fuel capitals—at least in the developed nations—sought long-term answers to one of the most demanding questions they had ever confronted: What would be their fate if the global oil, gas, and coal industries declined, losing their positions as the world's leading energy sources?

Throughout history, new fuels have emerged to replace once dominant fuels in a process generally discussed under the heading "energy transitions."[28] Such transformational change holds great opportunities for new regions closely tied to the emerging new energy source—think Houston in 1901. It also poses challenges to those regions closely tied to the old energy order—think Pittsburgh in 1901. If we are in the process of a long-term transition away from oil to other energy sources, one conclusion seems obvious: Those regions least deeply involved in all phases of the oil industry—for example, Calgary, Perth, and Los Angeles—should have the least difficult time adjusting to the decline of oil. Those with the longest, deepest involvement in the oil-related endeavors—Houston and Baton Rouge—should have the most difficulties adjusting to the rise of new fuels.

Yet a more complicated reality emerges from the cases on Pittsburgh and Houston. These two full-fledged energy capitals developed large, thick, and diverse industrial cores of energy-related economic activities, including steel production in the Pittsburgh region and natural gas and petrochemicals in Houston. A long era of sustained economic growth by these diversified cores shaped basic infrastructure, educational institutions, philanthropic foundations, skilled workforces, and aggressive civic elites. These inherited resources were available for use in diversifying the regional economies.

One little noted aspect of the case of Pittsburgh as a coal capital is instructive for all fossil-fuel capitals. Coal did not abruptly disappear from the national economy with the rise of oil and natural gas in the twentieth century. Instead, its level of production held relatively steady for almost fifty years before increasing in many parts of the world after the 1970s.[29] The spectacular growth of oil and natural gas in the mid-twentieth century came primarily from the surging demand for energy, not from the absolute decline in the demand for coal.

These past trends have obvious and important implications for the future of fossil fuels and for oil and natural gas capitals. These fuels will not suddenly disappear, as did a significant chunk of the oil and gas business in the great oil depression of the 1980s. Indeed, even the resulting bust proved to be a boon for Houston; as the industry consolidated operations in response to hard times, many functions previously performed elsewhere moved to the "energy capital of the world."

The future of fossil fuels—and of the regions closely tied to their production and processing—in the twenty-first century is by no means certain. Price and government policies in response to climate change will shape their futures. Given their vital role in the global economy and the magnitude of their use, it is unlikely that they will decline rapidly. Fossil-fuel capitals will have some time to adjust, although it is impossible to predict how much time.

The seven relatively successful energy capitals examined in this volume were well prepared by their histories to absorb coal- or oil-led development; the two regions cursed by oil were not. Houston and Los Angeles, for example, were not yet major cities near the turn of the twentieth century just before the discovery of oil, but they were the centers of dynamic regional economies tied into growing national markets via railroads and shipping. They were parts of a booming national economy not dominated by oil, yet receptive to its growth. A well-established federal system of government gave them the flexibility to respond to the challenges posed by oil in ways adapted to local conditions. Cultural attitudes embodied in laws encouraged individual participation in the new oil industry in these regions, which were open to the influx of both workers and capital. Those who had migrated to the regions before oil and those who came after generally shared the same language and cultural assumptions; in short, local/national interests, not foreign interests, controlled the development of oil.

There is a great divide between the experiences of Los Angeles and Houston (as well as Pittsburgh, Baton Rouge, Stavanger, Calgary, and Perth) and those of Tampico and Port-Gentil. The histories of other energy capitals not included in this volume will not necessarily show the same patterns as our case studies. A recently published book, *Oil Is Not a Curse: Ownership Structure and Institutions in Soviet Successor States* by political scientists Pauline Jones Luong and Erika Weinthal, compares the evolution of the independent oil-rich nations formed during the breakup of the Soviet Union in the early 1990s. They conclude that in these states (Russia, Azerbaijan, Kazakhstan, Turkmenistan, and Uzbekistan), oil is not necessarily a curse.[30] The authors make a historical argument that much of the literature on the oil curse errs by beginning the study of a region's development when oil is initially discovered, thereby slighting the long-term impact of the political institutions in place before oil. Once oil is discovered, they argue, the government's capacity to build an effective institutional framework for its development has been a key determinant of the balance of the positive and negative impacts of oil. The case of Gabon demonstrates how significant this could be in nations where colonial systems remained in place into the mid-twentieth century.

Their discussion of "ownership structure and institutions" encompasses the traditional debate on private versus public ownership, but it focuses on more complex issues. How does government use its powers to shape a working relationship with oil producers that encourages orderly development? How are oil revenues integrated into the nation's finances to produce a healthy, sustainable fiscal regime? How is private capital and innovation harnessed into a system of politics, law, regulation, and social attitudes capable of both promoting and regulating development over a long term, in the broad interest of the region and nation as a whole? The work of Luong and Weinthal informs the study of energy capitals by encouraging us to look harder at the range of choices made by governments in different regions at different times as they assert control over their own resources. Their long-term perspective leaves room for much-needed optimism by holding out the prospect that governments can learn to construct effective institutions to reduce the impacts of the oil curse.

Oil Is Not a Curse shows the value of looking beyond the study of individual nations or regions to broader, comparative groupings. The book puts forward interesting, if speculative conclusions about one such grouping—the nations of the former Soviet Union. Successful energy capitals that emerged in the developed world might be another grouping. Three others might be Latin American regions in addition to Tampico, African oil producers since the 1950s, and major producing regions in the Middle East. The case of Pittsburgh suggests that other promising research topics can be found in the experiences of regions that played critical roles in the rise or the decline of sources of energy; such areas would include regions in the recent past that have begun to emerge as centers of non-fossil-fuel-based energy.

In any place or time, putting energy capitals in the context of the nation and of the international economy will remain a major challenge. National energy, environmental, economic, and foreign policies create much of the institutional framework within which regional development goes forward. National attitudes and legal frameworks also shape assumptions about entrepreneurship and ownership in the energy industries. Beyond national perspectives, energy capitals also must be placed in the context of evolving global energy industries.

As we ponder our energy future in the early twenty-first century, we know too little about the process of change. One thing that we do know, however, is that fundamental changes in the global energy mix will have far-reaching impacts on both emerging and declining energy capitals. The nine historical case studies presented in this volume are a start in understanding these impacts, but we need broader analysis of a much wider array of cases. More study is needed of the struggles of less-developed nations to capture the benefits and minimize the costs of energy-led development. We also need to know more about the local impacts of fuels other than fossil fuels, especially the differences in the potential environmental impacts of their use on a much larger scale. Such studies would be improved by insights from disciplines other than history and political science, particularly from developmental economists and environmental scientists. In a world buffeted by concerns about energy, the study of the rise and fall of regions that have produced much of the world's energy in the past will prove useful in understanding the local impacts and global influences of future energy choices.

NOTES

Introduction

1. Timothy J. Healy, *Energy and Society* (San Francisco: Boyd and Fraser, 1976), 10–11, 44–45. See also Vaclav Smil, *Energy in Nature and Society: General Energetics of Complex Systems* (Cambridge, Mass.: MIT Press, 2008), 14.

2. Smil, *Energy in Nature and Society*, 203.

3. Power as distinct from energy is the rate of change of energy.

4. J. R. McNeill, *Something New under the Sun: An Environmental History of the Twentieth-Century World* (New York: W. W. Norton, 2000), 14. See also Smil, *Energy in Nature and Society*, 205.

5. Smil, *Energy in Nature and Society*, 206–27. See also Anthony N. Penna, *Nature's Bounty: Historical and Modern Environmental Perspectives* (Armonk, N.Y.: M. E. Sharpe, 1999), 234–35.

6. McNeill, *Something New under the Sun*, 15.

7. Alfred W. Crosby, *Children of the Sun: A History of Humanity's Unappeasable Appetite for Energy* (New York: W. W. Norton, 2006), xiv.

8. Smil, *Energy in Nature and Society*, 257.

9. Crosby, *Children of the Sun*, xiv. See also Healy, *Energy and Society*, 53.

10. McNeill, *Something New under the Sun*, 297.

11. Raquel Pinderhughes, *Alternative Urban Futures: Planning for Sustainable Development in Cities throughout the World* (Lanham, Md.: Rowman and Little field, 2004), 94–95.

12. Richard Stren, Rodney White, and Joseph Whitney, eds., *Sustainable Cities: Urbanization and the Environment in International Perspective* (Boulder: Westview Press, 1992), 24–25.

13. Pinderhughes, *Alternative Urban Futures*, 10–11.

14. Smil, *Energy in Nature and Society*, 307.

15. Ronald J. Johnston, *The American Urban System: A Geographic Perspective* (New York: St. Martin's Press, 1982), 304–5.

16. Thomas R. Detwyler and Melvin G. Marcus, eds., *Urbanization and Environment: The Physical Geography of the City* (Belmont, Calif.: Duxbury Press, 1972), 21.

17. Lisa Benton-Short and John Rennie Short, *Cities and Nature* (London: Routledge, 2008), 142–43.

18. Smil, *Energy in Nature and Society*, 260.

19. Anthony N. Penna, *The Human Footprint: A Global Environmental History* (New York: Wiley-Blackwell, 2010), 261.

20. Like wood, coal (also oil and natural gas) is a hydrocarbon containing carbon and hydrogen, but in different ratios. Energy is created by burning hydrogen and smoke and smog become the by-products of burning carbon. It is important, however, for an energy source to have a low carbon-to-hydrogen ratio. Coal possesses a better ratio of carbon to hydrogen than wood, and energy efficiencies therefore have increased since the increased use of coal. However, the acceleration of urbanization and the intense growth of industrialization meant much more widescale use of hydrocarbons relative to earlier times. See Penna, *Nature's Bounty*, 228–29.

21. Smil, *Energy in Nature and Society*, 307–8. See also Stren, White, and Whitney, eds., *Sustainable Cities*, 25.

22. Penna, *The Human Footprint*, 287.

23. Ibid., 308.

24. William Cronon, *Nature's Metropolis: Chicago and the Great West* (New York: W. W. Norton, 1991); Kathleen Brosnan, *Uniting Mountain & Plain: Cities, Law, and Environmental Change Along the Front Range* (Albuquerque: University of New Mexico Press, 2002).

25. The following are examples of useful energy histories relevant to the study of energy capitals: Ibrahim M. Al-But'hie and Mohammad A. Eben Saleh', "Urban and Industrial Development Planning as an Approach for Saudi Arabia: The Case Study of Jubail and Yanbu," *Habitat International* 26 (January 2002): 1–20; Saud Al-Oteibi, Allen G. Noble, and Frank J. Costa, "The Impact of Planning on Growth and Development in Riyadh, Saudi Arabia, 1970–1990," *GeoJournal* 29 (February 1993): 163–70; William Beaver, *Nuclear Power Goes On-Line: A History of Shippingport* (Westport, Conn.: Greenwood Press, 1990); Henry F. Bedford, *Seabrook Station: Citizen Politics and Nuclear Power* (Amherst: University of Massachusetts Press, 1990); Rómulo Betancourt, *Venezuela: Oil and Politics* (Boston: Houghton Mifflin, 1979); Brian Black, *Petrolia: The Landscape of America's First Oil Boom* (Baltimore: Johns Hopkins University Press, 2000); Howard I. Blutstein, *Venezuela: Politics in a Petroleum Republic* (New York: Praeger, 1984); Jonathan Brown, *Oil and Revolution in Mexico* (Berkeley: University of California Press, 1993); Craig E. Colten, ed., *Transforming New Orleans and Its Environs* (Pittsburgh: University of Pittsburgh Press, 2000); Michele Stenehjem Gerber, *On the Home Front: The Cold War Legacy of the Hanford Nuclear Site* (Lincoln: University of Nebraska Press, 2002); Mustapha Ben Hamouche, "The Changing Morphology of the Gulf Cities in the Age of Globalisation: The Case of Bahrain," *Habitat International* 28 (December 2004): 521–40; Tony Hodges, *Angola: Anatomy of an Oil State* (Bloomington: Indiana University Press, 2004); Augustine A. Ikein, *The Impact of Oil on a Developing Country: The Case of Nigeria* (New York: Praeger, 1990); Pauline Jones Luong and Erika Weinthal, "Prelude to the Resource Curse: Explaining Oil and Gas Development Strategies in the Soviet Successor States and Beyond," *Comparative Political Studies* 34, no. 4 (2001): 367–99; Paul Sabin, *Crude Politics: The California Oil Market, 1900–1940* (Berkeley: University of California Press, 2004); Jorge Salazar-Carrillo and Bernadette West, *Oil and Development in Venezuela during the 20th Century* (Westport, Conn.: Praeger, 2004); Myrna I. Santiago, *The Ecology of Oil: Environment, Labor, and the Mex-*

ican Revolution, 1900–1938 (Cambridge: Cambridge University Press, 2006); Lee Scamehorn, *High Altitude Energy: A History of Fossil Fuels in Colorado* (Boulder: The University of Colorado Press, 2002); Joel A. Tarr, ed., *Devastation and Renewal: An Environmental History of Pittsburgh and Its Region* (Pittsburgh: The University of Pittsburgh Press, 2003); Robert Vitalis, *America's Kingdom: Mythmaking on the Saudi Oil Frontier* (Stanford: Stanford University Press, 2007); James C. Williams, *Energy and the Making of Modern California* (Akron, Ohio: University of Akron Press, 1997): Martin V. Melosi, *Atomic Age America* (New York: Longmans, 2013).

26. See, for example, Saskia Sassen, *Cities in a World Economy*, 2nd ed. (Thousand Oaks, Calif.: Pine Forge Press, 2000); Sassen, *The Global City: New York, London, Tokyo* (Princeton, N.J.: Princeton University Press, 1991); Sassen, ed., *Global Networks, Linked Cities* (New York: Routledge, 2002); Arturo Almondoz, *Planning Latin American Capital Cities, 1850–1950*, Planning, History, and the Environment Series (London: Routledge, 2002); Mark Abrahamson, *Global Cities* (New York: Oxford University Press, 2004); Malcom Cross and Robert Moore, eds., *Globalization and the New City: Migrants, Minorities and Urban Transformation in Comparative Perspective* (New York: Palgrave, 2002); David L. A. Gordon, *Planning Twentieth Century Capital Cities* (London: Routledge, 2006).

27. Sassen, *Cities in a World Economy*, xvi.

28. Short discussions of many oil-producing regions are found in Augustine A. Ikein, *The Impact of Oil on a Developing Country: The Case of Nigeria* (New York: Praeger, 1990). See, also, Tony Hodges, *Angola: Anatomy of an Oil State* (Bloomington: Indiana University Press, 2001).

29. Among its various uses, some in the field of environmental justice have been using the term "sacrifice zones" to pinpoint communities placed at risk because of the siting of toxic facilities there. See Steve Lerner, *Sacrifice Zones: The Front Lines of Toxic Chemical Exposure in the United States* (Cambridge, Mass.: MIT Press, 2010).

PART I. Blessed by Fossil Fuels? Pittsburgh, Houston, Louisiana, and Los Angeles

1. Mark Walker, *The Salzburg Transaction: Expulsion and Redemption in Eighteenth-Century Germany* (Ithaca: Cornell University Press, 1982); and Gunther Barth, *Instant Cities: Urbanization and the Growth of San Francisco and Denver* (New York: Oxford University Press, 1975).

2. J. R. McNeill, *Something New under the Sun: An Environmental History of the Twentieth-Century World* (New York: Norton, 2000).

3. Daniel Yergin, *The Prize: The Epic Quest for Oil, Money and Power* (New York: Touchstone, 1991), 12.

CHAPTER 1. Pittsburgh as an Energy Capital: Perspectives on Coal and Natural Gas Transitions and the Environment

The authors would like to thank Chris Jones, Austin Mitchell, Edward K. Muller, and Werner Troesken for their helpful comments and Paula Levin for her aid in data processing.

1. Pittsburgh Regional Planning Association, *Region in Transition, Report of the Economic Study of the Pittsburgh Region* (Pittsburgh: University of Pittsburgh Press, 1963), 226.

2. The petroleum came from the oil fields in the northwest corner of the state, the nation's first major oil-producing area. In the 1860s oil was shipped down the Allegheny River on barges and other forms of river transport, with rail transport becoming increasingly important in the following decades. Pittsburgh became a petroleum warehouse and a refining center and 1866 it had eighty-six refineries. In the late nineteenth century, however, the refining industry played an increasingly diminished role in the city's economy because of railroad freight rate discrimination competition from Cleveland refineries, which had superior transport options for the crude. See A. Michael Sulman, "The Short Happy Life of Petroleum in Pittsburgh," *Pennsylvania History* 33 (January 1966): 50–69; Harold F. Williamson and Arnold R. Daum, *The American Petroleum Industry: The Age of Illumination 1859–1899* (Evanston, Ill.: Northwestern University Press, 1959), 55–60, 165–231. The environmental impacts of oil exploitation in northwestern Pennsylvania are dealt with by Brian Black, *Petrolia: The Landscape of America's First Oil Boom* (Baltimore: Johns Hopkins University Press, 2002).

3. For energy transitions see Vaclav Smil, *Energy Transitions: History, Requirements, Prospects* (New York: Praeger, 2010). For a study of the role of different energy sources in the American economy, see Sam H. Schurr and Bruce C. Netschert, *Energy in the American Economy 1850–1975: An Economic Study of Its History and Prospect* (Baltimore: Johns Hopkins University Press, 1960).

4. E. V. d'Invilliers, *Part I. Report on the Pittsburgh Coal Region, Annual Report of the Geological Survey of Pennsylvania for 1886* (Harrisburg: Board of Commissioners, 1887), 119–256.

5. Susan J. Tewalt et al., "Chapter C-A Digital Resource Model of the Upper Pennsylvania Pittsburgh Coal Bed, Monongahela Group, Northern Appalachian Basin Coal Region," U.S.G.S. Professional Paper 1675, http://pubs.usgs.gov/pp/p1625c/.

6. For discussions of coal in Pennsylvania, see Frederick Moore Binder, *Coal Age Empire: Pennsylvania Coal and Its Utilization to 1860* (Harrisburg: Pennsylvania Historical and Museum Commission, 1974); John N. Hoffman, "Major Economic Changes in the Mining and Distribution of Pennsylvania Bituminous Coal" (PhD dissertation, Pennsylvania State University, 1961); and Howard N. Eavenson, *The First Century and a Half of American Coal Industry* (Pittsburgh: privately printed, 1942), 138–226.

7. Howard N. Eavenson, "The Pittsburgh Coal Bed: Its Early History and Development," *American Institute of Mining Engineers, Transactions* 130 (Philadelphia: AIME, 1938): 55.

8. Eavenson, *The First Century,* table 49, 485–90; John J. Schanz Jr., "Historical Statistics of Pennsylvania's Mineral Industries, 1759–1955," Paper 1, Mineral Conservation Series, (State College: Pennsylvania State University, 1957); Hoffman, "Major Economic Changes," 64; and Wilbert G. Fritz and Theodore A. Veenstra, *Regional Shifts in the Bituminous Coal Industry with Special Reference to Pennsylvania* (Pittsburgh: Bureau of Business Research, 1935),

15–24. The four coal-mining counties composed a region where 77 percent of Pennsylvania's coal was mined although occupying only 14.7 percent of the state's land area.

9. Fritz and Veenstra, *Regional Shifts*, 65–66; U.S. Bureau of Mines, *Mineral Resources, 1924—Part II* (Washington, DC: GPO, 1925), 471.

10. National Coal Association, *Bituminous Coal Facts 1966* (Washington, DC: NCA, 1966), 60.

11. See V. Price Fishback, *Soft Coal, Hard Choice: The Economic Welfare of Bituminous Coal Miners 1890–1930* (New York: Oxford University Press, 1992).

12. Fritz and Veenstra, *Regional Shifts in the Bituminous Coal Industry*, 25–35.

13. Shera Λ. Moxley, *From Rivers to Lakes: Engineering Pittsburgh's Three Rivers, A Report from 3 Rivers 2nd Nature* (Pittsburgh: Carnegie Mellon University Studio for Creative Inquiry 2001), 16–17.

14. Pittsburgh was also an important manufacturer of railroad equipment such as rails, locomotives, and freight cars. For the impact of railroads on the Pittsburgh economy, see David Hounshell et al., "Economic Impact of Rail Transportation in Western Pennsylvania" (Pittsburgh: A Report to the National Park Service, 2005).

15. Hoffman, "Major Economic Changes," 109–12, 125; Pittsburgh Regional Planning Association, *Region in Transition*, 213–14.

16. Willard Glazier, "The Great Furnace of America," quoted in *Pittsburgh*, ed. Roy Lubove (New York: Franklin Watts, 1976), 22–29.

17. Kenneth Warren, *The American Steel Industry 1850–1970: A Geographical Interpretation* (London: Claredon Press, 1973), 15–28.

18. Edward K. Muller and Joel A. Tarr, "The Interaction of Natural and Built Environments in the Pittsburgh Landscape," in *Devastation and Renewal: An Environmental History of Pittsburgh and Its Region*, ed. Joel A. Tarr (Pittsburgh: University of Pittsburgh Press, 2003), 21.

19. John Ingham, *Making Iron and Steel: Independent Mills in Pittsburgh 1820–1920* (Columbus: Ohio State University Press, 1991), 47–95.

20. Kenneth Warren, *Triumphant Capitalism: Henry Clay Frick and the Industrial Transformation of America* (Pittsburgh: University of Pittsburgh Press, 1996), 21–268.

21. Warren, *The American Steel Industry*, 114–15, 135–36; Ingham, *Making Iron and Steel*, 140–51; and Edward Muller and Joel A. Tarr, "Pittsburgh's Industrial Corridors," in *Born of Fire: The Valley of Work*, ed. Barbara Jones (Pittsburgh: University of Pittsburgh Press, 2006), 15–35.

22. Edward K. Muller, "River City," in Tarr, ed., *Devastation and Renewal*, 41–63.

23. Warren, *The American Steel Industry*, 137.

24. Hoffman, "Major Economic Changes," 137.

25. Pittsburgh Regional Planning Association, *Region in Transition*, 229–33.

26. Coal was obtained primarily by room and pillar deep mining before the 1950s, when longwall mining and strip mining or surface mining became common. Both of these newer techniques produce large environmental damages but reduce the human risks in

coal mining. The volume of material involved in strip mining made its environmental impacts especially severe. For strip mining, see Mark Squillace, *Strip Mining Handbook*, chap. 2: "The Environmental Effects of Strip Mining," at http://sites.google.com/site/stripmininghandbook/chapter-2-1. For longwall mining, see The Center for Public Integrity, *The Hidden Costs of Clean Coal: The Environmental and Human Disaster of Longwall Mining*, at http://www.publicintegrity.org/investigations/longwall/pages/documents/.

27. The Surface Mining Control and Reclamation Act (SMCRA) of 1977 defined abandoned mine land (AML) as lands that were mined, left in an inadequate reclamation status, and abandoned before August 3, 1977, with no continuing reclamation responsibility by any individual or company under state or federal laws. Many sources deal with the environmental impacts of coal use. See, for example, "Coal Mining," at http://wiki.dickinson.edu/index.php/Coal_mining; "Environmental Effects of Coal Mining and Burning," *Wikipedia*, at http://en.wikipedia.org/wiki/Environmental_effects_of_coal; and "Environmental Impacts of Coal," at http://www.sourcewatch.org/index.php?title=Environmental_impacts_of_coal.

28. Some health authorities were reluctant to counsel the complete exclusion of mine acid drainage from streams because the acid waste had a germicidal effect on sewage pollution.

29. Nicholas Casner, "Acid Mine Drainage and Pittsburgh's Water Quality," in Tarr, ed., *Devastation and Renewal*, 89–109.

30. Ibid., 92–109.

31. Joel A. Tarr, *The Search for the Ultimate Sink: Urban Pollution in Historical Perspective* (Akron: University of Akron Press, 1996), 390–91.

32. "Inversion (meteorology)," *Wikipedia*, at http://en.wikipedia.org/wiki/Inversion_percent28meteorologypercent29.

33. See Eavenson, "The Pittsburgh Coal Bed," 20–1; John O'Connor, "The History of the Smoke Nuisance and of Smoke Abatement in Pittsburgh," *Industrial World* (March 14, 1913).

34. Angela Gugliotta, "How, When, and for Whom Was Smoke a Problem in Pittsburgh?" in Tarr, ed., *Devastation and Renewal*, 110–14.

35. A. B. Bellows, O. R. McBride, and A. A. Straub, *Some Engineering Phases of Pittsburgh's Smoke Problem*, vol. 6, *Mellon Smoke Investigation* (Pittsburgh: University of Pittsburgh, 1914), 12; Angela Gugliotta, "'Hell with the Lid Taken Off': A Cultural History of Air Pollution—Pittsburgh" (PhD Thesis, University of Notre Dame, 2004), 32–36, 58–59, 118, 122.

36. For a list of gas wells in the Pittsburgh region before 1885 see Francis C. Phillips, *Report on the Composition and Fuel-Value of Natural Gas, Part II, Annual Report, Geological Survey of Pennsylvania 1886* (Harrisburg: Board of Commissioners for the Geological Survey, 1887), 690–92.

37. The plants were the ironworks of Spang, Chafant and Company, the Isabella Blast Furnace, and the iron mills of Graff, Bennett & Company.

38. See "Glass: Pittsburgh as a Center," in Engineers' Society of Western Pennsylvania, *Pittsburgh* (Pittsburgh: The Society, 1930), 365–74.

39. "Report of a Committee, on Natural Gas, May 21, 1884," *Proceedings of Engineers' Society of Western Pennsylvania*, Pittsburg [*sic*], vol. 2, Sept. 1882–1884. The new well fields were located in Butler, Westmoreland, and Washington Counties as well as in the township of Tarentum in Allegheny County.

40. Westinghouse also patented a number of improvements in gas distribution technology including the use of double piping to avoid leaks. In 1886 he formed the Safety Appliance Co. to manufacturer gas meters and regulators. See Quentin R. Skrabec Jr., *George Westinghouse: Gentle Genius* (New York: Algora Publishing, 2007), 71–75.

41. For the early growth of the Philadelphia Company see George H. Thurston, *Pittsburgh's Progress, Industries and Resources* (Pittsburgh: A. A. Anderson and Son, 1886), 13. Thurston claimed that 3,000 families, 34 iron and steel mills, 60 glass factories, 300 small factories, and a hotel used natural gas by the late 1880s. By 1885, the Philadelphia Company piped in gas from 45 wells, had laid 331 miles of pipe, and supplied 2,637 homes. Stanley Paul Wagner, "Natural Gas Comes to Pittsburgh" (MA Thesis, University of Pittsburgh, 1947), 50–56. In 1890, *Brown's Directory of American Gas Companies* listed 78 gas companies in the Pittsburgh region. See *Brown's Director of American Gas Companies* (Duluth, Minn.: Harcourt Brace Jovanovich, 1890), 143–44.

42. E. V. d' Invilliers, *Report on the Pittsburgh Coal Region*, 18–19; "Natural Gas vs. Coal at Pittsburg [*sic*]," *National Labor Tribune*, 24 July 1886.

43. Gugliotta, "'Hell with the Lid Taken Off,'" 129–39.

44. "The City of Pittsburg [*sic*]," *Harper's Weekly*, February 27, 1892, 202–3. See the discussion of this article in Gugliotta, "'Hell with the Lid Taken Off,'" 143–47.

45. See, for instance, Thurston, *Pittsburgh's Progress* (pp. 6–8) published in 1886, where he repeats and reiterates optimistic comments about gas supply he originally made in 1876 and Thurston, *Allegheny County's Hundred Years* (Pittsburgh: A. A. Anderson and Son, 1888), 92–95, 202–9; Gugliotta, "'Hell with the Lid Taken Off,'" 128–31.

46. Regulations regarding natural gas pipeline construction and routes were specified by the state legislature in *No. 32: Laws of Pennsylvania, Session of 1885*, "An Act to Provide for the Incorporation and Regulation of Natural Gas Companies," 33–35.

47. Quoted in Wagner, "Natural Gas Comes to Pittsburgh," 42.

48. See, W. W. Thomson, *A Digest of the Acts of Assembly Relating to the General Ordinances of the City of Pittsburgh from 1804–1886* (Harrisburg, 1887), 369–725; Hiram Schock, comp. and ed., *Digest of the General Ordinances and Laws of the City of Pittsburgh to March 1, 1938* (Pittsburgh: The City, 1938), 728–37; see n. 1, p. 729, for the void of parts of the 1885 ordinance by the courts.

49. The Penn State Extension website reports that a Pennsylvania study found that the average volumes of water produced during shallow gas well drilling in western Pennsylvania was 25,000 gallons during drilling, 50,000 gallons during stimulation, and 150 gallons

per day during production. See, "Water Issues," Penn Sate Extension, at http://extension.psu.edu/naturalgas/issues/environmental/resources/water.

50. Pennsylvania Department of Environmental Protection, *Orphan Oil and Gas Wells and the Orphan Well Plugging Fund*, at http://www.portal.state.pa.us/portal/server.pt/community/abandoned_orphan_well_program/20292.

51. See the following acts within the *Laws of Pennsylvania*: Act No. 80, *Session of 1878*, 57 (plugging oil wells); Act No. 122, *Session of 1881*, 110–11 (plugging abandoned oil wells); Act No. 54, *Session of 1883*, 61–65 (oil pipeline regulations); Act No. 32, *Session of 1885*, 29–37 (for the incorporation and regulation of natural gas companies); Act No. 114, *Session of 1885*, 145–46 (to protect oil, gas, and water wells); Act No. 199, *Session of 1887*, 310–13 (right of gas companies to eminent domain); Act 114, *Session of 1891*, 122–23 (to prevent pollution of springs, water wells, and streams by water from abandoned oil and gas wells); and Act No. 322, *Session of 1921*, 912–14 (to regulate the drilling, operating, and abandoning of oil and gas wells).

52. For the history of the Sanitary Water Board, see http://www.portal.state.pa.us/portal/server.pt?open=514&objID=588469&mode=2. Issues".

53. See *Annual Report of the Pennsylvania Department of Health*, 1908 (Harrisburg) II: 486–88, Google Books, at com/books/about/Report.html?id=0xdNAAAAMAAJ.

54. *Collins v. Chartiers Valley Gas Co.*, 131 Pa. 143 (1890); *Nannie R. Collins v. Chartiers v. Gas Co.*, 139 Pa. 111 (1891).

55. J. M. T. Carpenter to J. N. Pew, Nov. 17, 1892, Pew Papers, Hagley Museum and Library, Wilmington; E. T. Bouser to the Peoples Natural Gas Company, July 19, 1900; E. Robbins to Peoples Natural Gas Company, Nov. 22, 1894, Pew Papers; and Moorhead and Head to Peoples Natural Gas Company, May 19, 1899, Pew Papers. The Pew Papers also contained "Proposal and Specifications for Drilling Wells," dated July 22, 1901.

56. The Peoples Natural Gas Company, "Proposal and Specifications for Drilling Wells," July 22, 1901, Pew Papers.

57. Wagner, "Natural Gas Comes to Pittsburgh," 73–74. By 1891, Carnegie's Edgar Thompson Works was using only coke for fuel and in 1893 the Philadelphia Company terminated its contract with the firm. See Edgar P. Allen, "Natural Resources of Pittsburgh," *Proceedings of the Engineers' Society of Western Pennsylvania* (1891) 7: 11–13; Gugliotta, "'Hell with the Lid Taken Off,'" 183–266, analyzes labor's position on the use of natural gas.

58. Gugliotta, "'Hell with the Lid Taken Off,'" 151–58.

59. *Proceedings of the Engineers' Society of Western Pennsylvania* 8 (1892): 42–43. The speaker was Dr. R. Stansbury Sutton, a prominent physician. Gugliotta, "'Hell with the Lid Taken Off,'" 200.

60. The best discussion of the role of the Ladies Health Protective Association and the consequent agitation for smoke control can be found in Gugliotta, "'Hell with the Lid Taken Off,'" 132–43.

61. Joel A. Tarr, "The Pittsburgh Survey as an Environmental Statement," in *Pittsburgh*

Surveyed: Social Science and Social Reform in the Early Twentieth Century, ed. Maurine W. Greenwald and Margo Anderson (Pittsburgh: University of Pittsburgh Press, 1996), 179–81.

62. Ibid., 374–404; Tarr, *The Search for the Ultimate Sink*, 241.

63. Samuel S. Wyer, *Gas Situation of the Philadelphia Company's Natural Gas Properties in West Virginia and Pennsylvania* (Columbus, Ohio: Samuel S. Wyer, July 7, 1920); John Gates Jr., "Domestic and Industrial Use of Natural Gas," *Gas Age* (May 1, 1919): 470–71; and Eugene D. Thoenen, *History of the Oil and Gas Industry in West Virginia* (Charleston, W.V.: Education Foundation, 1964), 361. The peak of gas production from Appalachian fields was in 1917. See Arion R. Tussing and Connie C. Barlow, *The Natural Gas Industry: Evolution, Structure, and Economics* (Cambridge: Ballinger Publishing, 1984), 32. In 1919 and 1920, a brief local boom occurred when drillers discovered gas near the city of McKeesport, fifteen miles from Pittsburgh, but the wells soon turned dry. See Meredith E. Johnson, *Geology and Mineral Resources, Topographic and Geologic Atlas of Pennsylvania No. 27, Pittsburgh Quadrangle* (Harrisburg: Bureau of Publications, 1929), 122–32; Joel A. Tarr, "There Will Be Gas," *Pittsburgh Post-Gazette*, 2 August 2009.

64. John G. Clark, *Energy and the Federal Government: Fossil Fuel Policies, 1900–1946* (Urbana: University of Illinois Press, 1987), 51–109.

65. See Public Service Commission of the Commonwealth of Pennsylvania, *Supply and Conservation of Natural Gas in the State of Pennsylvania, Proceedings at Pittsburgh, Pennsylvania, Jan. 8, 1919* (Harrisburg, 1919); F. F. Schauer (vice president and general manager, Equitable Gas), "A Resume of the History, Organization, Operation and Present Day Problems of the Equitable Gas Company Pittsburgh & West Virginia Gas Company, Philadelphia Oil Company" (unpublished manuscript, February 15, 1932), 6–7; Meredith E. Johnson, *Geology and Mineral Resources, Topographic and Geologic Atlas of Pennsylvania No. 27, Pittsburgh Quadrangle* (Harrisburg: Bureau of Publications, 1929), 110–11; and James H. Reed, "Pittsburgh and the Natural Gas Industry," in Greater Pittsburgh Chamber of Commerce, *Pittsburgh and the Pittsburgh Spirit: Addresses at the Chamber of Commerce, 1927–28* (Pittsburgh: Robert L. Forsythe, 1928), 126–27, 134. From 1922 to 1927 Equitable Gas manufactured coal gas that it mixed with natural gas to supplant decline natural gas supplies.

66. Johnson, *Geology and Mineral Resources, Pittsburgh Quadrangle*, 110. For glass, see Charles R. Fettke, *Glass Manufacture and the Glass Sand Industry of Pennsylvania*, #12, Pennsylvania Topographic and Geological Survey Commission (Harrisburg: L. L. Kuhn, 1919), 97.

67. Natural gas production in Pennsylvania (mostly western Pennsylvania counties) ranged from approximately 125,787,000 million cubic feet in 1920 to 101,951,00 in 1929. Production figures were sharply reduced during the Depression. See Schanz, "Historical Statistics of Pennsylvania's Mineral Industries, 1759–1955," 37.

68. Gugliotta, "'Hell with the Lid Taken Off,'" 501–9; Clark, *Energy and the Federal Government*, 145.

69. A national study done in 1931–33 found winter particulate levels remained among the worst in the nation but, as figure 1.2 shows, the general trend during the Depression

was for a decline in smoky days. See James E. Ives et al., *Atmospheric Pollution of American Cities for the Years 1931–1933, with Special Reference to the Solid Constituents of the Pollution*, Public Health Bulletin No. 224 (Washington, DC: GPO, 1936).

70. Tarr, *Search for the Ultimate Sink*, 232–33.

71. Joel A. Tarr and Carl Zimring, "The Struggle for Smoke Control in St. Louis: Achievement and Emulation," in *Common Fields: an Environmental History of St. Louis*, Andrew Hurley, ed. (St. Louis: Missouri Historical Society, 1997), 199–220.

72. Tarr, *Ultimate Sink*, 240–41, 246–47; Gugliotta, "'Hell with the Lid Taken Off,'" 595–97.

73. Tarr, *Ultimate Sink*, 243–51.

74. Clark, *Energy and the Federal Government*, 254–57, 349. Natural gas pipelines were extended to many parts of the country in the 1930s but not to the Pittsburgh area or other parts of Pennsylvania, New York, and New England. See John H. Herbert, *Clean Cheap Heat: The Development of Residential Markets for Natural Gas in the United States* (New York: Praeger, 1992), 91–92, and Pittsburgh and West Virginia Gas Company and Equitable Gas Company, *Annual Reports* 1940 and 1941 (Pittsburgh, 1940, 1941), 15–16 (1940), and 15–19 (1941).

75. Tarr, *Ultimate Sink*, 244–45. Other influential Pittsburgh businessmen besides Mellon, such as J. H. Hillman, had interests in both coal and competing fuels. Hillman, who owned Pittsburgh Coke and Chemical and controlled several coal, coke, and iron companies, served as president of the Texas Gas Transmission Corporation. In 1948, in return for gas supplies for his industrial interests, he made scarce steel piping available for the pipeline expansion of Texas Eastern, the firm that converted Little Inch and Big Inch to natural gas. See Christopher J. Castaneda and Joseph A. Pratt, *From Texas to the East: A Strategic History of Texas Eastern Corporation* (College Station: Texas A&M Press, 1993), 84–85.

76. Pratt, *From Texas to the East*, 252.

77. "Caution Urged on County Smoke Law," *Pittsburgh Press*, 21 December 1945; and "Coal Industry Opens Drive on Air Pollution," ibid., 3 February 1946. The coal industry created the Western Pennsylvania Conference on Air Pollution to study combustion equipment, fuels, prices, and methods to reduce smoke; it later evolved into Bituminous Coal Research.

78. For the story of the "Inch" lines see Castaneda and Pratt, *From Texas to the East*, 13–74.

79. Equitable Gas Company and Subsidiaries, *Annual Report* 1952, 19; Peoples Natural Gas Company, "Peak Storage Inventory, 1943–1957," *Annual Gas Sales Volume*, 20A.

80. Herbert, *Clean Cheap Heat*, 91–103.

81. The projected industry to manufacture treated coal (Disco) never became a success although one plant was constructed. To meet the ordinance's requirements, the Bureau of Smoke Prevention permitted a mixture of 50 percent high-volatile coal and 50 percent low-volatile coal to be used. See City of Pittsburgh Bureau of Smoke Prevention, *Report 1953* (Pittsburgh, 1954), 12. Discussions of Disco can be found in C. E. Lesher, "Disco: A Smokeless Fuel," and Caleb Davies Jr., "Tar and By-Products from the Disco Process," *Industrial and Engineering Chemistry* 33 (1941): 858–64.

82. The Bureau of Smoke Inspection reported that between 1948 and 1955, 4,173 domes-

tic natural gas conversion units, 1,283 commercial conversion units, and 1,743 domestic gas boilers were installed. In addition, 1,257 domestic mechanical coal stokers and 628 commercial stokers were installed. Domestic household and commercial interests installed new gas-consuming conversion units and boilers at a declining but still high rate through 1955, while stoker installations declined to almost zero by that date. See Bureau of Smoke Prevention, *Report 1955* (Pittsburgh, 1955), n.p. For Equitable sales increases, see Equitable Gas Company, *Annual Reports,* 1949 and 1956 (Pittsburgh: The Company, 1948 and 1956). For Peoples sales increases see Peoples Natural Gas Company, *Annual Gas Sales Volume,* 1943–1957. The Peoples compilation includes comparative fuel prices.

83. Peoples Natural Gas Company, *Annual Gas Sales Volume,* 251. For comparative costs of competing fuels, see Peoples Natural Gas Company, *Operating and Financial Statistics Years 1943–1972* (Pittsburgh, 1973), 23a. While there had been considerable discussion in both the 1941 and 1946 hearings about special provisions for the poor, cash subsidies of a limited amount were only provided for public assistance recipients. In enforcing the statute, the Bureau of Smoke Prevention attempted to educate low-income consumers as to the proper measures of firing smokeless coal rather than fining them for smoke violations.

84. Tarr, *Ultimate Sink,* 250.

85. Ibid., 262–82.

86. Sherie R. Mershon and Joel A. Tarr, "Strategies for Clean Air: The Pittsburgh and Allegheny County Smoke Control Movements, 1940–1960," in Tarr, ed., *Devastation and Renewal,* 163–66. The railroads hated the city smoke control ordinance and opposed any serious county regulation. In 1943 they used their political influence to secure exemption from a state enabling law that empowered Allegheny County to regulate smoke, but in 1947 Pittsburgh antismoke interests, including corporate leaders such as Richard King Mellon, were able to secure state authorization for county legislation.

87. For a study of the dieselization of American railroads, see Albert J. Churella, *From Steam to Diesel: Managerial Customs and Organizational Capabilities in the Twentieth-Century American Locomotive Industry* (Princeton: Princeton University Press, 1999).

88. In 1955 the Pennsylvania Railroad had 602 diesel units in service and only 18 steam engines. Pittsburgh Bureau of Smoke Prevention, *Reports, 1952* (p. 12), 1953 (pp. 14–18), 1954 (p. 20), and 1955 (pp. 16–20). Diesels could "smoke," but the bureau reported that no violations had been recorded for Pittsburgh railroads.

89. Pittsburgh Department of Public Health, Bureau of Smoke Prevention, "Pneumonia and Atmospheric Dust," *1954 Report* (Pittsburgh, 1954), 12, and Gugliotta, "'Hell with the Lid Taken Off,'" 404–15. For an earlier study that also found no significant relationship, see Ewald Tomakek and Edwin B. Wilson, "Pneumonia in Pittsburgh," *Proceedings of the American Philosophical Society* 53 (1924): 279–319.

90. Clay and Tarr (2011), "Coal, Smoke, and Death," working paper; Kenneth Y. Clay and Michael Greenstone, "The Impact of Air Pollution on Infant Mortality: Evidence from Geographic Variation in Pollution Shocks Induced by a Recession," *Quarterly Journal of Economics*

118 (August 2003): 1121–67, and Janet Currie and Matthew Neidell, "Air Pollution and Infant Health: What Can We Learn from California's Recent Experience?" *Quarterly Journal of Economics* 120 (2005): 1003–30.

91. Pittsburgh Bureau of Smoke Prevention, *Report on Stationary Stacks* (Pittsburgh, 1951), 13.

92. Roy Lubove, *Twentieth Century Pittsburgh: Government, Business, and Environmental Change* (New York: John Wiley, 1969), 16–41; Tarr, *Devastation and Renewal*, 173.

CHAPTER 2. The Energy Capital of the World? Oil-Led Development in Twentieth-Century Houston

The conceptualization of this chapter owes its genesis—and several key parts of the chapter's narrative on Houston—to our collaboration through our book *Energy Metropolis: An Environmental History of Houston and the Gulf Coast* (Pittsburgh: University of Pittsburgh Press, 2007). We also wish to thank two of our students, Tom McKinney and Terry Tomkins-Walsh, for work they completed on the study of Buffalo Bayou—also used in part in this chapter, and to Steven McDonald for material gathered on Houston's physical setting.

1. W. L. Fisher, J. H. McGowen, L. F. Brown Jr., and C. G. Groat, *Environmental Geologic Atlas of the Texas Coastal Zone—Galveston-Houston Area* (Austin: Bureau of Economic Geology, University of Texas at Austin, 1972), 1.

2. David G. McComb, *Galveston: A History* (Austin: University of Texas Press, 1986), 6. See also 7–8 and Houston Geological Society, *Geology of Houston & Vicinity, Texas* (Houston: Houston Geological Society, 1961), 3, 7; Robert R. Lankford and John J. W. Roger, comps., *Holocene Geology of the Galveston Bay Area* (Houston: Geological Society, 1969), 7, 1; Fisher et al., *Environmental Geologic Atlas,* 7.

3. Fisher et al., *Environmental Geologic Atlas,* 7; Jim Lester and Lisa Gonzalez, eds., *Ebb & Flow: Galveston Bay Characterization Highlights* (Galveston: Galveston Bay Estuary Program, 2001), 12; Joseph L. Clark and Elton M. Scott, *The Texas Gulf Coast: Its History and Development* (New York: Lewis Historical Publishing, 1955), 2: 14–16; Houston Geological Society, *Geology of Houston & Vicinity, Texas,* 3.

4. Fisher et al., *Environmental Geologic Atlas,* 7.

5. Ibid.; G. L. Fugate, "Development of Houston's Water Supply," *Journal of the American Water Works Association* 33 (October 1941): 1769–70.

6. Planning and Development Department, *Public Utilities Profile for Houston Texas* (Summer, 1994), III-15; Fugate, "Development of Houston's Water Supply," 1769–70. The geologic formations from which Houston obtains groundwater supplies are Upper Miocene, Pliocene, and Pleistocene in origin. See Nicholas A. Rose, "Ground Water and Relations of Geology to Its Occurrence in Houston District, Texas," *Bulletin of the American Association of Petroleum Geologists* 27 (August 1943): 1081.

7. See "Houston," *Twentieth Century Cities,* part 4 of Association of American Geogra-

phers, in *Contemporary Metropolitan America*, ed. John S. Adams (Cambridge, Mass.: Ballinger, 1976), 109, 121–24; Houston Chamber of Commerce, *Houston Facts '82*.

8. U.S. Environmental Protection Agency, *Heat Island Effect: Houston's Urban Fabric*, http://www.epa.gov/heatisland/pilot/houst_urbanfabric.html; U.S. EPA, *Heat Island Effect: Houston*, http://www.epa.gov/heatisland/pilot/houston.html.

9. Robert Thompson, "'The Air Conditioning Capital of the World': Houston and Climate Control," in Melosi and Pratt, eds., *Energy Metropolis*, 88–104.

10. Fisher et al., *Environmental Geological Atlas*, 1 and 7.

11. Espey, Huston and Associates, Inc., preparers, Archival Research: Houston-Galveston Navigation Channels, Texas Project—Galveston, Harris, Liberty and Chambers Counties, Texas, April 1993, 8, 10; McComb, *Galveston: A History*, 121–49; David Roth, "Texas Hurricane History" (National Weather Service, Lake Charles, Louisiana, 2004), http://www.srh.noaa.gov/lch/research/txhur.php.

12. Fisher et al., *Environmental Geologic Atlas*, 1.

13. Fisher et al., *Environmental Geologic Atlas*, 15, 20; Robert R. Stickney, *Estuarine Ecology of the Southeastern United States and Gulf of Mexico* (College Station: Texas A&M University Press, 1984), 247–80; Lester and Gonzalez, eds., *Ebb & Flow*, 9–11.

14. Joseph A. Pratt, *The Growth of a Refining Region* (Greenwich, Conn.: JAI Press, 1980), 35.

15. Marilyn M. Sibley, "Houston Ship Channel," *The Handbook of Texas, Online*, 1–3, www.tsha.utexas.edu/handbook/online/articles/view/HH/rhh11.html; Marilyn M. Sibley, *The Port of Houston: A History* (Houston: University of Texas Press, 1968), 102–45; Garvin Berry, "Promoters, Politicians Turned Shallow Bayou into Seaport," *Houston Business Journal* (November 22, 1893); Lynn M. Alperin, *Custodians of the Coast: History of the United States Army Engineers at Galveston* (Galveston: U.S. Army Corps of Engineers, 1977), 95–101.

16. Harold Willamson, Ralph Andreano, Arnold Daum, and Gilbert Klose, *The American Petroleum Industry: The Age of Energy, 1899–1959* (Evanston: Northwestern University Press, 1963). For the rise of oil in the southwestern United States, see Joseph A. Pratt, "The Ascent of Oil: The Transition from Coal to Oil in Early Twentieth Century America," in *Energy Transitions: Long-term Perspectives*, eds. Lewis J. Perelman, August W. Giebelhaus, and Michael D. Yokell (Boulder, Colo.: Westview Press, 1981): 9–34.

17. Barry J. Kaplan, "Houston: The Golden Buckle of the Sunbelt," in *Sunbelt Cities: Politics and Growth Since World War II*, eds. Richard M. Bernard and Bradley R. Rice (Austin: University of Texas Press, 1983), 197; Walter L. Buenger and Joseph A. Pratt, *But Also Good Business: Texas Commerce Banks and the Financing of Houston and Texas, 1886–1986* (College Station: Texas A&M University Press, 1986), 73, 109; Walter Rundell Jr., *Early Texas Oil: A Photographic History, 1866–1936* (College Station: Texas A&M University Press, 1977), 136; David G. McComb, *Houston: A History* (Austin: University of Texas Press, 1981), 78–79.

18. Frederick Law Olmsted, *A Journey through Texas, Or, A Saddle Trip on the Southwestern Frontier* (New York: Dix, Edwards and Co., 1857), 366.

19. For histories of a major law firm and a bank active in tying the region into the nation-

al economy, see Kenneth J. Lipartito and Joseph A. Pratt, *Baker & Botts in the Development of Modern Houston* (Austin: University of Texas Press, 1991); and Buenger and Pratt, *But Also Good Business.*

20. For a biography of one of the best-known migrants, "Mr. Houston," Jesse Jones, see Steven Fenberg, *Unprecedented Power: Jesse Jones, Capitalism, and the Common Good* (College Station: Texas A & M University Press, 2011).

21. John O. King, *Joseph Stephen Cullinan: A Study of Leadership in the Texas Petroleum Industry, 1897–1937* (Nashville: Vanderbilt University Press, 1970); Marquis James, *The Texaco Story: The First Fifty Years, 1902–1952* (New York: The Texas Company, 1953); James A. Clark and Michel T. Halbouty, *Spindletop* (New York: Random House, 1952); Craig Thompson, *Since Spindletop: The Human Story of Gulf's First Century* (Pittsburgh: Gulf Oil Corporation, 1951).

22. Hughes Tool, a factory that manufactured cutting-edge drill bits, remained one of the largest employers in the region for much of the twentieth century. Michael R. Botson Jr., *Labor, Civil Rights and the Hughes Tool Company* (College Station: Texas A & M University Press, 2005).

23. Joseph A. Pratt, "The Petroleum Industry in Transition: Antitrust and the Decline of Monopoly Control in Oil," *Journal of Economic History* 40, no. 4 (December 1980): 815–37.

24. King, *Joseph Stephen Cullinan.*

25. Henrietta M. Larson and Kenneth Wiggins Porter, *History of the Humble Oil & Refining Company: A Study in Industrial Growth* (New York: Harper and Brothers, 1959).

26. Pratt, *Growth of a Refining Region,* 36–42, 153–88.

27. For a brief history of the creation of Hughes Tool and the establishment of its large factory in Houston, see Botson, *Labor, Civil Rights, and the Hughes Tool Company,* 34–58.

28. "Buffalo Bayou," *The Handbook of Texas, Online* (Austin: Texas State Historical Association, December 4, 2002), 1, www.tsha.utexas.edu/handbook/online/articles/view/BB/rhb28.html. See also Marvin V. Melosi (with Thomas McKinney and Terry-Tomkins-Walsh), "Historical Significance of Buffalo Bayou, Houston, Texas," National Heritage Area Study (Denver: U.S. Department of Interior, U.S. National Park Service, 2005).

29. Melosi (with McKinney and Tomkins-Walsh), "Historical Significance of Buffalo Bayou," 52–56.

30. Marilyn McAdams Sibley, *The Port of Houston* (Austin: University of Texas Press, 1968); Lynn M. Alperin, *Custodians of the Coast: History of the United States Army Engineers at Galveston* (Galveston, Tex.: U.S. Army Corps of Engineers, 1977).

31. Marilyn McAdams Sibley, *The Port of Houston* (Austin: University of Texas Press, 1968); Lynn M. Alperin, *Custodians of the Coast: History of the United States Army Engineers at Galveston* (Galveston, Tex.: U.S. Army Corps of Engineers, 1977), 58.

32. Quoted in Beth Ann Shelton et al., *Houston: Growth and Decline in a Sunbelt Boomtown* (Philadelphia: Temple University Press, 1989), 15.

33. For a lively photographic history of one area refinery, see Barbara Wells, *Shell at Deer Park: The Story of the First Fifty Years* (Houston: Shell Oil Company, 1979).

34. Pratt, *The Growth of a Refining Region*, 3–7, 33. See also Joe Feagin, *Free Enterprise City* (New Brunswick, N.J.: Rutgers University Press, 1988), 62, 65–66; David G. McComb, *Texas: A Modern History* (Austin: University of Texas Press, 1989), 125: Kaplan, "Houston: The Golden Buckle of the Sunbelt," 197; Sibley, *The Port of Houston*, 161; James J. Parsons, "Recent Industrial Development in the Gulf South," *Geographical Review* 40, no. 1 (January 1950): 74. The statistics for 2008 are taken from Greater Houston Partnership, "Energy Capital Houston," vol. 8 (2008), 21.

35. General patterns of migration are discussed in Pratt, *Growth of a Refining Region*. See also, Bernadette Pruitt, "In Search of Freedom: Black Migration to Houston, 1914–1945," *The Houston Review of History and Culture* 3, no. 1 (Fall 2005): 48–57; Herbert Winthrop, *Negro Employment in Basic Industries—A Study of Racial Policies in Six Industries* (Philadelphia, Pa.: Trustees of the University of Pennsylvania, 1970); F. Ray Marshall, *The Negro and Organized Labor* (New York: Wiley, 1965); Robert D. Bullard, *Invisible Houston: The Black Experience in Boom and Bust* (College Station: Texas A & M University, 1987); Thomas Kreneck, *Mexican American Odyssey* (College Station: Texas A & M University Press, 2001).

36. Sethuraman Srinivasan, "The Struggle for Control: Technology and Organized Labor in Gulf Coast Refineries, 1913–1973" (PhD Dissertation, University of Houston, Department of History, 2001).

37. John Frey and H. Chandler Ide, eds., *A History of the Petroleum Administration for War, 1941–1945* (Washington, DC: Government Printing Office, 1946).

38. Greater Houston Partnership, "Energy Capital Houston," vol. 8 (2008), 47–53.

39. Christopher James Castaneda, *Regulated Enterprise: Natural Gas Pipelines and Northeastern Markets, 1938–1954* (Columbus: Ohio State University Press, 1993); Christopher James Castaneda and Joseph A. Pratt, *From Texas to the East: A Strategic History of Texas Eastern Corporation* (College Station: Texas A & M University Press, 1993); Christopher Castaneda and Clarence Smith, *Gas Pipelines and the Emergence of America's Regulatory State* (New York: Cambridge University Press, 1996); Frank Mangan, *The Pipeliners: The Story of El Paso Natural Gas* (El Paso: Guynes Press, 1978); David Raley, "Out of Gas: Tenneco in the Era of Natural Gas Regulation" (PhD Dissertation, University of Houston, Department of History, 2011).

40. Greater Houston Partnership, "Energy Capital Houston," vol. 8 (2008), 21.

41. Tyler Priest, *The Offshore Imperative: Shell Oil's Search for Petroleum in Postwar America* (College Station: Texas A & M University Press, 2007); Sibley, *Port of Houston*; Joseph L. Clark, *The Texas Gulf Coast: Its History and Development* (New York: The Lewis Historical Publishing, 1955); Joseph A. Pratt and Christopher J. Castaneda, *Builders: Herman and George R. Brown* (College Station: Texas A & M University Press, 1999).

42. The Offshore Technology Conference was created in 1969 and meets in Houston each May. In 2007, more than 67,000 attendees from offshore projects throughout the world gathered to discuss advances in the technology needed to find, drill, and produce offshore oil. Greater Houston Partnership, "Energy Capital Houston," vol. 8 (2008), 28.

43. One of the best sources on the energy crisis of the early 1970s remains Anthony Sampson, *The Seven Sisters* (New York: Bantam Books, 1976).

44. A good sense of the impact of the oil depression of the mid-1980s can be seen in Joseph M. Grant, *The Great Texas Bank Crash: An Insider's Account* (Austin: University of Texas Press, 1996); and in Ben F. Love, *Ben Love: My Life in Texas Commerce* (College Station: Texas A & M University Press, 2005), 218–51.

45. William D. Angel Jr., "The Politics of Space: NASA's Decision to Locate the Manned Spacecraft Center in Houston," *Houston Review of History and Culture* 6, Second Issue (1984): 63–81.

46. George Thomas Morgan and John O. King, *The Woodlands: New Community Development, 1964–1983* (College Station: Texas A & M University Press, 1987).

47. For a list of numerous relocations of oil-related companies to Houston in the recent past, see Greater Houston Partnership, "Energy Capital Houston," vol. 8 (2008), 21; Jon Ishii, "The Oil Industry and the Cities: Consolidation in the Oil Extraction Industry," *Houston Business: A Perspective on the Houston Economy,* Federal Reserve Bank of Dallas, April 1996.

48. This section is drawn from Martin V. Melosi, "Houston: Energy Capital," *New Geographies* 2 (2009): 97–102.

49. Planning and Development Department, *Public Utilities Profile for Houston, Texas* (Summer 1994), III-15.

50. City of Houston, *Public Water Supply System* (Houston: Department of Public Works and Engineering, 1948), 4.

51. Fugate, "Development of Houston's Water Supply," 1768–69.

52. "Report of Water Committee," *Annual Report, 1909,* 24.

53. "Report of Water Commissioner," *City Book of Houston, 1914* (Houston: The City, 1914), 106; "Mayor's Message," ibid., 80–81; *City Book of Houston, 1925* (Houston: The City, 1914), 41; Bud A. Randolph, "The History of Houston's Water Supply," *Texas Commercial News,* June 1927, 43.

54. Fugate, "Development of Houston's Water Supply," 1770–71.

55. "Spending $4,000,000 to $6,000,000 on Water System Urged," *Houston Chronicle,* 2 September 1937.

56. Fugate, "Development," 1772–74.

57. Alvord, Burdick, and Howson, "Report on an Adequate Water Supply for the City of Houston, Texas" (Chicago, February 1938), 1–3, 75–76.

58. "New Plan for Water Supply to Be Offered," *Houston Chronicle,* 3 February 1939.

59. J. M. Nagle, "Houston Gets Needed Water," *American City* 60 (February 1945): 77.

60. William W. McClendon, "The San Jacinto River Conservation and Reclamation District's Proposed Plan of Full Scale Development," *Slide Rule* (March 1945): 11; "Water Supply Dam to Be Built on San Jacinto," *Houston Post,* 14 July 1942; City of Houston, Utilities Department, *Engineering Report for Water Works Improvements,* January 17, 1944.

61. Water Department, *Report of Director for Year 1942,* 16; "Houston's Greater Water

Supply Near," *Houston* 23 (August 1952): 8–9; Water Supply and Conservation Committee, Houston Chamber of Commerce, *Water for the Houston Area* (Houston: Houston Chamber of Commerce, December 1954), 3; U.S. Department of the Interior, U.S. Geological Survey, "Characteristics of Water-Quality Data for Lake Houston, Selected Tributary Inflows to Lake Houston, and the Trinity River Near Lake Houston, August 1983–September 1990," *Water-Resources Investigations Report 99–4129* (Washington, DC: U.S. Department of Interior, 1999), 2, 4.

62. "Motorvista-Car Dealers and Driver Information for Houston, Texas," *Motorvista* (2007), http://motorvista.com/Dealers/Texas/Houston/.

63. Martin V. Melosi, "Community and the Growth of Houston," in *Effluent America: Cities, Industry, Energy, and the Environment*, ed. Martin V. Melosi (Pittsburgh: University of Pittsburgh Press, 2001), 194–95.

64. Tom Watson McKinney, "Superhighway Deluxe: Houston's Gulf Freeway," in Melosi and Pratt, eds., *Energy Metropolis*, 148–72.

65. See Diane C. Bates, "Urban Sprawl and the Piney Woods: Deforestation in the San Jacinto Watershed," in Melosi and Pratt, eds., *Energy Metropolis*, 173–84; Melosi, "Community and the Growth of Houston," 190–205.

66. Frank R. von der Mehden, ed., *The Ethnic Groups of Houston* (Houston: Rice University Studies, 1984). Gradual assimilation through education was accelerated during World War II, as women entered the refinery workforces in unprecedented numbers. See Gary J. Rabelais, "Humble Women at War: The Case of Humble's Baytown Refinery, 1942–1945," *The Houston Review of History and Culture* 2, no. 2 (Spring 2005): 33–36, 58.

67. Fredericka Meiners, *A History of Rice University: The Institute Years, 1907–1963* (Houston: Rice University Press, 1982); Joseph A. Pratt and Christopher J. Castaneda, *Builders* (College Station: Texas A & M University Press, 1999), 253–71.

68. Patrick Nicholson, *In Time: An Anecdotal History of the University of Houston* (Houston: Gulf Publishing, 1977); Ed Kilman and Theon Wright, *Hugh Roy Cullen: A Story of American Opportunity* (New York: Prentice-Hall, 1975).

69. For the early history of Houston philanthropy, see Thomas Henthorn, "Faith, Philanthropy, and Southern Progress: Social Policy and Urban Development in Houston, Texas, 1890–1930" (PhD Dissertation, Michigan State University, 2009). For a group portrait of many of these civic leaders, see Joseph A. Pratt, "8F and Many More: Business and Civic Leadership in Modern Houston," *The Houston Review of History and Culture* 1, no. 1 (Summer 2004): 2–7, 31–44.

70. See the following for examples of energy pollution studies with an industrial focus: Barbara L. Allen, *Uneasy Alchemy: Citizens and Experts in Louisiana's Chemical Corridor Disputes* (Cambridge, Mass.: MIT Press, 2003); Joan Norris Booth, *Cleaning Up: The Costs of Refinery Pollution* (New York: Council on Economic Priorities, 1975); Craig E. Colten, "Chicago Waste Lands: Refuse Disposal and Urban Growth, 1840–1990," *Journal of Historical Geography* 20 (April 1994): 124–42; Colten, *Industrial Waste in the Calumet Area, 1869–1970: An*

Historical Geography, Illinois Department of Energy and Natural Resources (1985); John T. Cumbler, "Whatever Happened to Industrial Waste? Reform, Compromise, and Science in Nineteenth Century Southern New England," *Journal of Social History* 29 (Fall 1995): 149–71; Neil Cunningham, Robert A. Kagan, and Dorothy Thornton, *Shades of Green: Business, Regulation, and Environment* (Stanford, Calif.: Stanford University Press, 2003); Hugh S. Gorman, "Manufacturing Brownfields: The Case of Neville Township, Pennsylvania, 1899–1989," *Technology and Culture* 38 (July 1997): 539–74; Gorman, *Redefining Efficiency: Pollution Concerns, Regulatory Mechanisms, and Technological Change in the U.S. Petroleum Industry* (Akron, Ohio: University of Akron Press, 2001); Andrew Hurley, "Creating Ecological Wastelands: Oil Pollution in New York City, 1870–1900," *Journal of Urban History* 20 (May 1994): 340–64; James E. Krier and Edmund Ursin, *Pollution and Policy: A Case Essay on California and Federal Experience with Motor Vehicle Air Pollution, 1940–1975* (Berkeley: University of California Press, 1977); Steve Lerner, *Diamond: A Struggle for Environmental Justice in Louisiana's Chemical Corridor* (Cambridge, Mass.: MIT Press, 2005); Martin V. Melosi, *Coping with Abundance: Energy and Environment in Industrial America* (New York: Knopf, 1985); Stephen Mosley, *The Chimney of the World: A History of Smoke Pollution in Victorian and Edwardian Manchester* (Cambridge: White Horse Press, 2001); Joseph A. Pratt, "Letting the Grandchildren Do: Environmental Planning during the Ascent of Oil as a Major Energy Source," *Public Historian* 2 (Summer 1980): 28–61; Joel A. Tarr and Bill Lamperes, "Changing Fuel Use Behavior and Energy Transitions: The Pittsburgh Smoke Control Movement, 1940–1950, a Case Study in Historical Analogy," *Journal of Social History* 14 (1981): 561–88; Jeffrey K. Stine and Joel A. Tarr, "At the Intersection of Histories: Technology and the Environment," *Technology and Culture* 39 (1998): 601–40; Gerald Markowitz and David Rosner, *Deceit and Denial: The Deadly Politics of Industrial Pollution* (Berkeley: University of California Press, 2002).

71. A pioneering of energy, growth, and environment in a region is James C. Williams, *Energy and the Making of Modern California* (Akron, Ohio: The University of Akron Press, 1997). An excellent account of the early politics of oil-led development in California is Paul Sabin, *Crude Politics: The California Oil Market, 1900–1940* (Berkeley: University of California Press, 2004).

72. Joseph A. Pratt, "Black Waters: Responses to America's First Oil Pollution Crisis," in *Essays in Public Works History*, no. 27 (Houston, Tex: APWA Press, July 2008); Joseph A. Pratt, "Growth or a Clean Environment? Responses to Petroleum-Related Pollution in the Gulf Coast Refining Region," *Business History Review* 52, no. 1 (Spring 1978): 1–29.

73. For a case study of the treatment of toxic wastes through time in Houston, see Kimberly A. Youngblood, "Voices of Doom: The Effects of a Grassroots Environmental Movement at the Brio Superfund Site," in Melosi and Pratt, eds., *Energy Metropolis*, 260–74. For a detailed description of Superfund sites in Harris County (where most of Houston is situated), see "Superfund Sites in Harris County," *Texas Commission on Environmental Quality*, 2011, http://www.tceq.texas.gov/remediation/superfund/sites/county/harris.html.

74. Pratt, *Growth of a Refining Region*, 72–75, 105. See also Shelton et al., *Houston: Growth*

and *Decline in a Sunbelt Boomtown*, 16–17; Parsons, "Recent Industrial Development in the Gulf South," 76–77; Randolph B. Campbell, *Gone to Texas: A History of the Lone Star State* (New York: Oxford University Press, 2003), 407–8; Kaplan, "Houston: Golden Buckle of the Sunbelt," 198; Feagin, *Free Enterprise City*, 66, 71; McComb, *Houston: A History*, 81, 128–29; Clark and Scott, *The Texas Gulf Coast*, 234; Warren Rose, *Catalyst of an Economy: The Economic Impact of the Port of Houston, 1958–1963* (Houston: Center for Research in Business and Economics, University of Houston, August 1965), 7; and several sections of Melosi and Pratt, eds., *Energy Metropolis*.

75. For issues on nonpoint pollution in Houston, see Martin V. Melosi, "Houston's Public Sinks: Sanitary Services from Local Concerns to Regional Challenges," in Melosi and Pratt, eds., *Energy Metropolis*, 134–47. See also Susan Smyer, "City of Houston Wastewater History," May 2008, http://documents.publicworks.houstontx.gov/documents/divisions/utilities/history_waste_water_operations.pdf.

76. Wendy Natt, "Sustainable Water Quality: Livable Houston Initiative," January 28, 2009, *Houston Tomorrow* http://www.houstontomorrow.org/initiatives/story/sustainable-water-quality/.

77. "Getting the Big Picture on Houston's Air Pollution," *NASA*, November 30, 2007, http://www.nasa.gov/vision/earth/everydaylife/archives/HP_ILP_Feature_03.html.

78. "Air Pollution 101," *Air Alliance Houston*, 2012, http://airalliancehouston.org/air_pollution_101/.

79. "A Closer Look at Air Pollution in Houston: Identifying Priority Health Risks," a summary of the *Report of the Mayor's Task Force on the Health Effects of Air Pollution*, 2007, https://sph.uth.tmc.edu/content/uploads/2011/12/UTReportrev.pdf.

80. Gorman, *Redefining Efficiency*.

81. See Feagin, *Free Enterprise City*, 61; see, also, Pratt and Castaneda, *Builders*, 157–91.

82. On the national level, see Robert Engler, *The Politics of Oil, Private Power and Democratic Directions* (New York: Macmillan, 1961). For Houston, see Feagin, *Free Enterprise City*. For Texas, see George Green, *The Establishment in Texas Politics: The Primitive Years, 1938–1957* (Westport, Conn.: Greenwood Press, 1979).

83. This pattern in the regulation of water pollution by oil in the United States is discussed in Pratt, "Black Waters."

84. Joseph A. Pratt, "A Mixed Blessing: Energy, Economic Growth, and Houston's Environment," in Melosi and Pratt, eds., *Energy Metropolis*, 21–51.

CHAPTER 3. Making a Lemon Out of Lemonade: Louisiana's Petrochemical Corridor

1. Joseph A. Pratt, *The Growth of a Refining Region* (Greenwich, Conn.: JAI Press, 1980); James C. Cobb, *Industrialization and Southern Society, 1877–1984* (Chicago: Dorsey Press, 1984); Keith Chapman, *The International Petrochemical Industry: Evolution and Location* (Cam-

bridge, Mass.: Blackwell, 1991); and Robert Lewis, "World War II Manufacturing and the Postwar Southern Economy," *Journal of Southern History* 73, no. 4 (2007): 837–66.

2. Louis C. Hunter, *Steamboats on the Western Rivers: An Economic and Technological History* (Cambridge, Mass.: Harvard University Press, 1949); and Paul Paskoff, *Troubled Waters: Steamboat Disasters, River Improvements, and American Public Policy, 1821–1860* (Baton Rouge: Louisiana State University Press, 2007).

3. For a discussion of the lower river floodplain, see Roger T. Saucier, *Recent Geomorphic History of the Pontchartrain Basin*, Coastal Studies Series, no. 9 (Baton Rouge: Louisiana State University, 1963).

4. D. O. Elliott, *The Improvement of the Lower Mississippi River for Flood Control and Navigation Vicksburg*, 3 vols. (Vicksburg, Miss.: U.S. Waterways Experiment Station, 1932); Robert Harrison, *Alluvial Empire* (Little Rock: U.S. Department of Agriculture, Economic Research Service, 1961), 1: 53–54; Craig E. Colten, *An Unnatural Metropolis: Wresting New Orleans from Nature* (Baton Rouge: Louisiana State University Press, 2005), esp. chap. 1; and Karen M. O'Neill, *Rivers by Design: State Power and the Origins of U.S. Flood Control* (Durham, N.C.: Duke University Press, 2006).

5. Harrison, *Alluvial Empire,* 159–60; and Charles A. Camillo and Matthew T. Pearcy, *Upon Their Shoulders: A History of the Mississippi River Commission* (Vicksburg: Mississippi River Commission, 2004), 138.

6. Harrison, *Alluvial Empire*, 161–63.

7. U.S. Army Corps of Engineers, *Floods and Flood Control on the Mississippi, 1973* (Washington, DC: USACE, 1973).

8. Lower Mississippi Region Comprehensive Study Coordinating Committee, *Lower Mississippi Region Comprehensive Study, Appendix E: Flood Problems* (Vicksburg: Lower Mississippi Region Comprehensive Study Coordinating Committee, 1974).

9. Ozarks Regional Commission, *Long Range Economic Impact of the 1973 Flood in Louisiana* (Baton Rouge: Research Associates, 1974), 28, 35–36, 84–88.

10. Extremely low water can also prompt navigation restrictions and disrupt industrial production.

11. Philip Shea, "The Spatial Impact of Government Decisions on the Production and Distribution of Louisiana Sugarcane, 1751–1972" (PhD Dissertation, Michigan State University, 1974), 90; and John B. Rehder, *Delta Sugar: Louisiana's Vanishing Plantation Landscape* (Baltimore: Johns Hopkins University Press, 1999).

12. Rehder, *Delta Sugar*, 127 and 299; and Elizabeth Vaughan, "Louisiana Sugar: A Geohistorical Perspective" (PhD Dissertation, Louisiana State University, 2003).

13. Elizabeth Vaughan, "Louisiana Sugar."

14. Raymond E. Sanafelt, "The Baton Rouge–New Orleans Petrochemical Industrial Region: A Functional Region Study" (PhD Dissertation, Louisiana State University, 1977); and Shea, "Spatial Impact," 159–60. Sugarcane Acreage from the U.S. Census of Agriculture, 1910–2007 (Washington, DC: USDA).

15. Sanafelt, "Baton Rouge–New Orleans," 22.

16. Ralph W. and Muriel E. Hidy, *Pioneering in Big Business: History of Standard Oil Company (New Jersey)* (New York: Harper and Brothers, 1955), 420, and John L. Loos, *Oil on Stream! A History of Interstate Oil Pipe Line Company, 1909–1959* (Baton Rouge: Louisiana State University Press, 1959), esp. chap. 1 and 7.

17. Henrietta M. Larson and Kenneth W. Porter, *History of Humble Oil & Refining Company: A Study in Industrial Growth* (New York: Harper and Brothers, 1959), chap. 21; and Charles S. Popple, *Standard Oil Company (New Jersey) in World War II* (New York: Standard Oil Company, 1952), 242–43.

18. Louisiana Division of Employment Security, *Oil and Jobs in Louisiana* (Baton Rouge: Louisiana Division of Employment Security, 1960); and Lewis "World War II Manufacturing."

19. Chapman, *International Petrochemical Industry*, 74.

20. Scott A. Hemmerling, "Environmental Equity in Southeast Louisiana: Oil, People, Policy, and the Geography of Industrial Hazards" (PhD Dissertation, Louisiana State University, 2007).

21. Robert N. McMichael, "Plant Location Factors in the Petrochemical Industry in Louisiana" (PhD Dissertation, Louisiana State University, 1961), 41; and U.S. Department of the Interior, Bureau of Mines, *Petroleum Refineries in the United States* (Washington, DC, 1962), 1. The Mississippi River petrochemical complex generally is treated as part of the larger Gulf Coast production and refining region. See Pratt, *The Growth of a Refining Region*; and Chapman, *The International Petrochemical Industry*.

22. Illinois Central Gulf Railroad Company, *Available Rail-River Industrial Sites: Along the Illinois Central Gulf Railroad and the East Side of Mississippi River in Louisiana* (Chicago: Industrial Development Department, Illinois Central Railroad, 1974).

23. Public Affairs Research Council, *Accelerating Louisiana's Economic Growth* (Baton Rouge: Public Affairs Research Council, 1977), 132–33.

24. Sanafelt, "Baton Rouge–New Orleans"; Chapman, *International Petrochemical Industry*, chap. 5; Hemmerling, "Environmental Equity"; and Elaine G. Yodis and Craig E. Colten, *Geography of Louisiana* (Boston: McGraw-Hill, 2007), chapter 10. U.S. Department of Energy, *Petroleum Refineries in the U.S. and U.S. Territories* (Washington, DC, 1979), 2.

25. Louisiana Wildlife and Fisheries Commission, *Eighth Biennial Report, 1958–59* (New Orleans: Louisiana Wildlife and Fisheries Commission, 1960), 170–71.

26. A second petrochemical complex developed near the southwest Louisiana city of Lake Charles. It enjoyed the same state policies and access to crude. Yet, it did not have access to a navigable waterway like the Mississippi River.

27. Sanafelt, "Baton Rouge–New Orleans," 49.

28. Louisiana Division of Employment Security, *Oil and Oil Jobs in Louisiana*; William R. Freudenburg and Robert Gramling, *Oil in Troubled Water: Perceptions, Politics, and the Battle over Offshore Drilling* (Albany: State University of New York Press, 1994), chap. 2; and Jacques

M. Henry and Carl L. Bankston III, *Blue Collar Bayou: Louisiana Cajuns in the New Economy of Ethnicity* (Westport, Conn.: Praeger, 2002), chap. 3.

29. Loren C. Scott, *Finding Permanent Solutions to Louisiana's Recurring Crisis* (Baton Rouge: Council for a Better Louisiana, 1987), 16–19.

30. For a discussion of Outer Continental Shelf development, see U.S. Department of the Interior, Minerals Management Service, *Deepwater Gulf of Mexico 2002: America's Expanding Frontier* (New Orleans: U.S. Department of the Interior, Minerals Management Service, Gulf of Mexico OCS Region, 2002).

31. Louisiana Department of Natural Resources, *Louisiana Petrochemical Industry Assessment* (Baton Rouge: Louisiana Department of Natural Resources, 1983); Scott, *Finding Permanent Solutions*; and Alan G. Pulsipher, *Accounting for Socioeconomic Change from Offshore Oil and Gas: Cumulative Effects on Louisiana's Coastal Parishes* (Baton Rouge: Louisiana State University, Center for Energy Studies, 2006).

32. Pulsipher, *Accounting for Socioeconomic Change*, 15.

33. See Craig E. Colten, "The Rusting of the Chemical Corridor," *Technology and Culture* 47, no. 1 (2006): 95–101.

34. Loren C. Scott, *The Energy Sector: Still a Giant Economic Engine for the Louisiana Economy* (Baton Rouge: Louisiana Mid-Continent Oil and Gas Association, 2002).

35. Craig E. Colten, "Too Much of a Good Thing: Industrial Pollution in the Lower Mississippi River," in *Transforming New Orleans and Its Environs: Centuries of Change*, ed. Craig E. Colten (Pittsburgh: University of Pittsburgh Press, 2000), 141–59. For an expanded discussion of brine pollution see Hugh S. Gorman, *Redefining Efficiency: Pollution Concerns, Regulatory Mechanisms, and Technological Change in the U. S. Petroleum Industry* (Akron, Ohio: University of Akron Press, 2001), esp. chap. 2 and 7.

36. Louisiana Department of Wildlife and Fisheries, *Third Biennial Report, 1948–1949* (Baton Rouge: Louisiana Department of Wildlife and Fisheries, 1949), 352–53.

37. New Orleans Association for Commerce, *A Statement of Facts Concerning Resources of the New Orleans Region for Chemical and Allied Industries* (New Orleans: New Orleans Association for Commerce, 1942), 12.

38. Louisiana Department of Wild Life and Fisheries, *Third Biennial Report* (Baton Rouge: Louisiana Department of Wild Life and Fisheries, 1948–49), 368–69.

39. U.S. Public Health Service, *Summary Report on Water Pollution: Southwest–Lower Mississippi Drainage Basins* (Washington, DC: USPHS, Water Pollution Division, 1951).

40. M. L. Eddars, L. R. Kister, and Glenn Scarcia, *Water Resources of the New Orleans Area, Louisiana*, Geological Survey Circular 374 (Washington, DC: U.S. Geological Survey, 1956); and Louisiana Stream Control Commission, *Proceedings of Meetings of the Louisiana Stream Control Commission* (Baton Rouge, May 15, 1958–September 22, 1966), see especially April 5, 1960.

41. Louisiana Wild Life and Fisheries Commission, *Ninth Biennial Report 1960–1961* (New Orleans, 1962), 197–98.

42. U.S. Department of the Interior, Federal Water Pollution Control Administration, *Endrin Pollution in the Lower Mississippi River Basin* (Dallas: USDOI, Federal Water Pollution Control Administration, 1966).

43. Colten, "Too Much of a Good Thing," 153.

44. Robert H. Harris and Edward M. Brecher, "Is the Water Safe to Drink?" in three parts: *Consumer Reports* 39 (June 1974): 436–42, (July 1974): 538–42, and (August 1974): 623–27. See also, T. A. DeRouen and J. E. Diem, "New Orleans Drinking Water Controversy: A Statistical Perspective," *American Journal of Public Health* 65, no. 10 (1975): 1060–62.

45. U.S. Environmental Protection Agency, *Industrial Pollution of the Lower Mississippi River in Louisiana* (Dallas: USEPA, 1972); and Colten, "Too Much of a Good Thing."

46. Executive Office of the President, Office of Emergency Planning, *Hurricane Betsy, August 27–September 10, 1965* (Washington, DC: Executive Office of the President, Office of Emergency Planning, 1965), 15–18.

47. "Cause of Shell Blasts Sought," *New Orleans Times Picayune*, 23 June 1979, 1–2; "Explosion Hit Town Like a Hurricane," *New Orleans Times Picayune*, 6 May 1988, 1 and 3. "One Worker Killed and Several Hurt in Blast at Louisiana Refinery," *New York Times* 25 December 1989 at http://www.nytimes.com/1989/12/25/us/one-worker-killed-and-several-hurt-in-blast-at-louisiana-refinery.html?pagewanted=1].

48. Susan L. Cutter and John P. Tiefenbacher, "Chemical Hazards in Urban America," *Urban Geography* 12, no. 5 (1991): 417–30.

49. U.S. Environmental Protection Agency, *1998 Toxics Release Inventory: Louisiana* (Dallas: USEPA, 1998); and U.S. Environmental Protection Agency, *2000 Toxics Release Inventory: Louisiana* (Dallas: USEPA, 2000).

50. See Joel B. Goldsteen, *Danger All Around: Waste Storage on the Texas and Louisiana Gulf Coast* (Austin: University of Texas Press, 1993), chap. 7 and 8; Barbara L. Allen, *Uneasy Alchemy: Citizens and Experts in Louisiana's Chemical Corridor Disputes* (Cambridge, Mass.: MIT Press, 2003), chap. 1; R. V. Rohli, S. A. Hsu, B. W. Blanchard, and R. L. Fontenot, "Short-Range Projection of Tropospheric Ozone Concentrations and Exceedances in Baton Rouge, Louisiana," *Weather and Forecasting* 18 (2003): 371–83; and Scott Frickel and James R. Elliott, "Tracking Industrial Land Use Conversion: A New Approach for Studying Relict Waste and Urban Development," *Organization and Environment* 21 (2008): 128–47.

51. A discussion of citizen concern with pollution in Louisiana is in Craig E. Colten, "Contesting Pollution in Dixie: The Case of Corney Creek," *Journal of Southern History* 72, no. 3 (August 2006): 605–34.

52. Hemmerling, "Environmental Equity in Southeast Louisiana," 256.

53. Interview with William Fontenot (retired community liaison, Louisiana State Attorney General's Office), February 6, 2010. See also Craig E. Colten, "Environmental Protection in Louisiana: An Historical Paradox," *Southern Studies*, n.s. 12 (Spring/Summer 2005): 75–92. A key exposé of the environmental justice situation is Louisiana Advisory Committee to the U.S. Commission on Civil Rights, *The Battle for Environmental Justice in Louisiana:*

Government, Industry, and the People (Baton Rouge: Louisiana Advisory Committee to the U.S. Commission on Civil Rights, 1993).

54. John C. Rodrigue, *Reconstruction in the Cane Fields: From Slavery to Free Labor in Louisiana's Sugar Parishes 1862–1880* (Baton Rouge: Louisiana State University Press, 2001); and Reheder, *Delta Sugar*.

55. McMichael, "Plant Location Factors," 77. See also, Timothy J. Minchin, *Forging a Common Bond: Labor and Environmental Activism during the BASF Lockout* (Gainesville: University of Florida Press, 2003).

56. See Hemmerling, "Environmental Equity," especially chap. 6.

57. Interview with Fontenot, February 6, 2010; Barbara Allen, *Uneasy Alchemy*, 36–43; Colten, "Environmental Protection in Louisiana." For discussions of the emergence of the environmental justice movement see Eileen McGurty, *Transforming Environmentalism: Warren County, PCBs, and the Origins of Environmental Justice* (New Brunswick, N. J.: Rutgers University Press, 2007); and Barbara Allen, "Cradle of a Revolution: The Industrial Transformation of Louisiana's Lower Mississippi River," *Technology and Culture* 47, no. 1 (2006): 112–19.

58. Minchin, *Forging a Common Bond*.

59. Charles A. Flanagan, "Mapping the Other Truth in the Shintech Case: Emancipatory Mapping for Environmental Justice in South Louisiana" (PhD Dissertation, Louisiana State University, 2005); Gerald Markowitz and David Rosner, *Deceit and Denial: The Deadly Politics of Industrial Pollution* (Berkeley: University of California Press, 2002), chap. 8; Abigail Blodgett, "An Analysis of Pollution and Community Advocacy in 'Cancer Alley': Setting an Example for the Environmental Justice Movement in St. James Parish Louisiana," *Local Environment* 11, no. 6 (2006): 647–61; and Revathi Hines, "African Americans' Struggle for Environmental Justice and the Case of the Shintech Plant: Lessons Learned from a War Waged," *Journal of Black Studies* 31, no. 6 (2001): 777–89.

60. The most extensive accounting of this effort is Steve Lerner, *Diamond: A Struggle for Environmental Justice in Louisiana's Chemical Corridor* (Cambridge, Mass.: MIT Press, 2005).

61. Cutter and Tiefenbacher, "Chemical Hazards in Urban America."

62. Dow Chemical Company, *A Diamond in the Sugar Bowl: A History of Dow Chemical Company's Louisiana Operations, 1956–2006* (Plaquemine, La.: Dow Chemical Company, 2006), 54–55.

63. "Mississippi River Closed to Gulf of Mexico," *New Orleans Times Picayune*, 24 July 2008, http://www.nola.com/news/index.ssf/2008/07/coast_guard_closes_almost_20_m .html.

64. Scott, *The Energy Sector*.

CHAPTER 4. Los Angeles, the Energy Capital of Southern California

Some of this material was previously published in "The Nature and Business of War: Drilling for Oil in World War II Los Angeles," in *Cities in Nature: Urban Environments of the American West, Essays in Honor of Hal K. Rothman*, ed. Char Miller, in press; and "Black Gold and the Beach: Offshore Oil, Beaches and Federal Power in Southern California," *Journal of the West* 44, no. 1 (2005).

1. U.S. Census Bureau, *2008 Economic Census* (Washington, DC: U.S. Department of Commerce). Measured by gross receipts, oil refineries in the Los Angeles metropolitan statistical area account for 45 percent of California's refining capacity. Important oil refining centers include Vernon and El Segundo as well as Long Beach and other Los Angeles area communities.

2. Gerald White, "California Oil Boom of the 1860s: The Ordeal of Benjamin Silliman, Jr.," *University of Wyoming Publications* 32 (1966): 2–3.

3. Ibid., 5–6.

4. "Abstracts of Articles of Incorporation for California Oil Companies on File in Alameda," Linda Vista Oil, Articles of Incorporation, 1900, and California Standard Oil Company, Articles of Incorporation, 1899, in "Abstracts of Articles of Incorporation for California Oil Companies on File in Alameda," Bancroft Library, Berkeley, Calif.

5. William R. Freudenberg and Robert Gramling, *Oil in Troubled Waters: Perceptions, Politics, and the Battle over Offshore Drilling* (New York: SUNY Press, 1994), 15, 72–73; Daniel Yergin, *The Prize: The Epic Quest for Oil, Money & Power* (Reed Business Information, 1991), 25, 28–9; Nancy Quam-Wicham, "Cities Sacrificed on the Altar of Oil," *Environmental History* 3, no. 2 (April 1998): 191.

6. Ralph Andreano, "The Structure of the California Petroleum Industry, 1895–1911," *Pacific Historical Review* 39, no. 2 (1970): 186.

7. Quam-Wicham, "Cities Sacrificed on the Altar of Oil," 191.

8. The *Los Angeles Times* contains many descriptions of the chaos and filth that accompanied oil development. Another indication of the extent and persistence of oil problems comes from the repeated efforts of city leaders to reduce the noise, pollution, and other consequences of oil development by ordinance. City ordinances, lawsuits, or other legal means to curb oil nuisances were recorded in 1901, 1926, 1929. See "Measures to Stop the Oil Nuisance," *Los Angeles Times*, 3 May 1901; "War to Bar Oil Leaks in Sea Starts," *Los Angeles Times*, 9 November 1926; "Oil Pollution Evil Banished," *Los Angeles Times*, 21 March 1929, ProQuest.com.

9. For a description of the impact of oil booms on Huntington Beach, see Jim Combs, "Another Oil Boom Brings Chaos to Huntington Beach," *Fortnight* 18, no. 6 (March 16, 1953): 11–13. *Fortnight* reported that most Huntington Beach residents were so money-struck that they welcomed even the chaos and filth. Nancy Quam-Wickham argues convincingly that many of these problems arose from the unusually high natural gas pressures overlying most of the oil deposits in the Los Angeles Basin. For vivid descriptions of the consequences, see Quam-Wickham, "Cities Sacrificed on the Altar of Oil," 192–94.

Oil wells lowered residential property values, something oil men acknowledged explicitly when they sought to transform oil fields into residential subdivisions. For example, in

1905 several oil companies asked the city council to force "holdouts" to give up their oil wells so that the majority could maximize profits from a residential development in the oil field, "Obsequies of Oiley Oozers," *Los Angeles Times*, 19 November 1905. A strike in 1917, ten miles from downtown, left even the "prominent oil operators" among home owners dismayed and "feel[ing] that they would rather make money out of oil in some other locality," "Oil Spoils Country Places," *The Oil Age* 8, no. 3 (March 1917): 17.

10. Fred W. Viehe, "Black Gold Suburbs: The Influence of the Extractive Industry on the Suburbanization of Los Angeles, 1890–1930," *Journal of Urban History* 8, no. 3 (1981): 3–26.

11. "That Oil Ordinance," *Los Angeles Times*, 25 January 1897, ProQuest.com.

12. Freudenberg, *Oil in Troubled Waters*, 17.

13. For one example of an oil company purchasing land to block another's access, see "News of Southern Counties: Stop Drilling on Tidelands," *Los Angeles Times,* 16 December 1927, ProQuest.com.

14. "Origin and History of Tidelands Case," *Congressional Digest* 27, no. 10 (1948): 233; "Conflicting State and Federal Claims of Title in Submerged Lands of the Continental Shelf," *Yale Law Review* 56 (1947): 356.

15. Guy W. Finney to John Anson Ford, 21 August 1936, JAFP, Huntington Library. Finney was one of the organizers of this movement.

By the mid-1930s, advocates promoted public beaches as a balm to quell the radicalization of the young. J. Spencer Smith of the American Shore and Beach Preservation Association campaigned for public beaches with dire warnings that the "inaccessibility of hundreds of miles of the best strands on the Atlantic and Pacific coasts as well as the Great Lakes and the Gulf of Mexico has done much to stir up unrest, especially among the younger generation." This, he argued was all the more worrisome given the "many conditions today which drive one to communistic tendencies." "Extract given from Speech of J. Spencer Smith, President of the American Shore and Beach Preservation Assoc at Long Beach, 25 Sept 1936 Association of the Pacific Coast Convention," Communication from Alfred J. Barnes, Box A625, Los Angeles City Archives.

16. C. C. Young, "'Save the Beaches' Is Plea Issued by Gov. C. C. Young," *Venice Evening Vanguard,* 4 January 1928.

17. Ibid.

18. Sarah S. Elkind, *How Local Politics Shape Federal Policy: Business, Power and the Environment in Twentieth Century Los Angeles* (Chapel Hill: University of North Carolina Press, 2011), 28.

19. J. C. Rendler to Los Angeles City Council, 11 July 1930, Los Angeles City Archives, Box A453, Communication #6094; C. C. Young, "'Save the Beaches,'" *Venice Evening Vanguard,* 4 January 1928; "Tideland Oil," *Business Week,* December 2, 1944, 32; "A. A. Newton to Report on Harbor Probe," *Venice Evening Vanguard,* 20 February 1936.

20. "Council Asked to Ban Beach Oil Drilling," *Los Angeles Times*, 26 February 1943; "Attack on Oil Drilling Move Proves Futile," *Los Angeles Times,* 17 April 1943, ProQuest.com.

21. One of the main objections that arose to oil development in the Venice–Del Mar field was the uneven distribution of benefits in the field itself. Oil companies did not lease all the properties in a field, frustrating many property owners whose neighbors negotiated leases first. One proposed solution was the community lease, in which a group of homeowners on a block leased their land as a block and shared royalties. Venice city councilmember B. M. Hansen proposed just this in early 1930, arguing that it would maximize oil production, increase efficiency within the field, and more equally distribute the benefits of oil development. B. M. Hansen to Los Angeles City Council, 6 February 1930, City Archives Box A434, Communication #1167; Mrs. R. Blum to Los Angeles City Council, 5 February 1930, City Archives Box A435, Communication #1319; Bertha S. Edwards to Los Angeles City Council and Planning Commission, 10 February 1930, City Archives Box A435, Communication #1307. Hansen's proposals would have allowed the majority of homeowners on any block to approve or reject oil drilling for the whole block. Community leases increased production by slowing the rate of production and, in the process, conserving the natural gas pressures necessary for extracting oil from the ground.

22. George Acret, "Summary of the Venice Oil Situation," City Archives Box A434, Communication #879; Spencer H. Horner to Councilman E. Webster, "Resolution" 8 February 1930, City Archives Box A435, Communication #1308.

23. "Hearing Slated on Oil Drilling," *Los Angeles Times,* 18 January 1932; "That Oil-Drilling Scheme," *Los Angeles Times,* 19 January 1932, ProQuest.com.

24. "Oil Drillers' Hearing Rests," *Los Angeles Times,* 12 December 1931; "Oil Test Voted Near Park Site," *Los Angeles Times,* 30 December 1931. Venice was already casting its shadow across the Los Angeles City Field: the damage bond for drilling companies in Venice was only $5,000.

25. "Unwise Oil Drilling Plan," *Los Angeles Times,* 2 January 1932, ProQuest.com. Opposition statements are found in "Yankwich Protests Oil Plans," *Los Angeles Times,* 14 January 1932; "Hearing Slated on Oil Drilling," *Los Angeles Times,* 18 January 1932; "Women Add Protest on Oil Drilling," *Los Angeles Times,* 19 January 1932; "Oil Aftermath Clean-Up Asked," *Los Angeles Times,* 20 January 1932; "Clamour Raised at Oil Hearing," *Los Angeles Times,* 22 January 1932, all from ProQuest.com; "Against Drilling in Residential Districts," *The Municipal League Bulletin* 9 (20 January 1932): 3–5.

26. "Oil Aftermath Clean-Up Asked," *Los Angeles Times,* 20 January 1932, ProQuest.com.

27. Marshall Stimson to Ed Ainsworth (*Times* editor), 9 February 1944, Box 1, Marshall Stimson Collection, San Marino, Calif.: HEH.

28. "Oil Drilling Scotched," *Los Angeles Times,* 30 January 1932; "Randall Would Restrict Areas for Oil Drilling," *Los Angeles Times,* 30 January 1932, both from ProQuest.com.

29. "California Petroleum Situation—December 1942," 8 February 1943, in "Statistics: California Situation," Petroleum Administration for the War Records, Box 138 (Laguna Niguel, Calif.: National Archives). According to this memo, in December 1942, California

drillers had 18,500 wells producing 774,000 barrels of oil per day. This was the largest output since November 1929.

30. "Minutes of the Annual Meeting of Shareholders of Burnoel Petroleum Corporation," 31 July 1942, in Burnoel Petroleum Company Papers, San Marino, Calif.: Huntington Library.

31. Howard Kegley, "California Oil News," *Los Angeles Times*, 21 February 1942; "Gilmore Area Addition Seen," *Los Angeles Times*, 9 October 1942, both from ProQuest.com. The A. F. Gilmore Company was part of a local business empire that included a dairy farm, oil company, real estate, and sports facilities including the city's first baseball arena. The A. F. Gilmore Company survives today as the owner and operator of an upscale shopping mall and what it bills as "the original Farmer's Market" at Fairfax Boulevard and Third Street. For Shell Oil's involvement at Gilmore Island, see "Oil Drilling Vote Deferred," *Los Angeles Times*, 27 June 1942.

32. Proposed Gilmore Strip Oil Drilling Debated," *Los Angeles Times*, 18 June 1942, ProQuest.com. The *Los Angeles Times* listed the Wilshire District Civic League, the Miracle Mile Association, and an "oil man" named Byron D. Seaver among the opposition.

33. "Gilmore Island Oil Drilling Opponents Ask for More Time," *Los Angeles Times*, 4 June 1942; "Proposed Gilmore Strip Oil Drilling Debated," *Los Angeles Times*, 18 June 1942; "Gilmore Island Oil Drilling Plea Supported by Knox," *Los Angeles Times*, 27 March 1942; "O. P. M. Urges Permit to Drill for Oil on Gilmore 'Island,'" *Los Angeles Times*, 28 May 1942, all from ProQuest.com.

34. "Mayor Vetoes Oil Ordinance," *Los Angeles Times*, 22 September 1942, ProQuest.com.

35. "Ruling on Elysian Park Oil Drilling Deferred," *Los Angeles Times*, 11 August 1943, ProQuest.com.

36. "Oil Drill Law Backed," *Los Angeles Times*, 12 June 1943, ProQuest.com.

37. D. G. Springer (Save Your Homes Association) to Stimson, 8 March 1943, Box 1, Marshall Stimson Collection, HEH.

38. Stimson to Ickes, 19 June 1943, Box 1, Marshall Stimson Collection, HEH.

39. D. G. Springer (Save Your Homes Association) to Stimson, 8 March 1943.

40. Stimson to Ickes, 19 June 1943. Mayor Fletcher Bowron lobbied to impose rules that would have forced oil companies to close wells drilled to meet the wartime emergency as soon as peace returned. The industry insisted that this would make the wells too uneconomical, an argument that seemed to lend force to assertions that the oil companies were more interested in profit than patriotism. See "Other City Drilling Programs Proposed," *Petroleum World* 40 (August 1943): 69. See also Bowron to city council, 7 July 1943, "In the Matter of the Proposed Ordinance," broadside, Box 1, Marshall Stimson Collection, HEH.

41. "Attack on Oil Drilling Move Proves Futile," *Los Angeles Times*, 17 April 1943, ProQuest.com.

42. Stimson to Ickes, 19 June 1943. See also, "Council Defeats Move to Repeal Drilling Action," *Los Angeles Times*, 21 July 1943, ProQuest.com.

43. "Oil Drilling Law Backed," *Los Angeles Times,* 12 June 1943, ProQuest.com.

44. Bruce Murchison, as quoted in "Notes Taken at Public Hearing on the Oil Drilling Ordinance, Thursday Afternoon, July 1, 1943," 3, in Box 56, Bowron Papers, Tuscaloosa: University of Alabama Libraries.

45. "J. O. Smith," "Notes Taken at Public Hearing on the Oil Drilling Ordinance, Thursday Afternoon, July 1, 1943," 4, in Box 56, Bowron Papers, Tuscaloosa: University of Alabama Libraries.

46. Carrington King, "PAW Gives California the Green Light," *Petroleum World* 40 (April 1943): 18. For the secretary of the Navy's position, see Martin Van Couvering, "Searching for Oil in the City When Los Angeles Was Young—and Now!" *Petroleum World* 40 (March 1943): 56.

47. Carrington King, "PAW Gives California the Green Light," *Petroleum World* 40 (April 1943): 18. For the secretary of the Navy's position, see Martin Van Couvering, "Searching for Oil in the City When Los Angeles Was Young—and Now!" *Petroleum World* 40 (Mar 1943): 56.

48. "Mayor Asks Council's Views on Oil Drilling," *Los Angeles Times,* 7 July 1943; "Mayor Proposes Safeguards for City Oil Drilling," *Los Angeles Times,* 15 July 1943, both from ProQuest.com; Ickes to Bowron, 19 January 1943, 3, in Box 56, Bowron Papers, Tuscaloosa: University of Alabama Libraries; "'Every Field and Pool Must Be Drilled,'—PAW," *Petroleum World* 40, no. 8 (August 1943): 21.

49. "Oil 'Politics' Laid to Mayor," *Citizen News,* 30 August 1944, Clippings, Bowron Papers, Tuscaloosa: University of Alabama Libraries.

50. As quoted in "Larger Reserve of Oil Urged at Council Hearing," 24 June 1943, ProQuest.com. For readers' letters on the subject of urban drilling, see R. G. Winter, "Why the Delay"; Mrs. A. L. Gribling, "Wells Favored"; Kenneth J. McIntosh, "Why Limit It"; Olga Dickenson, "Produce It Quickly"; and P. J. Yorba, "Permission Favored," all in *Los Angeles Times,* 14 February 1944, ProQuest.com.

51. "Oil Drilling in City Backed," *Los Angeles Times,* 14 February 1944. The *Los Angeles Times* filled its editorial page of the February 14 issue with letters on the oil question; only two of the dozen or so letters opposed urban drilling. Marshall Stimson wrote one of these.

52. Ora E. Knight, "Nonsense to . . . Fuss," *Los Angeles Times,* 14 February 1944, clippings, Bowron Papers, Tuscaloosa: University of Alabama Libraries.

53. Continental Oil Co. executive, as quoted in "Oil Well Showdown," *Daily World,* 21 June 1946, Clippings, Bowron Papers, Tuscaloosa: University of Alabama Libraries.

54. Oklahoma City had far worse problems containing oil drilling in residential neighborhoods. There, oil wells spread from unincorporated lands outside the city, through residential areas, and eventually onto the grounds of the state capitol. See Morris P. Moore, "Zoning against Oil Wells," *American City* (Sept 1930): 157–58; David S. Robertson, "Oil Derricks and Corinthian Columns: The Industrial Transformation of the Oklahoma State Capital Grounds," *Journal of Cultural Geography* 16, no. 1 (1996): 17–35.

55. Greg Hise, "'Nature's Workshop,' Industry and Urban Expansion in Southern California," *Journal of Historical Geography* 27, no. 1 (2001): 75–76.

56. Ibid., 74–92.

57. Merry Ovnick, *Los Angeles: The End of the Rainbow* (Los Angeles: Balcony Press, 1994), 105.

58. Delores Nason McBroome, "'All Men Up and No Man Down': Black Angelenos Confront Refracted Racism, 1900–1940," in *City of Promise: Race and Historical Change in Los Angeles,* ed. Martin Schiesl and Mark M. Dodge (Claremont, Calif.: Regina Books, 2006), 59–61.

59. Douglas Flamming, *Bound for Freedom: Black Los Angeles in Jim Crow America* (Berkeley: University of California Press, 2005), 17–54, 65–67. Historians of Los Angeles have completed a number of nuanced studies of race relations in the region, but among the best are Flamming, *Bound for Freedom* as well as Becky M. Nicolaides, *My Blue Heaven: Life and Politics in the Working-Class Suburbs of Los Angeles, 1920–1965* (Chicago, Ill.: University of Chicago Press, 2002); Josh Sides, *L. A. City Limits: African American Los Angeles from the Great Depression to the Present* (Berkeley: University of California Press, 2003); Mark Wild, *Street Meeting: Multiethnic Neighborhoods in Early Twentieth Century Los Angeles* (Berkeley: University of California Press, 2005).

60. Flamming, *Bound for Freedom,* 83.

61. Ibid., 4–5, 90, 310.

62. Barbara Milkovich, "What Price 'Progress' in Huntington Beach? A Preservationist's Perspective on the Oil and Real Estate Booms of the 1920s and 1980s," *Journal of Orange County Studies* 2 (1989): 28.

63. Los Angeles County population in 1910 was 504,131; in 1920, 936,455; and in 1930, 2,208,492.

64. Hise, "Nature's Workshop," 86.

65. Melville C. Branch, "Oil Extraction, Urban Environment, and City Planning," *Journal of the American Institute of Planners* 38, no. 3 (1972): 143.

66. "Origin and History of Tidelands Case," *Congressional Digest* 27, no. 10 (1948): 233; "Conflicting State and Federal Claims of Title in Submerged Lands of the Continental Shelf," *Yale Law Review* 56 (1947): 356–57.

67. E. J. Preston leased a section of offshore lands near Huntington Beach in the 1930s. A competing company blocked access to his pier. His efforts to seek relief through the California Legislature stymied; he filed for a federal lease and began disputing the state's title to the land in question. See E. J. Preston, "Speech Opposing Senate Joint Resolution no. 48 as Delivered by E. J. Preston before the Judiciary Committee of the United States Senate," Washington, DC, 7 February 1946, in Arnold Papers, Box 124, Pa-Pu folder, San Marino, Calif.: Huntington Library.

68. Downey, *Truth about the Tidelands,* 16–18. This resolution, Senate Joint Resolution 208, began life as Senate Bill 2164 in April 1937. The issue was reintroduced as a joint resolution in August of the same year. There is some dispute over whether the idea originat-

ed with Nye, Franklin Roosevelt, or Harold Ickes. See "Who Owns Offshore Oil Lands," *Business Week,* September 11, 1937, 48–49. According to Ernest R. Bartley, Ickes asked Nye to pursue the naval oil reserve proposal because Nye's home state had no oil resources and therefore no local oil lobby to influence Nye's decision. See Ernest R. Bartley, *The Tidelands Oil Controversy: A Legal and Historical Analysis* (Austin: University of Texas Press, 1953), 101.

69. "Tideland Oil Bill Foes Fight House Showdown," *Los Angeles Times,* 23 May 1938.

70. Downey, *Truth about the Tidelands,* 26. The bill was called a "quitclaim" after the legal process by which one person renounces title to property in favor of another's claim. In this case, advocates of state ownership wanted the federal government to renounce its claim in favor of the states.

71. For more on air pollution politics in Los Angeles, see Sarah Elkind, "Los Angeles Nature: Urban Environmental Politics in the Twentieth Century," in *City, Country, Empire: New Directions in Environmental History,* ed. Jeffry M. Diefendorf and Kurk Dorsey (Pittsburgh: University of Pittsburgh Press, 2005); and Scott Hamilton Dewey, *Don't Breathe the Air: Air Pollution and U.S. Environmental Politics, 1945–1970* (College Station: Texas A & M Press, 2000). In many American cities, the combination of residential segregation and real estate prices has concentrated poor and minority populations in neighborhoods with very poor air quality. In Los Angeles, the persistence of incorporated municipalities such as Vernon that are zoned almost exclusively for industry, exacerbated this problem by protecting industries from local public activism, at least until air pollution control was effectively regionalized. Nevertheless, accidents of geography muddy the correlation between air quality and race and class in the Los Angeles area; many relatively wealthy communities in the San Gabriel Valley experience some of the region's worst smog, while some more working-class neighborhoods benefit from coastal winds that blow smog inland. On recent challenges to Vernon's status as an incorporated municipality, see Adam Nagourney, "Plan Would Erase All-Business Town," *New York Times,* 1 March 2011. For an excellent discussion of housing segregation and environmental injustice, see Andrew Hurley, "The Social Biases of Environmental Change in Gary, Indiana, 1945–1980," *Environmental Review* 12, no. 4 (1988): 1–19.

PART II. Distant yet Central? Perth, Calgary, and Stavanger

1. Daniel Yergin, *The Quest: Energy, Security, and the Remaking of the Modern World* (New York: Penguin, 2011), 270.

CHAPTER 5. Scoping Perth as an Energy Capital

1. All population figures are from the Australian Bureau of Statistics, Historical Population Figures, http://www.abs.gov.au/ausstats/abs@.nsf/cat/3105.0.65.00.

2. R. E. N. Twopenny, *Town Life in Australia* (Elliot Stock, 1883), 169 quoted in C. T. Stannage, *People of Perth: A Social History of Western Australia's Capital City* (Perth: City of Perth, 1979), 85.

3. R. Woodall and G. A. Travis, "Gold Production to 1930," in *Mining in Western Australia*, ed. R. T. Prider (Nedlands: UWA Press, 1979), 58.

4. This and the preceding paragraph are based on Jenny Gregory, "Perth" in *Historical Encyclopedia of Western Australia*, ed. Jenny Gregory and Jan Gothard (Crawley: UWA Press, 2009).

5. This and the preceding paragraphs are based on information in Jenny Gregory, *City of Light: Perth since the 1950s* (Perth: City of Perth, 2003).

6. Hon Christian Porter, Western Australian Treasurer, 2012 Budget Speech, WA Hansard, 17 May 2012.

7. Alan Bonds, "Lynn, Robert John (1873–1928)," *Australian Dictionary of Biography*, vol. 10 (Melbourne: Melbourne University Press, 1986), 182.

8. Garrick Moore, "Coal," in *Historical Encyclopedia of Western Australia*, ed. Gregory and Gothard (Crawley: UWA Press, 2009).

9. Louise Boylen and John McIlwraith, *Power for the People: A History of Gas and Electricity in Western Australia* (Perth: State Energy Commission of Western Australia, 1994), 64–70.

10. There have been considerable developments in alternative sources of power, but they are beyond the scope of this chapter.

11. Reserve Bank of Australia, Inflation Calculator, http://www.rba.gov.au/calculator/annualPreDecimal.html.

12. Terry Filkin, "West Australian Petroleum Pty Ltd: The Early Years 1952–1967," in *Private Enterprise, Government and Society: Studies in Western Australian History*, ed. Frank Broeze (Centre for Western Australian History, University of Western Australia, 1992), XIII, 117. As a child I well remember my grandfather's depression after losing money when the value of WAPET shares plunged.

13. ACIL Tasman, "Nation Builder: How the North West Shelf Project Has Driven Economic Transformation in Australia," Economics Policy Strategy Paper prepared for the NWSV participants: BHP Billiton Petroleum (North West Shelf) Pty Ltd., BP Developments Australia Pty Ltd., Chevron Australia Pty Ltd., Japan Australia LNG (MIMI) Pty Ltd., Shell Development (Australia) Pty Ltd, Woodside Energy Ltd (Operator), October 2009.

14. ACIL Tasman, "Nation Builder," 6–10.

15. Ibid., 6.

16. Ibid., 24–25.

17. ACIL Tasman, "Nation Builder," pie chart and table, 29–30.

18. The southwest of Western Australia is the largest alumina region in the world and produces 15 percent of global supply, thus the alumina industry is concomitantly the largest consumer of natural gas in the state. Alcoa is the largest producer of alumina in Western Australia.

19. ACIL Tasman, "Nation Builder," pie chart, 31.

20. "Future of LNG," Special Report, *Weekend Australian* (April 24–25, 2010): 3.

21. "The Battle for James Price Point," 30 May 2012, *newmatilda.com*, newmatilda .com/2012/05/30/kimberleys-gas-port-protest-quashed-0.

22. "EPA Ticks Woodside's James Price Point LNG Hub," *The Australian,* 17 July 2012.

23. Perth Mint Australia, "History," http://www.perthmint.com.au/visit_the_mint_ the_perth_mint_history.aspx.

24. Richard G. Hartley, "Manufacturing," in Gregory and Gothard, eds., *Historical Encyclopedia of Western Australia* (Crawley: UWA Press, 2009), 549–51. See also http://members .westnet.com.au/wundowietc/wci.html.

25. Hartley, "Mining Technology and Mineral Processing," in Gregory and Gothard, eds., *Historical Encyclopedia,* 591.

26. Rodney Tiffen, *Scandals: Media, Politics & Corruption in Contemporary Australia* (Sydney: UNSW Press, 1999), 26.

27. John Phaceas, "BHP about to Quit Briquette Plant," *Sydney Morning Herald,* 8 August 2005; Andrew Trouson, "BHP Dispute Hits $1.7bn," *The Australian,* 5 December 2007.

28. See Sue Graham-Taylor, *50 Years of Fuelling Western Australia* (Perth: BP, 2005) for a detailed discussion.

29. Lenore Layman, "Development Ideology in Western Australia, 1933–1965," *Australian Historical Studies* 20, no. 79 (1982): 246–51.

30. Chris Brown, general manager Strategic Development BHP Iron Ore, "BHP's New Direct Reduced Iron Project," speech to CIP (Chemical Industry and Professions), February 21, 1996, recorded by Chemlink, http://www.chemlink.com.au/cipbhp.htm.

31. Gordon Stephenson and J. A. Hepburn, *Plan for the Metropolitan Region of Perth* (Perth: Government Printing Office, 1955), 52, 60.

32. Ibid., 69 and 130.

33. Ibid., 82, 136, 140.

34. Ibid., 70.

35. Ibid., 235.

36. Ibid., 69.

37. The Acts of Parliament Western Australia, "Oil Refinery Industry (Anglo-Iranian Oil Company Limited) Act, March 1952" (Perth: Government Printer, 1952), 258–62, cited in Sarah Brown, "Surveying Our Past and Building Our Future: An Environmental History of an Australian Suburb," *Limina: A Journal of Historical and Cultural Studies* 13 (2007): 25.

38. The general manager of the Oil Refinery wrote that he "definitely did not want an 'oil town'—it is essential that workers' homes be interspersed between various industrial projects and other avocations," General Manager of the Anglo-Iranian Oil Company to the Chairman of the State Housing Commission, letter, 10 March 1952, State Housing Commission File 666/52, quoted in C. F. Makin, "Social Differentiation and the Concept of Community in a Western Australian Township" (MA Thesis, University of Western Australia, 1962), 58. This point was also made by Feilman: "Although the impetus to the development of the township came from the building of the Refinery, Medina, or any other of the neigh-

bourhood units, should not become an 'oil town'—only a proportion of the houses are for Refinery employees," quoted in *Sydney Morning Herald,* 25 April 1955. Cited in Brown, "Surveying Our Past," fn. 25, 32.

39. Urban Air Pollution in Australia, www.environment.gov.au/atmosphere/airquality/publications/urban-air/pubs/chapter1.pdf/.

40. Graham-Taylor, *50 Years of Fuelling Western Australia,* 47.

41. Peter J. Hurley, William L. Physick, Ashok K. Luhar, and Mary Edwards, "The Air Pollution Model (TAPM) Version 3, Part 2: Summary of Some Verification Studies," CSIRO Atmospheric Research Technical Paper, No. 72 (Canberra: Commonwealth Scientific and Industrial Research Organisation, 2005).

42. Western Australian Land Authority (Landcorp), Hope Valley Wattleup Redevelopment Project Master Plan, December (Perth: WA Land Authority, 2004); Appeals Report on Hope Valley—Wattleup Redevelopment Project (EPA Bulletin 1133), Final report 9/8/04 (Perth: Environmental Protection Authority, 2004).

43. Landcorp, Latitude 32 Planning website, http://www.latitude32planning.com.au/Project-Overview/.

44. Landcorp, "Timetable Announced for End to Wattleup Tenancies," Media Release, 5 April 2005.

45. Tucker7287, online forum, ABC, 3 October 2005, http://www2b.abc.net.au/4corners/forum/archives/archive107/newposts/63/topic63578.shtm.

46. David, online forum, ABC, 4 October 2005.

47. Marin, online forum, ABC, 3 October 2005.

48. Town of Kwinana, http://www.kwinana.wa.gov.au/About-Kwinana/Industry.aspx.

49. http://www.latitude32.com.au/.

50. Kwinana Industries Council, http://www.kic.org.au/kia.asp.

51. Department of Foreign Affairs and Trade, "Western Australia," fact sheet, June 2012, Canberra, Department of Foreign Affairs and Trade.

52. City of Perth, Annual Report, 2010–2011, 27.

CHAPTER 6. At Arm's Length: Energy and the Construction of a Peripheral Prairie Petrometropolis

1. "Canada's Power Shifts Westward," *Calgary Herald,* 22 March 2010, A-10.

2. Classic literature in this vein includes John Richards and Larry Pratt's *Prairie Capitalism: Power and Influence in the New West* (Toronto: McClelland and Stewart, 1979); Edward Shaffer, "Class and Oil in Alberta," in *Oil and Class Struggle,* eds. Petter Nore and Terisa Turner (London: Zed Press, 1980); Edward Shaffer, *Canada's Oil and the American Empire* (Edmonton: Hurtig Publishers, 1983); David H. Breen, *Alberta's Petroleum Industry and the Conservation Board* (Edmonton: The University of Alberta Press, 1993).

3. Andy Merrifield, *Metromarxism: A Marxist Tale of the City* (New York: Routledge, 2002), 1–2.

4. See, for example, Ken Cruickshank and Nancy B. Bouchier, "Blighted Areas and Obnoxious Industries: Constructing Environmental Inequality in an Industrial Waterfront, Hamilton, Ontario, 1890–1960," *Environmental History* 9 (July 2004): 464–96; Mike Davis, *Ecology of Fear: Los Angeles and the Imagination of Disaster* (New York: Metropolitan Books, 1998); Andrew Hurley, *Environmental Inequalities: Class, Race, and Pollution in Gary, Indiana, 1945–1980* (Akron, Ohio: University of Akron Press, 1996); Thomas Lambert and Christopher Boerner, "Environmental Inequality: Economic Causes, Economic Solutions," *Yale Journal on Regulation* 14 (1997): 195–234; Joel A. Tarr, ed., *Devastation and Renewal: An Environmental History of Pittsburgh and Its Region,* (Pittsburgh: University of Pittsburgh Press, 2003); Robert Gottlieb, *Forcing the Spring: The Transformation of the American Environmental Movement* (Washington, DC: Island Press, 1993).

5. Bruce Podobnik, *Global Energy Shifts: Fostering Sustainability in a Turbulent Age* (Philadelphia: Temple University Press, 2006), 29–37; Eric Hobsbawm, *Industry and Empire: The Birth of the Industrial Revolution* (New York: The New Press, 1999), 87–88; C. Knick Harley, "Coal Exports and British Shipping, 1850–1913," *Explorations in Economic History* 26 (1989): 312.

6. Vaclav Smil, *Energy Transitions: History, Requirements, Prospects* (Santa Barbara, Calif.: Praeger, 2010), 75.

7. Martin V. Melosi and Joseph A. Pratt, "Energy Capital and Opportunity City: Houston in the Twentieth Century," 34–35 (forthcoming in *Energy Capitals* edited collection).

8. Some of the city's more prominent oil figures like Robert A. Brown Jr., organizer of Federated Petroleums, and Stanley Milner, founder of the Chieftain companies, were born in Calgary. Others including Frank McMahon of Pacific Petroleums and Westcoast Transmission fame and Dome Petroleum's Jack Gallagher were from other parts of Canada. And individuals like Ted Link, leader of the Imperial Oil team responsible for the 1947 strike, J. C. Anderson, founder of Anderson Exploration, and Ralph F. Will, the first president of Alberta Gas Trunk Line, were American; see Canadian Petroleum Hall of Fame, http://www.canadianpetroleumhalloffame.ca/index.html.

9. See Shaffer, "Class and Oil in Alberta," 260–61.

10. Richards and Pratt, *Prairie Capitalism,* 61–63; Breen, *Alberta's Petroleum Industry,* 10–15, 154.

11. Shaffer, *Canada's Oil,* 34–45; Robert D. Bott, *Evolution of Canada's Oil and Gas Industry* (Calgary: Canadian Centre for Energy Information, 2004), 17–18.

12. Hearings held in 1931 established that energy equivalent to twenty-five thousand tons of coal *per day* was being flared in Turner Valley; see Breen, *Alberta's Petroleum Industry,* 37, 51, 70.

13. Ibid., 216–18, 237.

14. Breen, "Calgary: The City and the Petroleum Industry," 55–71.

15. Imperial Oil Calgary Refinery Fonds, Glenbow Museum and Archives, Calgary.

16. City of Calgary Electric System Inventory Fonds, City of Calgary Corporate Records and Archives, 1–2.

17. City of Calgary Electric System Inventory Fonds, 2–3; Enmax, http://www.enmax .com/Corporation/About+Enmax/Our+Company/History.htm.

18. TransAlta Utilities, http://www.transalta.com/about-us/history.

19. S. J. Davies to City Commissioners, "Ref: Generating Plant," September 24, 1959, "Fuel—Stan Davies, 1959," Electric System Series III, box 21, file 54, bay 224, shelf 003, City of Calgary Corporate Records and Archives; H. G. Acres and Company, "Report on Thermal Power Generation, March 1959," 4, Electric System Series III, box 24, file 96, bay 224, shelf 03, City of Calgary Corporate Records and Archives.

20. Imperial was acutely aware of its problematic public image in Alberta. In the wake of its Leduc discoveries, it was keen to lower expectations in central Alberta, a region it felt harbored "strong socialist leanings," and to teach the rules of the oil game as played in southern Alberta. One early public relations booklet outlining talking points for Imperial employees in the Edmonton area and addressed the dominant role played by private enterprise and the alienation of subsurface mineral rights. Imperial framed the latter as a factor "beyond its control." In fact, industry benefited immensely from the provincial government's control of these rights, particularly the Right-Of-Entry Arbitration Act of 1947. This forced property owners to allow industry access to their land following an arbitration process that could not be appealed. In contrast, landowners owned subsurface mineral rights in Manitoba, most of Saskatchewan, and Texas; see Imperial Oil, "Public Relations Aspects of the Leduc Oil Discovery," Imperial Oil Archives, Series 16, External Affairs, Research & Analysis, Business Library, accession number 80–0005, box 013, file 22, Glenbow Museum and Archives, Calgary.

21. H. H. Hewetson, Imperial's president, invoked national security as prime reason for exploiting Alberta oil as early as 1948; see "National Aspects of Oil Development," address by H. H. Hewetson, President, Imperial Oil, Ltd., December 6, 1948; Imperial Oil Archives, Series 16, External Affairs, Research & Analysis, Business Library, accession number 80–0005, box 012, file 21, Glenbow Museum and Archives, Calgary.

22. See Carl O. Nickle, "Results of Survey May Be Known in May," *Calgary Herald,* 28 April 1951; Neill C. Wilson and Frank J. Taylor, *The Building of Trans Mountain, Canada's First Oil Pipeline Across the Rockies* (Vancouver: Trans Mountain Oil Pipe Line Company, 1954), xi. A public relations booklet published by Trans Mountain in 1952 emphasized the economic rather than the national security benefits of the pipeline; see "Oil Across the Rockies: Trans Mountain Oil Pipe Line Company," Series 16, Imperial Oil Archives, External Affairs, Research and Analysis, Business Library, accession number 80–0005, box 003, file 04, Glenbow Museum and Archives, Calgary.

23. Richards and Pratt note that Howe pressured Manning to allow the export of gas. In fact the premier was more than willing to participate when properly reassured that a real emergency was at hand; Ernest Manning to C. D. Howe, January 13, 1951, Charles Wil-

son to Howe, February 16, 1951, Howe to Manning, February 21, 1951, Manning to Howe, March 17, 1951; MG27 III B20 Vol. 31, 8–2-1, Pipelines, Fuel, 1950–1951, File No. 23, Howe Papers, National Archives, Ottawa. For their part, American coal suppliers contested Anaconda's claim that natural gas was essential for the processing of strategic minerals, especially manganese. They noted Anaconda had in the past used coal and oil to fire steam plant for electrolysis, developed an oil/gas dual fuel capability, and was using natural gas for steam, not metallurgy. National security, they argued, was a pretext contrived to allow the Montana Power Corporation to import Canadian natural gas in order expand the domestic heating market at a time when other American natural gas producers were diverting their own reserves for defense manufacturing; "In the Matter of the Proposed Importation of Natural Gas by Montana Power Company Pursuant to Applications Pending before the Federal Power Commission in Dockets No. G-1712 and G-1717: Memorandum Reviewing Evidence Setting Forth Reasons Why Importation of Canadian Gas Should Not Be Permitted," October 19, 1951, MG27 III B20 Vol. 31, 8–2-1, Pipelines, Fuels 1951–1952, File No. 22, 1–23, Howe Papers, National Archives, Ottawa.

24. W. L. Dack, "Gas Pipeline Picture Due for Changes Soon? Washington Hearings May Decide One Phase of the Problem, But Cost Factors Are Raising Some Questions on the All-Canadian Route," *The Financial Post*, 14 March 1953.

25. Richards and Pratt, *Prairie Capitalism,* 68.

26. In attacking the Montreal plan, Imperial spoke not only on behalf of the Canadian oil industry but of the national interest. It decried "unusual government intervention," a choice of phrase that underscored years of effort by both federal and provincial governments to foster a regulatory environment conducive to Imperial and the majors and negotiate the export of resources to the United States. Imperial claimed that if a Montreal pipeline made economic sense, the same oil industry that had bankrolled the Inter-Provincial and Trans-Mountain pipelines would have built it. Of course, Imperial and the other majors were the majority investors in those enterprises; see "Markets for Canada's Oil and the National Interest," October 11, 1960, Imperial Oil Archives, Series 16, External Affairs, Research & Analysis, Business Library, accession number 80–0005, box 002, file 30, Glenbow Museum and Archives, Calgary.

27. Shaffer, *Canada's Oil,* 156–61.

28. S. J. Davies to City Commissioners, "Reference—Fuel Thermal Power Plant," June 29, 1959; Davies to Albert Bishop, August 3, 1959; William D. Kirkland to A. Bishop, September 18, 1959; "Fuel—Stan Davies," Electric System Series III, box 21, file 54, bay 224, shelf 003, City of Calgary Corporate Records, Archives; H. G. Acres and Company, "Report on Thermal Power Generation," March 1959.

29. Calgary Power, *Annual Report 1964,* 4.

30. Calgary Power, *Annual Report 1967,* 4; *Annual Report 1968,* 4.

31. "Home Oil Company Papers, 1936–1972," Glenbow Museum and Archives, Calgary; Richards and Pratt, *Prairie Capitalism,* 83–84.

32. Richards and Pratt, *Prairie Capitalism,* 240–41; Kevin Taft, *Shredding the Public Interest: Ralph Klein and 25 Years of One-Party Government* (Edmonton: The University of Alberta Press and Parkland Institute, 1997), 41–49. One of the government's chief instruments to this end was the Calgary-based Alberta Energy Company (AEC), created in 1973 as a public-private corporation designed expressly to compete in nonconventional oil and gas. In a similar vein, AGTL, renamed Nova in 1980, reinvented itself as a diversified petrochemical and industrial concern.

33. Avison Young, *Calgary Office Market Report, Spring 2008,* 1.

34. Even so, the provincial government rejected conventional industrial diversification, asserting that it did not want the large population associated with a manufacturing economy; see Martin Douglas, "In Alberta, Help Isn't Just Around the Corner," *New York Times,* 19 June 1983, A2.

35. John F. Burns, "Dome Spurs Canadian Debate," *New York Times,* 16 April 1987; Alan Bayless, "Bid for Bow Valley, a Healthy Oil Firm, Tests Canada Ban on Foreign Takeovers," *Wall Street Journal,* 10 August 1987, 1.

36. Deborah C. Sawyer, "Dome Petroleum Limited," *The Canadian Encyclopedia,* http://www.thecanadianencyclopedia.com/index.cfm?PgNm=TCE&Params=A1ARTA0002341.

37. Larry Pratt, *Energy: Free Trade and the Price We Paid* (Edmonton: Parkland Institute, 2001), 24–29. In 1991, 27 percent of the production revenue of Alberta oil and gas companies went to the provincial government as royalties. As revised by the Alberta government, companies did not have to pay royalties on the first CAD $1 million of oil from each new exploration well and the first $400,000 from each new development well, a move that benefited smaller companies; see Suzanne McGee, "Alberta Launches Incentive Program that Analysts Say Will Boost Drilling," *Wall Street Journal,* 8 November 1991.

38. Pratt, *Energy,* 24–29; Taft, *Shredding the Public Interest,* 46.

39. Pratt, *Energy,* 26–27.

40. Calgary Transit carried over 94 million passengers in 2009. Inaugurated in 1981, the C-Train, the city's light rail system, boasted two lines with 46 kilometers of track, 26 stations, and 157 vehicles carrying around 250,000 people per day by the late 2000s, dwarfing Houston's METROrail system. After Toronto's streetcar system, Calgary's C-Train carried more passengers than any other analogous system in the United States and Canada: see Calgary Transit, "LRT Technical Data," http://www.calgarytransit.com/html/technical _information.html, http://www.calgarytransit.com/html/statistics.html, and American Public Transportation Association Ridership Report, Fourth Quarter 2009, http://www.apta .com/resources/statistics/Documents/Ridership/2009_q4ridership_APTA.pdf.

41. John Hubbell and Dave Colquhoun, "Light Rail Transit in Calgary: The First 25 Years," paper presented at the 2006 Joint International Light Rail Conference, St. Louis, Missouri, April 8–12, 2006, 8–10.

42. Tamsin Carlisle, "Calgary Becomes Outpost on High-Tech Frontier," *Wall Street Journal,* 24 March 1998, 1.

43. James Brooke, "In the Catbird Seat (or the Hot Seat, Like Seattle?)," *New York Times,* 9 June 2000, A-4. On the other hand, Mary Valentich describes deep-rooted sexism within Calgary's municipal government; see "Calgary's Resistance to Changing from 'Alderman' to 'Councillor,'" *Theory in Action* 2, no. 1 (January 2009): 66–85.

44. Enmax, "Wind Generation," http://www.enmax.com/Corporation/Clean+Power/ Generation+Facilities/Wind+Generation/default.htm. Enmax's efforts to develop its own generating plant supplied an ironic footnote in the history of municipal utility ownership, highlighting the diverging interests of Calgary-based power producers and the city-owned distributor. Decades after Calgary had abandoned its own generating plant, it had become heavily dependent on coal-fired electricity produced in central Alberta largely by Calgary-based ATCO and TransAlta, with the traditional giant controlling about 50 percent of the market by 2008. With the deregulation of Alberta's electricity market between 1996 and 2000, Enmax was keen to compete in generation and sought to develop its own large plant close to Calgary. But its plans were complicated by the passage in November 2009 of Bill 50. This enabled the provincial government to assume authority for transmission infrastructure from the Alberta Utilities Commission. Critics charged the bill set the stage for the publicly financed expansion of the Alberta electrical grid, furthering the dominance of the Edmonton generating complex. Others argued the bill prepared the ground for massive electricity exports to the United States; see Enmax, "More About Bill 50," http://www.enmax.com/Corporation/Bi1150/Questions+and+Answers/ FAQs.htm; Dina O'Meara and Renata D'Aliesio, "Controversial Edmonton-Calgary Power Line Pushes Ahead," *Calgary Herald,* 16 July 2009; Renata D'Aliesio, "Alberta Government's Bill 50 'Looming in the Shadows,'" *Calgary Herald*, 18 August 2009; Steve Snyder, TransAlta President and CEO to Edmonton Chamber of Commerce, "Breaking the Triple-E Equation," November 19, 2008; http://suppliers.transalta.com/transalta/webcms .nsf/AllDoc/593AB61B256C8C4E87257157004F9B1D?OpenDocument.

45. Calgary Transit, http://calgarytransit.com/index.html.

46. Enmax, "Downtown District Energy Centre," http://www.enmax.com/ Corporation/Clean+Power/OtherProjects/District+Energy/Calgary+District+Energy +Centre.htm.

47. Industry Development Branch, Alberta Finance & Enterprise, *Alberta Chemical Operations,* August 2009; Government of Alberta, "Chemicals and Petrochemicals," http:// www.alberta-canada.com/industries/861.html.

48. Mike Poehlmann, "Hub Oil Accident," 12th Annual IPEIA Conference, Banff, Alberta, February 5–7, 2008; www.ipeia.com/Misc_Docs/2008%20Mike%20Poehlmann.pdf.

49. TransAlta, "Plants in Operation," http://www.transalta.com/facilities/plants -operation; Capital Power Corporation, "Our Operations," http://www.capitalpower.com/ About/OurOperations/Pages/Pages/default.aspx; ATCO Power, "Our Facilities," http:// www.atcopower.com/OurFacilities/NorthAmerica/; Government of Alberta, Energy, Electricity, http://www.energy.alberta.ca/Electricity/682.asp.

50. W. F. Donahue, E. W. Allen, and D. W. Schindler, "Impacts of Coal-Fired Power Plants on Trace Metals and Polycyclic Aromatic Hydrocarbons (PAHs) in Lake Sediments in Central Alberta, Canada," *Journal of Paleolimnology* 35 (2006): 111–28.

51. Hugh McCullum, *Fuelling Fortress America: A Report on the Athabasca Tar Sands and U.S. Demands for Canada's Energy* (Edmonton: Canadian Center for Policy Alternatives, Parkland Institute, Polaris Institute, 2006), 49. See also Paul Chastko, *Developing Alberta's Oil Sands: From Karl Clark to Kyoto* (Calgary: University of Calgary Press, 2004).

52. Andrew Willis, "Tough Talks over Syncrude Stake," *The Globe and Mail*, 29 October 2009, http://www.theglobeandmail.com/globe-investor/investment-ideas/streetwise/tough-talks-over-syncrude-stake/article1343382/.

53. Andrew Nikiforuk, *Tar Sands: Dirty Oil and the Future of a Continent* (Vancouver: Greystone Books, 2008), 120–21.

54. Sam Kean, "Eco-Alchemy in Alberta: The Oil of the Future Has Serious Reclamation Challenges Right Now," *Science* 326 (November 20, 2009): 1052–55; Matt Price, *11 Million Litres a Day: The Tar Sands' Leaking Legacy* (Toronto: Environmental Defense, 2008), 1–23. Disposing of the sulfur by-product of synthetic fuel processing was one of the lesser-known problems of rapid tar sands development. In the early 2000s, Syncrude's plant at Fort McMurray produced 1,700 tons a day. Too costly to ship in a world market saturated with byproduct sulfur, it accumulated in two sixty-foot-high pyramids; see Alexei Barrionuevo, "A Chip off the Block Is Going to Smell Like Rotten Eggs: Sulfur Is Piling Up in Alberta: Millions of Tons Nobody Needs or Can Get Rid Of," *Wall Street Journal*, 4 November 2003, A-1.

55. Shawn McCarthy, "In Mackenzie Valley, Frustration and a Sense of Foreboding," *Globe and Mail*, 30 August 2009, http://www.theglobeandmail.com/report-on-business/industry-news/energy-and-resources/in-mackenzie-valley-frustration-and-a-sense-of-foreboding/article1269987/.

56. In 2008, Canada exported about 912 million barrels of oil (2.5 million barrels per day) and about 3.5 trillion cubic feet of gas to the United States. This constituted 90 percent of U.S. gas imports; see U.S. Energy Information Agency, "Canada-Oil," http://www.eia.doe.gov/emeu/cabs/Canada/Oil.html; "Canada-Natural Gas," http://www.eia.doe.gov/cabs/Canada/Natural Gas.html.

57. The larger of the two, Bridgewater Bank, was formed in 1997 and has around $3 billion in mortgages and over $400 million in guaranteed investment certificates, http://www.bridgewaterbank.ca/cps/rde/xchg/SID-53ED365C-736A777F/bridgewater/web/16.htm; DirectCash Bank is a limited-purpose virtual bank designed to provide ATM services, http://www.dcbank.ca/aboutus.shtml.

58. For example, Calgary had the second-highest concentration of head offices or regional headquarters after Toronto, but employed only 4 percent of Canada's finance, insurance, and real estate (FIRE) workforce; see William J. Coffey and Richard G. Shearmur, "Employment in Canadian Cities," in *Canadian Cities in Transition: Local Through Global Perspectives*, eds. Trudi Bunting and Pierre Filion (New York: Oxford University Press, 2006),

256–57. By 2008, the energy sector comprised the Calgary's second-largest source of GDP, worth around CAN $10 billion or 14.5 percent of the city's economy, behind only FIRE, worth $15.5 billion. On the other hand, manufacturing was worth only $6 billion, 8.6 percent of a $52.4 billion economy. In contrast, manufacturing was much more important to the Houston area, which in 2009 produced $63 billion in goods in an economy worth around $325.5 billion; see Calgary Economic Development, http://www.calgaryeconomic development.com; City of Houston, "Houston Facts and Figures," http://www .houstontx.gov/abouthouston/houstonfacts.html; Greater Houston Partnership, "Advanced Manufacturing," http://www.houston.org/economic-development/industry -sectors/advanced-manufacturing/index.aspx.

59. Carola Hoyos, "PetroChina Takes Back Energy Top Slot," *Financial Times*, 26 January 2010, http://www.ft.com/cms/s/0/305c6878–09e9–11df-8b23–00144feabdc0.html. In 2012, the four largest energy companies with significant levels of Canadian ownership—Suncor, Enbridge, TransCanada, and Canadian Natural—ranked 20th, 29th, 31st, and 36th, respectively among the world's top 50 largest energy firms by market capitalization. Imperial ranked 28th; see *PFC Energy 50: The Definitive Annual Ranking of the World's Largest Listed Energy Firms*, January 2013, http://www.pfcenergy.com/pfc50.aspx.

60. In late 2012, the China National Offshore Oil Corporation purchased Nexen for CAN $15.1 billion; Laura Payton, "Government OK's foreign bids for Nexen, Progress Energy," CBC News December 7, 2012, http://www.cbc.ca/news/business/story/2012/12/07/ cnooc-nexen-takeover.html.

61. Adam W. Legge, "2010 Calgary Economic Outlook: A Charlie Brown Economy," October 6, 2009, Calgary Economic Development, 11, www.calgaryeconomicdevelopment .com/2010%20Calgary%20outlook%20Report%20FINAL.pdf.

62. In 2009, Canada's reserves of natural gas ranked 21st in the world, with about 58 trillion cubic feet out of total world reserves of around 6,300 trillion cubic feet. At the 2008 rate of consumption of 6 trillion cubic feet a year, these reserves were sufficient only for around another decade. Matters with oil were even starker. Canada is often claimed to possess the world's second-largest oil reserves in the world, with around 179 billion proven barrels. But 95 percent (170 billion barrels) is in the form of tar sand. With conventional oil accounting for only 39 percent of daily oil production (2008) of around 3.35 million barrels a day, sustaining output will require the massive expansion of the tar sands; U.S. Energy Information Administration, "Country Analysis Briefs: Canada," http://tonto.eia.doe.gov/ country/country_energy_data.cfm?fips=CA, and http://www.eia.doe.gov/emeu/cabs/ Canada/Oil.html.

CHAPTER 7. Oil Shocks in an Oil City: The View from Stavanger, Norway, 1973–2008

1. Rolf Danielsen, "Den store krise," in *Stavanger på 1800-tallet,* ed. Rolf Danielsen (Stavanger: Dreyer Bok, 1975), 281–316.

2. Anders Haaland and Helge W. Nordvik, "Industribyen tar form," in *Stavanger mellom sild og olje: Hermetikkbyen 1900–1940*, ed. Rolf Danielsen (Stavanger: Dreyer Bok, 1988), 1: 49–167. Also see generally Gunnar Nerheim and Bjørn S. Utne, *Smedvig: A Story of Canning, Shipping and Contract Drilling from Western Norway* (Stavanger: Peder Smedvig, 1992).

3. Gunnar Nerheim, Lars Gaute Jøssang, and Bjørn S. Utne, *I vekst og forandring. Rosenberg Verft 100 år, 1896–1996* (Stavanger: Kværner Rosenberg a.s., 1995).

4. Gunnar Nerheim, Trygve Brandal, and Edgar Hovland, *Klart det lønner seg å samarbeide: Sparebank 1 SR-Bank, 1976–2001* (Stavanger: SR Bank, 2001), 49–51; Lars Gaute Jøssang, Svein Ivar Langhelle, and Olav Tysdal, *Sandneshistorien—fra husklynge*, bind 1 (Bergen: Fagbokforlaget, 2010), 223–60.

5. Gunnar Nerheim, "The Offshore Drilling Business, 1950–1990: Some Development Patterns," in *Management, Finance and Industrial Relations in the Maritime Industry*, ed. Simon P. Ville and David M. Williams (Ghent: International Maritime Economic History Association, 1994), B 14; Proceedings Eleventh International Economic History Congress, Milano, September 1994 (Milan: Università Bocconi, 1994), 137–49; Joseph A. Pratt, Tyler Priest, and Christopher Castaneda, *Offshore Pioneers: Brown & Root and the History of Offshore Oil and Gas* (Houston: Gulf Professional Publishing, 1997).

6. Stig S. Kvendseth, *Giant Discovery: A History of Ekofisk through the First 20 Years* (Stavanger: Phillips Petroleum Company Norway, 1988), 11–17; William C. Wertz, *Phillips: The First 66 Years* (Bartlesville, Okla.: Phillips Petroleum Co., 1988), 161–62.

7. Kvendseth, *Giant Discovery*, 23–31; Tore J. Hanisch and Gunnar Nerheim, *Norsk oljehistorie*, bind 1 (Oslo: Leseselskapet, 1992), 119–23, 189–97.

8. Gunnar Nerheim, *Norsk oljehistorie: En gassnasjon blir til* (Oslo: Leseselskapet, 1996), 2: 76–78.

9. Nerheim, Brandal, and Hovland, *Klart det lønner seg å samarbeide*, 85.

10. M. A. Adelman, *The World Petroleum Market* (Baltimore: The Johns Hopkins University Press, 1972); Peter R. Odell, *Oil and World Power: Background to the Oil Crisis* (London: Penguin Books Ltd., 1974); Stephen G. Rabe, *The Road to OPEC: United States Relations with Venezuela, 1919–1976* (Austin: University of Texas Press, 1982); Abbas Alnasrawi, *OPEC in a Changing World Economy* (Baltimore: The Johns Hopkins University Press, 1985); James M. Griffin and David J. Teece, *OPEC Behavior and World Oil Prices* (London: George Allen and Unwin, 1982); Mohammed E. Ahrari, *OPEC: The Failing Giant* (Lexington: The University of Kentucky Press, 1986); Ian Skeet, *OPEC: Twenty-Five Years of Prices and Politics* (Cambridge: Cambridge University Press, 1988); Steven A. Schneider, *The Oil Price Revolution* (Baltimore: The Johns Hopkins University Press, 1983).

11. Walter L. Buenger and Joseph A. Pratt, *But Also Good Business: Texas Commerce Banks and the Financing of Houston and Texas, 1886–1986* (College Station: Texas A&M University Press, 1986), 320–2i.

12. *Calgary in Fact 1978/1979* (Calgary: City of Calgary, Department of Business Development, 1978), 1; *Calgary in Fact 1981/1982* (Calgary: City of Calgary, Department of Business

Development, 1981), 1; "Calgary Reaches 560, 618," *The Sunday Calgary Albertan,* 25 May 1980.

13. *Calgary in Fact 1981/1982,* 19.

14. J. Joseph Fitzgerald, *Black Gold with Grit: The Alberta Oil Sands* (Vancouver: Grays Publishing, 1978), 181–95; Tom Kennedy, *Quest: Canada's Search for Arctic Oil* (Edmonton: Reidmore Books, 1988).

15. Nerheim, *Norsk oljehistorie,* 2: 76–78.

16. Gunnar Nerheim, "The Condeep Concept: The Development and Breakthrough of Concrete Gravity Platforms," *History of Technology* 16 (1994): 15–34.

17. Elf Aquitaine began building up a permanent organization from 1974. French oil expertise was imported from headquarters in Paris or other Elf subsidiaries throughout the world. The number of employees in Stavanger increased from 133 persons in 1974 to 1,381 in 1985. Mobil as operator of the Statfjord field built up a local organization in Stavanger, hiring more than a hundred employees each year during the late 1970s. Nerheim, Brandal, and Hovland, *Klart det lønner seg å samarbeide,* 110–20.

18. Hanisch and Gunnar Nerheim, *Norsk oljehistorie,* 1: 352–72.

19. *Statistisk Årbok 1975* (Oslo: Central Bureau of Statistics of Norway, 1975), 289; *Statistisk Årbok 1985* (Oslo: Central Bureau of Statistics of Norway, 1985), 332.

20. P. Cross, "Tracking the Business Cycle: Monthly Analysis of the Economy at Statistics Canada, 1926–2001," *Canadian Economic Observer,* December 2001, 14, no. 12, 16.

21. "1982 a Bleak Year," *Calgary Herald,* 23 December 1982; "More Going Under," *Calgary Herald,* 7 July 1982.

22. "The Needs a Break," *The Calgary Sunday Sun,* 18 October 1981.

23. "It's a Trickle: Flood of Newcomers Dries Up," *Calgary Herald,* 25 November 1982; "It's a Buyer's Market for Office Space," *Calgary Herald,* 27 August 1983.

24. Torbjørn Eika, "En oljesmurt økonomi—Med harelabb over 35 års konjunkturhistorie," *Samfunnsspeilet,* no. 5–6 (2007) (Oslo: Central Bureau of Statistics of Norway, 1985).

25. Nerheim, Brandal, and Hovland, *Klart det lønner seg å samarbeide,* 100–115.

26. Andreas Benedictow, "Norsk økonomi—en konjunkturhistorie," *Samfunnsspeilet,* no. 5–6 (2006) (Oslo: Central Bureau of Statistics of Norway, 1985).

27. *Faktaheftet 1990: Den Norske Kontinentalsokkelen,* utgitt av OED (Oslo: Ministry of Petroleum and Energy, 1990), 98–100.

28. Årsmelding, Arbeidsformidlingen i Rogaland 1988 (Oslo: Statens Forvaltningstjeneste, 1988); and Årsmelding, Arbeidsformidlingen i Rogaland 1990 (Oslo: Statens Forvaltningstjeneste, 1990).

29. Houston, *Economic Overview,* no. 6 (1989): 5; Barton A. Smith, *Houston's Economy Gains in the National Race: A Survey of Local Economics and the Real Estate Market* (Houston: University of Houston Center for Public Policy, 1990).

30. The recession in the United States in the early 1990s was characterized by a prolonged period of weak expansion and limited job growth. The price of crude oil did not in-

crease to any large degree until after the Iraqi invasion of Kuwait in August 1990, followed by the Gulf War in 1991. When the war to liberate Kuwait ended, crude oil prices entered a period of steady decline until 1994, when the price cycle turned again. The United States economy was strong and the Asian Pacific region was booming. Asian consumption accounted for all but three hundred thousand barrels of that gain.

31. Strong economic growth in the United States, combined with a 1995 turnaround in oil and natural gas extraction, led to new employment growth in Houston. The job market in Houston in 1997 was the strongest since 1976. "New jobs, tight labor markets, huge profits from energy, a housing boom, rising land prices and strong consumer demand combine to provide a buoyant economic atmosphere long missing from Houston," "Houston Economy Heats Up," *Houston Business: A Perspective on the Houston Economy* (Dallas: Federal Reserve Bank of Dallas, August 1997), 35. In the fall of 1997 companies had problems recruiting the people or the rigs they needed. In fall 1997 geologists and geophysicists were in great demand again, and drilling contractors had problems recruiting skilled drilling employees. L. M. Sixel, "It's a Job Seeker's Market as Local Businesses Expand, Move or Open—And Need People," *Houston Chronicle,* 26 October 1997.

32. In 1996, Houston dominated employment in the American oil industry with more than 56,000 upstream jobs. That was nearly four times the oil employment in Dallas, its nearest competitor. Houston also dominated the headquarters and producer sectors where crucial industry decisions are made. While overall industry employment shrank by 27.3 percent after 1987, Houston was the only city to gain in upstream oil jobs. The decline of domestic oil fields and low oil prices forced the oil industry to concentrate its activities, and both oil producers and service companies chose to do this in Houston, Dallas, and New Orleans. Bill Gilmer, "Oil and the Houston Economy Today," *Houston Business: A Perspective on the Houston Economy* (Dallas: Federal Reserve Bank of Dallas, January 2000); "Oil-Related Employment: Long Term Adjustment in Nine Cities," *Houston Business: A Perspective on the Houston Economy* (Dallas: Federal Reserve Bank of Dallas, September 1998), n.p.; Jun Ishii, "The Oil Industry and the Cities: Consolidation in the Oil Extraction Industry," *Houston Business: A Perspective on the Houston Economy* (Dallas: Federal Reserve Bank of Dallas, April 1996), n.p.

33. "Calgary Reaps a Renaissance," *The Globe and Mail,* 3 January 1996; "Head Offices Flock to Calgary," *Calgary Herald,* 12 June 1997.

34. William K. Carrol, "Westward Ho? The Shifting Geography of Corporate Power in Canada," *Journal of Canadian Studies* 36, no. 4 (Winter 2001/2002): 118–42.

35. "It's Calgary All the Way: Edmonton's Economy Lags Even Further behind Its More Dynamic Rival," *Alberta Report,* 13 June 1994.

36. "Provincial Trends in GDP," *Canadian Economic Observer,* June 1998, 3.1.

37. Nerheim, Brandal, and Hovland, *Klart det lønner seg å samarbeide,* 338–44.

38. "Asian Flu and Oil Glut Weaken Outlook for Houston," *Houston Business: A Perspective on the Houston Economy* (Dallas: Federal Reserve Bank of Dallas, March 1998), n.p.; and

"Weak Commodity Prices Take Toll on Gulf Coast Economy," *Houston Business: A Perspective on the Houston Economy* (Dallas: Federal Reserve Bank of Dallas, March 1999), n.p.

39. "Calgarians Endure the Pains and Gains of Spiralling Growth," *National Post,* 19 December 1999.

40. Between 1996 and 2001 the population of Alberta grew by more than 10 percent and passed the three million mark. Population in Calgary increased by 16 percent between 1996 and 2001. People were moving from British Columbia to Alberta, as well as from Saskatchewan, Ontario, and the Maritime Provinces. The economic growth in the province was the strongest since the early 1980s. Calgary outperformed all other major Canadian cities on most economic indicators. "Calgary's Prosperity at Risk," *Calgary Herald,* 27 September 2001; "Calgary's Growth Unmatched," *Calgary Herald,* 13 March 2002.

41. Torbjørn Eika, "Det svinger i norsk økonomi," *Samfunnsspeilet,* no. 5–6 (2008) (Oslo: Central Bureau of Statistics of Norway, 1985).

42. Stig E. Omre, "Regional konjunkturutvikling: Sterkest konjunkturnedgang for Agder-Rogaland," *Samfunnsspeilet,* no. 2 (2003) (Oslo: Central Bureau of Statistics of Norway, 1985).

43. "10.000 arbeidsplasser vil ryke," *Stavanger Aftenblad,* 16 October 1999; "Pilene peker nedover i Rogaland," *Stavanger Aftenblad,* 15 November 2000.

44. L. M. Sixel, "Job Figures Not as Good as Estimated," *Houston Chronicle,* 8 March 2003.

45. Charles Savage, "A Painful Past Breeds Caution: Despite Leap in Oil Prices, Firms Still Wary in Houston," *Boston Globe,* 17 June 2004.

46. Simon Romero, "Soaring Oil Prices, But No New Boom in Houston," *New York Times,* 14 July 2004.

47. L. M. Sixel, "Houston Adds Jobs But Can't Keep Up," *Houston Chronicle,* 17 December 2005.

48. Kristina Shevory, "Square Feet; Houston, the Oil Town, Is Sharing in a Boom," *New York Times,* 14 March 2007.

49. J. Wilkinson, "Provincial Growth in 2002," *Canadian Economic Observer,* May 2003, 3.7; Philip Cross and Geoff Bowlby, "The Alberta Economic Juggernaut: The Boom on the Rose," *Canadian Economic Observer,* September 2006, 3.1.

50. Philip Cross, "Canada's Economic Growth in Review," *Canadian Economic Observer,* April 2005, 3.6. Production from oil sands increased from 444,000 barrels a day in 1996 to 778,000 barrels a day in 2003.

51. "Rising Oil, Rising Towers," *Business in Calgary,* September 2005, 45; "Residential Real Estate Remains Hot," *Business in Calgary,* October 2005, 61.

52. "Forget Boom, Calgary Is Exploding," *Calgary Herald,* 8 April 8, 2005; "Building Boom Sees 30 % Jump over 2004," *Calgary Herald,* 2 November 2005.

53. P. Cross, "Year End Review: Westward Ho!," *Canadian Economic Observer,* April 2007, 3.7.

54. Torbjørn Eika, "En oljesmurt økonomi—Med harelabb over 35 års konjunktur-historie," *Samfunnsspeilet,* no. 5–6 (2007) (Oslo: Central Bureau of Statistics of Norway, 1985).

55. Joel Kotkin, "How Houston Will Weather the Recession," *Forbes Magazine,* October 3, 2009, n.p.

56. L. M. Sixel, "Houston's Economy Leads Pack. Job Growth Off a Bit, But Area Well Ahead of Nation," *Houston Chronicle,* 27 March 2008.

57. L. M. Sixel, "Houston's Jobless Rate Exceeds the State's," *Houston Chronicle,* 21 August 2009.

58. Paul M. Hohenberg and Lynn Hollen Lees, *The Making of Modern Europe 1000–1994* (Cambridge, Mass.: Harvard University Press, 1995), 59–73.

PART III. Cursed by Oil? Tampico and Port-Gentil

1. Richard M. Auty, *Sustaining Development in Mineral Economies: The Resource Curse Thesis* (London: Routledge, 1993); Jeffrey D. Sachs and Andrew M. Warner, "Natural Resource Abundance and Economic Growth," National Bureau of Economic Research Working Paper 5398, December 1995, at http://www.nber.org/papers/w5398.pdf; and Michael L. Ross, *The Oil Curse: How Petroleum Wealth Shapes the Development of Nations* (Princeton: Princeton University Press, 2012).

2. Daniel Yergin, *The Prize: The Epic Quest for Oil, Money and Power* (New York: Touchstone, 1991), 666–67, 730–33.

CHAPTER 8. Tampico, Mexico: The Rise and Decline of an Energy Metropolis

1. Martin V. Melosi and Joseph A. Pratt, *Energy Metropolis: An Environmental History of Houston and the Gulf Coast* (Pittsburgh: University of Pittsburg Press, 2007).

2. Carlos González Salas, *Nuevas Crónicas de Tampico* (Ciudad Victoria: Universidad Autónoma de Tamaulipas, 2007), 285.

3. Instituto Mexicano del Transporte, *Integración modal y regional en el sistema portuario Tampico-Altamira* (Sanfandilla, Querétaro: Secretaría de Comunicaciones y Transportes, 1999), 30.

4. Anna Zalik, "Liquefied Natural Gas and Fossil Capitalism," *Monthly Review,* November 2008, http://www.monthlyreview.org/081130zalik.php.

5. "Mexico: Ports and Terminals," Central Intelligence Agency, *The CIA World Factbook,* http://www.cia.gov/library/publications/the-world-factbook/notesanddefs.html?countryName=Mexico&countryCode=mex&regionCode=na#2120.

6. J. A. Spender, *Weetman Pearson First Viscount Cowdray, 1856–1927* (London: Cassell, 1930), 188.

7. Myrna Santiago, *The Ecology of Oil: Environment, Labor, and the Mexican Revolution, 1900–1938* (Cambridge: Cambridge University Press, 2006), 25–26.

8. Howard T. Fisher and Marion Hall Fisher, *Life in Mexico: The Letters of Fanny Calderón*

de la Barca, with New Material from the Author's Private Journals (New York: Anchor Books, Doubleday, 1970), 619–20.

9. Thomas L. Rogers, *Mexico, Si Señor* (Boston: Mexican Central Railway Co., Ltd, 1893), 374.

10. Charles Hamilton, *Early Day Oil Tales of Mexico* (Houston: Gulf Publishing, 1966), 19; Román Piña Chan, "El desarrollo de la tradición huasteca," in *Huaxtecos y Totonacos: Una antología histórico-cultural*, ed. Lorenzo Ochoa (Mexico City: Consejo Nacional para la Cultura y las Artes, 1990), 172.

11. Desmond Young, *Member for Mexico: A Biography of Weetman Pearson, First Viscount Cowdray* (London: Cassell, 1966), 127.

12. PanAmerican Petroleum and Transport Company, *Mexican Petroleum* (New York: PanAmerican Petroleum and Transport Company, 1922), 52; Lorenzo Meyer, *México y los Estados Unidos en el conflicto petrolero, 1917–1942* (Mexico City: El Colegio de México, 1981), 46; Richard O'Connor, *The Oil Barons: Men of Greed and Grandeur* (Boston: Little, Brown, 1971), 159.

13. Meyer, *México*, 21.

14. Hamilton, *Early Day*, 140; Carlos González Salas, *Tampico es lo azul* (Mexico City: Miguel Angel Porrúa and Universidad Autónoma de Tamaulipas, 1990), 135.

15. Hamilton, *Early Day*, 22, 125.

16. Marcial E. Ocasio-Melèndez, *Capitalism and Development: Tampico, Madero 1878–1924* (New York: Peter Lang, 1998), 163–67.

17. Meyer, *México*, 76.

18. O'Connor, *The Oil Barons*, 171.

19. PanAmerican Petroleum, *Mexican Petroleum*, 3.

20. Meyer, *México*, 21.

21. Ocasio Meléndez, *Capitalism,* 200; Santiago, *The Ecology of Oil*, 118–20.

22. Hamilton, *Early Day*, 160–61, 187; PanAmerican, *Mexican Petroleum,* 172.

23. S. Lief Adleson, "Historia social de los obreros industriales de Tampico, 1906–1919" (PhD Dissertation, El Colegio de México, 1982), 390.

24. PanAmerican, *Mexican Petroleum*, 11–12, 50; Adleson, "Historia social," 39.

25. There were two exceptions, 1918 and 1928, when Mexico consumed 19 and 21 percent of its production, respectively. Meyer, *México*, 21.

26. PanAmerican, *Mexican Petroleum*, 172, 181.

27. Santiago, *The Ecology of Oil*, 126.

28. Chamber of Commerce to Municipal President, 1920, Archivo Histórico del Ayuntamiento de Tampico, Exp. 38, no. 32.

29. Santiago, *The Ecology of Oil*, 129, 142–44.

30. Ocasio Meléndez, *Capitalism*, 220.

31. Ibid., 218–19; Aurelio Regalado, "Nuevo Inventario: Historia de la Ciudad: El servicio de 'taxi aéreo' en el auge petrolero," *El Sol de Tampico*, 14 April 2008.

32. Aurelio Regalado, "Nuevo Inventario: Historia de la Ciudad: Los primeros dos aeropuertos y la muerte de George H. Rihl," *El Sol de Tampico*, 16 April 2008.

33. Juan Manuel Torrea and Ignacio Fuentes, *Tampico: Apuntes para su historia* (Mexico City: Editorial Nuestra Patria, 1942), 308; Verna Carleton Millan, *Mexico Reborn* (Cambridge, Mass.: The Riverside Press, 1939), 216.

34. González Salas, *Nuevas Crónicas*, 114, 289.

35. *El Luchador*, 19 June 1915; Adleson, "Historia Social," 110–11.

36. Alejandro Salín, Secretary of the Tampico Labor Board to the Labor Board, November 16, 1917, Archivo General de la Nación (AGN), Departamento del Trabajo (DT), Caja 169, Exp 40; Interview with Superintendent Coxon, 1918?, Doheny Mexican Collection, Labor, File I, Item #2952; Inspector Cayetano Pérez Ruíz to the Department of Labor, September 18, 1920, AGN/DT, Caja 220, Exp 6.

37. *Excélsior*, 10 June 1921; Torrea y Fuentes, *Tampico*, 409–10; S. Lief Adleson, "Coyuntura y conciencia: Factores convergentes en la fundación de los sindicatos petroleros de Tampico durante la década de 1920," in *El trabajo y los trabajadores en la historia de México*, eds. Elsa Cecilia Frost, Michael C. Meyer, and Josefina Zoraida Vásquez (Mexico City: El Colegio de México, 1979), 647–51; Ocasio Meléndez, *Capitalism*, 238.

38. Jonathan C. Brown, "The Structure of the Foreign-Owned Petroleum Industry in Mexico, 1880–1938," in *The Mexican Petroleum Industry*, eds. Jonathan C. Brown and Alan Knight (Austin: The University of Texas Press, 1992), 16.

39. *Fanal* (March–April 1937): 8.

40. *El Mundo*, 2 June 1930.

41. "Americans Safe in Tampico Ruins," *New York Times*, 27 September 1933; Aurelio Regalado Hernández, "Nuevo Inventario: Historia de la Ciudad: Imágenes del 33, de Miguel Carranza Rivas," *El Sol de Tampico*, 18 July 2008.

42. Meyer, *México*, 296–97.

43. George W. Grayson, *The Politics of Mexican Oil* (Pittsburgh: University of Pittsburg Press, 1980), 24–25; Santiago, *The Ecology of Oil*, 350–53.

44. Fredda Jean Bullard, *Mexico's Natural Gas: The Beginning of an Industry* (Austin: The University of Texas Press, 1968), 85.

45. Meyer, *México*, 363; Grayson, *The Politics*, 28.

46. Grayson, *The Politics*, 36.

47. René Guzmán García, *Sucesos, Personajes, y Símbolos de Ciudad Madero* (Ciudad Madero: Imprenta Libre Comercio, 2008), 86.

48. González Salas, *Nuevas Crónicas*, 233, 270.

49. *Enciclopedia Regional Ilustrada del Estado de Tamaulipas* (Mexico City: Fernández Editores, 1956), 13, 48.

50. Grayson, *The Politics*, 41, 62.

51. Forrest Jones, "Buried Treasure," *Latin Trade*, August 2004, 38–39.

52. "Más de 200 comercios afectados por apagones en Altamira," *Metrópoli Tamauli-*

pas, August 30, 2009, http://metropolitamaulipas.com/index.php?news=69503; "Ecología hace 'cacería de brujas' con pepenadores," *Metrópoli Tamaulipas,* 21 August 2009, http://metropolitamaulipas.com/index.php?news=69417.

53. R. Stringer, I. Labunska, and K. Brigden, *Informe: Organochlorine and Heavy Metal Contaminants in the Environment around Primex, Altamira, Mexico* (Exeter, UK: Greenpeace Research Laboratories, Department of Biological Sciences, University of Exeter, March 2001), 2, 11.

54. "Transnacionales matan las lagunas de Altamira," Servicios Informativos y Publicaciones del Sureste, August 12, 2009, http://www.sipse.com/noticias/imprimir?8910.

55. "Industrias niegan contaminar lagunas," *Entorno a Tamaulipas,* August 13, 2009, http://www.entornoatamaulipas.com/noticias/templates/nota.aspx?articleid=15489&zoneid=13; "Ejidatarios exigen terminar con la contaminación," and "El municipio es culpable," *Entorno a Tamaulipas,* August 15, 2009, http://www.entornoatamaulipas.com/noticias/templates/nota.aspx?articleid=15566&zoneid=13.

56. Instituto Mexicano del Transporte, *Integración modal,* xv, 6, 31.

57. *Enciclopedia Regional,* 49.

CHAPTER 9. Port-Gentil: From Forestry Capital to Energy Capital

1. BBC, "Gabon Rampage After Poll Results," September 3, 2009.

2. BBC, "Gabon Locks Down City Amid Riots," September 3, 2009.

3. Bloomberg, "Gabon Riots Continued Overnight in Port-Gentil," Bloomberg News, September 4, 2009.

4. France 24, "French Oil Groups Pulls Foreign Staff out of Riot-Hit Port-Gentil," France 24, September 5, 2009.

5. Reuters, "Total Removes Staff from Riot-Hit Gabon Oil Hub," Reuters, September 5, 2009.

6. Al Jazeera, "Gabon Riots Follow Bongo Poll Win," Al Jazeera, September 3, 2009.

7. Le Monde, "Les émeutes de Port-Gentil auraient fait plusieurs dizaines de morts selon une figure de l'opposition," Le Monde, September 9, 2009,

8. Terry Lynn Karl, *The Paradox of Plenty: Oil Boom and Petro-States* (Berkeley: University of California Press, 1997).

9. "Moyen Congo" was a French colony that included both present-day Gabon and neighboring Congo-Brazzaville. For Emile Gentil, see David Gardinier and Douglas Yates, *Historical Dictionary of Gabon,* 3rd ed. (Metuchen, N.J.: Scarecrow Press, 2006), 262.

10. Ibid, 156–57.

11. E. Heiser, *Émile Gentil, 1866–1914* (Paris: Notice Biographiques, 1976).

12. Marcel Souzy, *Les coloniaux français illustres* (Lyon: B. Arnaud, 1941).

13. Jean-Marie Hombert and Louis Perrois, *Cœur d'Afrique: Gorilles, cannibales et Pygmées dans le Gabon du Paul du Chaillu* (Paris: CNRS, 2007): 250–51.

14. G. Boussougou, *Organisation de l'espace dans le nord de l'île Mandji* (unpublished thesis, Université de Bordeaux III, 1981), 5.

15. Michel Mbadinga, "Elf et Port-Gentil," *Networks and Communication Studies* 14, no. 3–4 (2000): 267–68.

16. Gardinier and Yates, *Historical Dictionary*, 247.

17. Mbadinga, "Elf et Port-Gentil," 269.

18. Ibid.

19. Edward Morel, *The British Case in the French Congo* (London: Heinemann, 1903).

20. Gardinier and Yates, *Historical Dictionary*, 90–91.

21. Michael Reed, "*La Coupe Familiale* at Ndjolé," in *Culture, Ecology, and Politics in Gabon's Rainforests*, ed. Michael Reed and James Barnes (Lewiston, N.Y.: Edwin Mellen Press, 2003): 215–40.

22. IMF, "Gabon/Wood: Metric Tons," *International Financial Statistics* (January 2010).

23. John Heilbrunn, "Oil and Water? Elite Politicians and Corruption in France," *Comparative Politics* 37, no. 3 (April 2005): 277.

24. Pierre Péan, *Affaires Africaines* (Paris: Fayard, 1983).

25. Gardinier and Yates, *Historical Dictionary*, 49–50.

26. For an eyewitness account of the coup by the American ambassador at that time, see Charles and Alice Darlington, *African Betrayal* (New York: David McKay, 1968). For a critical analysis of Foccart's influence, read Pean, *Affaires Africaines*.

27. Mbadinga, "Elf et Port-Gentil," 275.

28. Douglas A. Yates, *The French Oil Industry and the Corps des Mines in Africa* (Trenton/Asmara: Africa World Press, 2009): 72.

29. J. Bouquerel, "Port-Gentil, centre économique du Gabon," *Les Cahiers d'Outre-Mer*, no. 79 (1967): 247–74.

30. Douglas A. Yates, *The Rentier State in Africa: Oil-Rent Dependency and Neocolonialism in the Republic of Gabon* (Trenton/Asmara: Africa Wold Press, 1996), 56–57.

31. Gardinier and Yates, *Historical Dictionary*, 261.

32. Ibid., 39.

33. Ibid., 262.

34. Bessora [nom de plume, Sandrine Ngon Nguéma], *Pétroleum* (Paris: Denoël, 2004), 127.

35. Ibid., 89.

36. This has been a topic of several (unpublished) theses and dissertations by Gabonese scholars living in France: Albert Enongo Bikoro, *L'impact des firmes multinationals sur l'économie gabonaise: L'exemple d'Elf Gabon* (DEA thesis, Université d'Amiens, 1979); Clément Ndong, *L'impact de la rente minière au Gabon* (PhD dissertation, Université de Paris X-Nanterre, 1980); Cyrille Biban Endamane, *Analyse du role et de l'avenir des firmes multinationals dans les pays du Tiers-monde: Elf-Gabon et l'économie du pétrole* (PhD dissertation, University of Paris, 1997); cf. Marcelin Ngomo Ondo Enguie, *Les enjeux économiques et géopolitiques du*

pétrole du Gabon (des années 1950 à 2005) (PhD dissertation, Université Montesquieu-Bordeaux IV, 2010), 1.

37. Ibid., 43, 112.

38. Marc Mvé Bekale, *Gabon: La Postcolonie en Débat* (Paris: L'Harmattan, 2003), 41.

39. Bessora, *Pétroleum*, 112.

40. Gardinier and Yates, *Historical Dictionary*, 290.

41. Ibid., 6–8.

42. British Petroleum, *BP Statistical Review of World Energy* (2010), www.bp.com.

43. Francois Ngolet, "Ideological Manipulations and Political Longevity: The Power of Omar Bongo in Gabon since 1967," *African Studies Review* 43, no. 2 (September 2000).

44. For first-person accounts by actors involved see Loïk Le Floch-Prigent, *Affaire Elf: Affaire d'Etat* (Paris: Le Cherche Midi, 2001); and Pierre Lethier, *Argent Secret: L'Espion de l'Affaire Elf Parle* (Paris: Albin Michel, 2001).

45. Douglas A. Yates, "French Puppet, Chinese Strings: Sino-Gabonese Relations," in *Crouching Tiger, Hidden Dragon? Africa and China,* ed. Kweku Ampiah and Asnusha Naidu (Scottsville, South Africa: University of KwaZulu-Natal Press, 2008).

46. Ibid., 209.

47. Jonas Moulenda, "Silence, on pollue à Obangué," www.brainforest.org.

48. Ibid.

49. Bessora, *Pétroleum*, 276.

50. Pierre Eric Mbog Batassi, "Gabon: Port-Gentil bénéficie d'un vaste plan de développement," March 7, 2010, Afrik.com.

CONCLUSION: Comparative Perspectives on Energy Capitals

1. The concept of the resource curse or the oil curse has been applied to numerous countries with varying results. For an introduction to the literature and several examples of its application, see J. D. Sachs and A. M. Warner, "The Curse of Natural Resources," *European Economic Review* 45, no. 4–6 (2001): 827–38; Mahmoud A. El-Gamal and Amy Meyers Jaffe, *Oil, Dollars, Debt, and Crisis: The Global Curse of Black Gold* (Cambridge: Cambridge University Press, 2007); Tony Hodges, *Angola: Anatomy of an Oil State* (Bloomington: University of Indiana Press, 2001); Aldo Musacchio, Eric Werker, and Jonathan Schlefer, "Angola and the Resource Curse," Case # N9–711–016 (Boston: Harvard Business School Publishing, 2010); David White, "The 'Resource Curse' Anew: Why a Grand World Bank Oil Project Has Fast Run into the Sand," *Financial Times*, 23 January 2006, 13; Carrie Ferman and Benjamin C. Esty, "The Chad-Cameroon Petroleum Development and Pipeline Project (A) and (B)," Cases 9–202–010 and 9–202–012 (Boston: Harvard Business School Publishing, 2001).

2. More than forty years ago, the geographer Brian J. L. Berry used the memorable phrase "cities as systems within systems of cities" as a reminder that regional development takes place within broader networks of economic activities. His phrase remains suggestive

when grappling with the concept of energy capitals. Brian J. L. Berry, "Cities as Systems within Systems of Cities," *Papers in Regional Science* 13, no. 1 (January 1964): 147–63.

3. Karen R. Merrill, *The Oil Crisis of 1973–1974: A Brief History with Documents* (Boston: Bedford/St. Martins, 2007); Daniel Yergin, *The Prize: The Epic Quest for Oil, Money, and Power* (New York: Simon and Schuster, 1991), 561–781. The best-written account of OPEC's rise remains Anthony Sampson, *The Seven Sisters: The Great Oil Companies and the World They Have Shaped* (New York: Viking Press, 1975).

4. Ricardo Soares de Oliveira, *Oil and Politics in the Gulf of Guinea* (New York: Columbia University Press, 2007), 19–158; Gerard Prunier and Rachel Gisselquist, "The Sudan: A Successful Failed State," in *State Failure and State Weakness in a Time of Terror*, ed. R. I. Rotberg (Washington, DC: Brookings Institution, 2003), 101–27.

5. Juan Carlos Boue, *Venezuela: The Political Economy of Oil* (Oxford: Oxford University Press, 1993); Terry Lynn Carl, *The Paradox of Plenty: Oil Booms and Petro-States* (Berkeley: University of California Press, 1997); and Jorge Salazar-Carrillo and Bernadette West, *Oil and Development in Venezuela during the Twentieth Century* (Westport, Conn.: Praeger, 2004).

6. For example, see Ricardo Soares de Oliveira, *Oil and Politics,* 302–27.

7. Macartan Humphreys, Jeffrey D. Sachs, and Joseph E. Stiglitz, eds., *Escaping the Resource Curse* (New York: Columbia University Press, 2009); Pauline Jones Luong and Erika Weinthal, *Oil Is Not a Curse: Ownership Structure and Institutions in Soviet Successor States* (Cambridge: Cambridge University Press, 2010). This book is an interesting addition to discussion of which nations might be susceptible to the oil curse and why.

8. Augustine A. Ikein, *The Impact of Oil on a Developing Nation: The Case of Nigeria* (New York: Praeger, 1990); Ike Okonta and Oronto Douglas, *Where Vultures Feast: Shell, Human Rights, and Oil* (New York: Verso, 2003).

9. In the early 1970s, I found this to be a comparison of great potential value, and I worked for a year on a dissertation that sought to explain why oil-led development became a long-term impetus for growth in the Houston region but not in the area around Tampico. I finally gave up and focused exclusively on the Houston-Beaumont area. A quarter of a century later, Myrna Santiago published her extraordinary book on the impact of the early Mexican oil industry in and around Tampico. Clio works in mysterious ways.

10. Myrna Santiago, *The Ecology of Oil: Environment, Labor, and the Mexican Revolution, 1900–1938* (Cambridge: Cambridge University Press, 2006); Jonathan C. Brown, "The Structure of the Foreign-Owned Petroleum Industry in Mexico, 1880–1938," in *The Mexican Petroleum Industry*, ed. Jonathan C. Brown and Alan Knight (Austin: The University of Texas Press, 1992).

11. Lorenzo Meyer, *Mexico and the United States in the Oil Controversy, 1917–1942* (Austin: The University of Texas Press, 1977); George W. Grayson, *The Politics of Mexican Oil* (Pittsburgh: University of Pittsburgh Press, 1980).

12. "Topping" refers to a basic refining process that removes a portion of some of the lighter streams from crude oil so that the remainder can be sold as residual fuel oil. For

more complex refining processes, much of the crude produced in Mexico was shipped to U.S. refineries. For an interesting view from the inside out of the Mexican oil industry in the early 1920s, see Pan American Petroleum Transport and Transport Company, *Mexican Petroleum* (New York: Pan American Petroleum and Transport Company, 1922). A detailed comparison of the company's refineries at Tampico and Destrehan, Louisiana, is presented on pages 169–84.

13. Jonathan C. Brown, "Why Foreign Oil Companies Shifted Their Production from Mexico to Venezuela in the 1920s," *The American Historical Review* 90, no. 2 (1985): 362–85; Brown, *Oil and Revolution in Mexico* (Berkeley: University of California Press, 1993).

14. Santiago, *The Ecology of Oil.*

15. This observation seems particularly important to me in light of my experience teaching many Mexican American students during almost thirty years at the University of Houston. These sons and daughters of working grandparents or parents from Mexico often discuss the lack of educational opportunities in Mexico for less privileged students.

16. For a good introduction to the oil curse literature, see Ricardo Soares de Oliveira, *Oil and Politics in the Gulf of Guinea* (New York: Columbia University Press, 2007).

17. Richard Gordon and Thomas Stenvoil, "Statoil: A Study in Political Entrepreneurship," case study prepared for The Changing Role of National Oil Companies in International Energy Markets Conference, James A. Baker Institute for Public Policy, Rice University, Houston, Texas, March 2007.

18. For a good history of Alberta's oil industry, see David H. Breen, *Alberta's Petroleum Industry and the Conservation Board* (Edmonton: University of Alberta Press, 1993); also, John Richards and Larry Pratt, *Prairie Capitalism: Power and Influence in the New West* (Toronto: McClelland and Stewart, 1979).

19. For an interesting history of Alberta's oil sands, see Paul Chasko, *Developing Alberta's Oil Sands: From Karl Clark to Kyoto* (Calgary: University of Calgary Press, 2004).

20. Jenny Gregory, *City of Lights: A History of Perth since the 1950s* (Perth: University of Western Australia Press, 2003).

21. "The Gorgon LNG Project," *The Australian Business Journal* (August 23, 2012), http://www.australianbusinessjournal.com.au/the-gorgon-project/.

22. Joseph A. Pratt, "Letting the Grandchildren Do It: Environmental Planning during the Ascent of Oil as a Major Energy Source," *Public Historian* 2 (1980): 28–61; Hugh S. Gorman, *Redefining Efficiency: Pollution Concerns, Regulatory Mechanisms, and Technological Change in the U.S. Petroleum Industry* (Akron: The University of Akron Press, 2001).

23. Joseph Schumpeter, *Capitalism, Socialism, and Democracy* (New York: Harper Perennial, 2008, reprint).

24. Thanks to the History of the Urban Environment series edited by Joel Tarr and Martin Melosi at the University of Pittsburgh Press, the reader has the opportunity to make a broad comparison of the environmental histories of Pittsburgh and Houston. See Joel A.

Tarr, ed., *Devastation and Renewal: An Environmental History of Pittsburgh and Its Region* (Pittsburgh: University of Pittsburgh Press, 2003); and Melosi and Pratt, eds., *Energy Metropolis*.

25. For a broad comparison of environmental issues in Los Angeles and Houston, see William Deverell and Greg Hise, eds., *Land of Sunshine: An Environmental History of Metropolitan Los Angeles* (Pittsburgh: University of Pittsburgh Press, 2007); and Melosi and Pratt, eds., *Energy Metropolis*.

26. Steve Lerner, *Diamond in the Rough: A Struggle for Environmental Justice in Louisiana's Chemical Corridor* (Cambridge, Mass.: MIT Press, 2005); Craig E. Colten, "The Rusting of the Chemical Corridor," *Technology and Culture* 47, no. 1 (2006): 95–101.

27. Craig C. Colten, "Too Much of a Good Thing: Industrial Pollution in the Lower Mississippi River," *Transforming New Orleans and Its Environs: Centuries of Change*, ed. Craig C. Colten (Pittsburgh: University of Pittsburgh Press, 2000), 141–59. The collection of essays in *Transforming New Orleans* provides a useful contrast to the anthologies cited above on the environmental histories of Pittsburgh, Houston, and Los Angeles.

28. Vaclav Smil, *Energy Transitions: History, Requirements, Prospects* (New York: Praeger, 2010); Peter A. O'Connor, "Energy Transitions," *The Pardee Papers* no. 12 (November 2010).

29. For an overview of shifts in energy demand in the United States, see Sam H. Schurr and Bruce C. Netschert, *Energy in the American Economy, 1850–1975: An Economic Study of Its History and Prospect* (Baltimore: Johns Hopkins University Press [Resources for the Future], 1960).

30. Pauline Jones Luong and Erica Weinthal, *Oil Is Not a Curse*, 322–36.

CONTRIBUTORS

Kathleen Brosnan is the Paul and Doris Eaton Travis Chair of Modern American History at the University of Oklahoma. She previously taught at the University of Tennessee and the University of Houston (UH). At UH, she served as associate director of the Center for Public History and associate dean of the College of Liberal Arts and Social Sciences. Dr. Brosnan's first book, *Uniting Mountain and Plain: Cities, Law, and Environmental Change along the Front Range* (2002), examines the integration of diverse and distant hinterlands into the Denver-based regional economic system following the discovery of gold. Dr. Brosnan is the editor of the four-volume *Encyclopedia of American Environmental History* (2010) and the coeditor of a collection of essays, *City Dreams, Country Schemes: Utopian Visions of the Twentieth-Century American West* (2011). Dr. Brosnan is currently finishing a book manuscript, "Napa Nature: An Environmental History of America's Premier Wine Region."

Karen Bradley Clay is associate professor of economics and public policy at the Heinz College at Carnegie Mellon University and a faculty research fellow at the National Bureau of Economic Research. She received her BA (1988) in economics and mathematics with highest honors from University of Virginia and her PhD (1994) in economics from Stanford University. Prior to joining Carnegie Mellon, Professor Clay was a faculty member in the Department of Economics at University of Toronto. Her book, *The Evolution of a Nation: How Geography and Law Shaped the American States,* coauthored with Daniel Berkowitz, was published with Princeton University Press in 2011.

Craig Colten is the Carl O. Sauer Professor of Geography at Louisiana State University and the director for Human Dimensions at the Water Institute of the Gulf. He spent more than a decade in government and the private sector conducting research on the history and geography of hazardous waste, when he coauthored *The Road to Love Canal: Managing Industrial Waste before the EPA* (1996). Upon returning to the academy, he expanded the scope of his inquiry to urban environmental topics and has since produced the award-winning *An Unnatural Metropolis: Wresting New Orleans from Nature* (2005) and *Perilous Place, Powerful Storms: Hurricane Protection in Coastal Louisiana* (2009), among other books and articles. Currently he is working on a historical geography of water resources in the American South.

Matthew N. Eisler is a lecturer of Science, Technology, and Society in the Department of Engineering and Society at the University of Virginia. Dr. Eisler's research interests include the politics and discourse of research and development relating to materials sciences and energy and power source systems since the Second World War. He explored these issues in his first book, *Overpotential: Fuel Cells, Futurism, and the Making of a Power Panacea* (2012). He is currently working on projects assessing the history of the lithium ion rechargeable battery and "sustainable innovation" discourse in global context.

Sarah Elkind is a professor of history at San Diego State University. Her publications include *How Local Politics Shape Federal Policy: Business, Power, and the Environment in Twentieth Century Los Angeles* (2011), which investigates the local government practices that gave business groups legitimacy in local governance and surprising influence in national policy; *Bay Cities and Water Politics: The Battle for Resources in Boston and Oakland* (1998), which won the Abel Wolman Award for best book in public works history in 1998; *Public Works and Public Health: Reflections on Urban Politics and Environment, 1880–1925* (1999); "Public Oil, Private Oil: The Tidelands Oil Controversy, World War II and the Control of the Environment," in *The Way We Really Were: The Golden State in the Second Great War*, edited by Roger Lotchin; and "Building a Better Jungle: Growth, Reform and Public Works in American Cities, 1880–1930," *Journal of Urban History* (1997).

Jenny Gregory is Winthrop Professor of History, head of the School of Humanities, and director of the Centre for Western Australian History at the University of Western Australia (UWA). She has written or edited ten books and more than seventy articles, book chapters, and proceedings. Current projects include research into international trends in Australian urban planning in the postwar era and on the impact of the loss of urban heritage places. She is currently vice president of the National Trust (WA) after years as president and then chair of council (1998–2010). In 2010 she was appointed a member of the Order of Australia for her service to history and the community. She was inaugural president of the History Council of WA (2001–2006) and is a fellow of the Royal Historical Society, London.

Martin V. Melosi, the Hugh Roy and Lillie Cranz Cullen University Professor and director of the Center for Public History at the University of Houston (UH), is one of the nation's leading scholars in urban and environmental history. Melosi received the Distinguished Service Award from the American Society for Environmental History (ASEH) in 2009 and the Esther Farfel Award from UH in 2005, the highest honor for a UH faculty member. Dr. Melosi is the general editor for the History of the Urban Environment Series of the University of Pittsburgh Press (with Joel A. Tarr), and he served on the Scientific Committee for Postgraduate Studies on Urbanism at the University of Geneva and has been president of the ASEH, the National Council on Public History, the Public Works Historical Society, and the Urban History Association. He has written or edited nineteen books and more than ninety proceedings, articles, and book chapters. Melosi's current project is "An Island Not So Far: Staten Island, New Yorkers, and Fresh Kills."

Gunnar Nerheim is professor of economic history at the University of Stavanger, Norway. His main fields of interest are economic history, business history, history of technology, and urban history. He was the CEO of the Norwegian Museum of Science and Technology in Oslo from 1995 to 2005 and director at the Stavanger Museum from 2006 until 2008, before assuming his present position, where one of his main tasks is to manage all externally financed historical research projects. For more than two decades he has conducted research and published books on different aspects of the Norwegian energy history. His extensive work includes the two-volume history of oil in Norway. The first volume was published in 1992 and the second volume in 1996. More recently, he served as the primary editor of a two-volume history of the city of Sandnes (2010).

Joseph Pratt, the Cullen Professor of History and Business at the University of Houston (UH), is a leading historian of the petroleum industry. He taught at UC Berkeley, the Harvard Business School, and Texas A&M University before coming to UH in 1986. At UH he has served as chair of the History Department and acting dean of the College of Liberal Arts and Social Sciences. He currently directs the Energy and Sustainability minor at UH. Professor Pratt has been a consultant for the PBS miniseries on the oil industry, *The Prize*, and for the American Experience documentary on the Trans-Alaska Pipeline. He is the editor of the Oil and Regional History Series for the Texas A&M University Press. He is also the founder of the Houston History Project and editor of *Houston History,* a magazine of popular history. Dr. Pratt is the author or coauthor of numerous books and articles. His research is primarily in energy history and the history of the Houston region and his most recent publication is *Exxon: Transforming Energy, 1973–2005* (2013).

Myrna Santiago received her PhD in history from the University of California, Berkeley, where her specialization was Latin America. After stints as lecturer at UC Berkeley and Mills College, she joined St. Mary's College of California, a small Catholic liberal arts school in the Bay Area, in 1998. She is an associate professor and teaches courses on Latin American history and world history. Her book, *The Ecology of Oil: Environment, Labor, and the Mexican Revolution, 1900–1938* (2006), received the Elinor Melville Book Prize from the Conference on Latin American History and the Bryce Wood Book Award from the Latin American Studies Association. She is currently working on questions of workers' health and safety rights.

Joel Tarr is the Richard S. Caliguiri University Professor of History and Policy at Carnegie Mellon University, where he has been on the faculty since 1967. His research interests include urban history, the development of urban technological systems, and the urban environment. He is currently exploring the environmental effects of traditional natural gas development in Pennsylvania. A multiple prize-winning author and editor of books such as *Technology and the Rise of the Networked City in Europe and America* (edited with Gabriel Dupuy); *The Search for the Ultimate Sink: Urban Pollution in Historical Perspective*; *Devastation and Renewal: An Environmental History of Pittsburgh and Its Region*; and *The Horse in the City: Living Machines in the Nineteenth Century* (coauthor Clay McShane), Tarr received from the Society for the History of Technology its highest award, the Leonardo da Vinci Medal, presented to an individual who has made an outstanding contribution to the history of technology. He is coeditor with Martin V. Melosi of the University of Pittsburgh's the History of the Urban Environment Series.

Douglas Yates, a tenured professor of Anglo-American law at the public University of Cergy-Pontoise, also teaches political science at the American University of Paris and the American Graduate School in Paris. He has spent twenty years researching the oil industry in Africa, with a special concentration on the Republic of Gabon. His books include *The Rentier State in Africa: Oil-Rent Dependency and Neocolonialism in the Republic of Gabon* (1996), *Oil Policy in the Gulf of Guinea*, coedited with Rudolf Traub-Merz (2003), and *The French Oil Industry and the Corps des Mines in Africa* (2009). His latest book is *The Scramble for African Oil* (2012). As a consultant and presenter for the U.S. government and numerous nongovernmental organizations, Yates has traveled to South Africa, Morocco, Gabon, Mauritania, Senegal, Burkina Faso, Togo, Benin, Cameroon, Nigeria, Guinea, and Tanzania.

INDEX

acid, as pollution from coal mining, 12, 202n28

Affaires Africanines (Péan), 175

Africa, 184. *See also* Gabon

African Americans, 87

Agondjo-Okawe, Pierre-Louis, 172–73

agriculture, 35, 60, 164; declining role in Louisiana, 58–59, 63–65; effects of refineries' arrival on, 63, 67; industry's relation to, 70, 75; in lower floodplain of Mississippi River, 58; water usage by, 46–47; workers leaving, 38–39

air conditioning, Houston's dependence on, 32

air pollution: in Baton Rouge, 72; effects of, 18, 27–28; in Houston, 53–54; in Kwinana, 106–8; in Los Angeles, 89, 227n71; in Pittsburgh, 5, *13,* 13–14, 18, *19,* 27–28, 205n69; Pittsburgh's efforts to address, 15, 26–27, 190–91; sources of, 5, 19, 53, 93, 106–8. *See also* environmental damage; pollution

air pollution controls, 206n77, 207n86; effects of, 26–27, *27,* 28, 29; in Los Angeles, 89; in Pittsburgh, 2, 14, 18, 20–22, *21,* 26–27; in St. Louis, 20. *See also* regulations

Aitken, Max (Lord Beaverbrook), 117

Alberta, 122, 234n34, 241n40; economy of, 137, 234n37; electricity generation in, 120, 235n44; energy production distant from Calgary, 124; energy reserves in, 115, 118; multinationals recruited to, 92; oil industry in, 115–20, 131, 140, 189, 234n37; political elites of, 92, 114. *See also* Calgary; Edmonton

Alexander, Titus, 87

Allegheny River, steel mills along, 10

Altamira, as energy city, 156–57

Anglo-Iranian Oil Company (later British Petroleum), 93, 98, 104–6, 229n38

Asia, 138; oil consumption in, 137, 239n30

Atchafalaya River, 62

Athabasca tar sands, Canadian efforts to develop, 121

Australia, 100, 106; revenue from energy industry, 101–2. *See also* Perth; Western Australia

automobiles, 4, 115, 153; air pollution from, 53, 89; Houston's sprawl from, 48–50

Barnett, Colin, 102

Barrow Island, natural gas from, 100, 102

BASF, public opposition to, 73

Baton Rouge, 3, 58, 60, 72; considered flood proof, 62, 65. *See also* petrochemical corridor, Louisiana's

Beaumont, Spindletop oil discovery in, 35–36

Bennett, R. B., 117

Berry, Brian J. L., 247n2

BHP Billton, 102, 104, 110

biofuel, sugarcane considered for, 65, 75

Bongo, Ali, 159–60, 178–79

Bongo, Omar, 168, 172–74, 176, 178

Bow River, hydroelectric power from, 117–18

Bowron, Fletcher, 83–85, 224n40

Breen, David, 115–17

Britain, 113, 129

British Continental Shelf, oil on, 129

Brown, George, 42

Brown, Robert A., Jr., 120–21

Brown, Sarah, 106

Bru, Roland, 165–66

business, investment in petroleum-related infrastructure, 40 business elite, 112; in Calgary, 231n8; as factor in development of energy capitals, 2, 4, 91; in Gabon, 169, 178–79; in Houston, 2, 35, 42, 51–52; influence on pollution regulations, 22–23, 54–55; of natural gas transmission companies, 41; in Pittsburgh, 10–11, 22–23, 206n75; in Tampico, 154–55

Calderón de las Barca, Fanny, 149

Calgary: business elite in, 114, 231n8; difference from other energy capitals, 113–14; distance from energy production, 126, 188; economy of, 123, 135, 140, 236n58, 241n40; effects of energy prices on, 131–32, 138–39; effects of free trade in energy on, 122–23; electricity generation for, 120, 124, 235n44; as energy capital, 91–92, 112–15, 125–26, 137, 141–42; energy production distant from, 114, 124; growth of, 121, 241n40; independent oil companies in, 120–23, 125–26; oil industry based in, 111–12, 116–17, 125–26, 236n58; population of, 135, 241n40; relations with international oil companies, 190; urban ecology of, 123, 234n40

California, 78; environmental movement in, 77; management of offshore oil, 88; oil drilling along beaches of, 80–82, 222n15; ownership of petroleum rights in, 88; regulation of oil industry in, 191; WWII demand for oil from, 83–85. See also Los Angeles

Canada, 122; exports of oil and natural gas from, 125–26, 232n23, 233n26, 236n56; natural gas reserves of, 237n62; oil industry in, 111–12, 115–16, 118–20; oil reserves of, 237n62; regional politics in, 111, 114–15; regulating oil industry, 116, 118, 121, 233n26; state ownership of Petro-Canada, 121–23

cancer: in Louisiana's petrochemical corridor, 71, 73; pollution from industries in Kwinana and, 108–9

capital, financial, 96; accumulated from processing crude oil, 38; for development of energy industries, 35, 52, 124, 157, 185–86; as factor in development of energy capitals, 4, 91, 143, 167; foreign investment In Mexico, 144, 185; invested in coal mining, 98; invested in oil industry, 39–40, 50, 182–83; oil companies trying to recoup, 115–16

carbon dioxide, produced by tar sands processing, 125

Cárdenas, Lázaro, 156, 185

Carnegie, Andrew, 9–10, 14

chemicals industry, 54, 191; importance in Houston's economy, 37–40. See also petrochemical industry

China, and Gabon, 176–77

Ciudad Madero, 148, 156–57, *157*

class, social, 96, 153, 227n71, 249n15; Alberta's elite worried about working-class politics, 92, 114; in Gabon, 164–65

climate: Calgary's, 123; Houston's, 31–32, 54; Tampico's, 148–49

climate change, 57, 158

coal, 99, 118; as cheap fuel, 9; cleaner-burning, 21–23, 206n81, 207n83; coke produced from, 9–11; compared to other energy sources, 198n20; consumption of, 8, *11*, 20, 21–22; for electricity generation, 99, 120, 124, 235n44; environmental effects of, 11–14; importance of, 1–2, 5, 9, 113; importance for railroads, 8, 26; industry shifting from natural gas back to, 18–19; natural gas *vs.*, 24–25, 120; Pittsburgh's seam of, 6–7; Pittsburgh's transition to natural gas from, 2, 14–15, 26–29, 193–94; smokeless, 25; transition away from, 6, 22, 33, 98–99

coal companies, 7; responsibility for damages, 12, 17

coal gas, 205n65

coal industry, 23; effects of smoke control regulations on, 22, 29; involved in air pollution control, 20–21, 206n77

coal mining, 7, 200n8, 201n26; companies' responsibility for damages from, 12, 17; environmental effects of, 12, 17, 202nn27–28; loss of jobs in, 7, 15

coal-oil industry, in Pittsburgh, 5

coal prices, 24, *24,* 98

coke: environmental effects of production, 12–13; for iron and steel industries, 9–11, 204n57; produced from coal, 9–11

Cold War, pollution ignored in, 82

colonialism: effects on Gabon, 164–65, 169, 178, 187; effects on Port-Gentil, 144, 162–67; French, 165, 245n9; in Gabon, 159–62

commercial-residential energy use: of coal, 11, *11*, 24–25; of natural gas, 14–15, 19, 24–25, *25*, 100; smoke emissions from, 20–22, 27; transition away from coal in, 22, 26–27

conservation, 130, 134; waste by oil industry *vs.*, 116, 119

corruption: in Gabon, 173, 175–76; in oil industry, 154–55, 175–76; of "successful failed states," 184

cotton, 37–38

Court, Charles, 100, 105

Cullen, Hugh Roy, 51

Cullinan, Joseph, 36

dangers: of industrial accidents, 71, 73, 103–4, 124; from natural gas, 15; from oil drilling under residential neighborhoods, 79

Davis, Morgan, 42

De Gaulle, Charles, 165

Depression: effects on energy industry, 22, 96, 155; pollution ignored in, 53, 82; reduced coal consumption in, 8, 20, 205n69

Diaz, Porfirio, 185

Diefenbaker, John, 120

Doheny, Edward L., 48, 149–51, 154–55

Dumas, Russell, 104–5

Durand-Reville, Luc, 164

Ecology of Oil, The (Santiago), 186

Edmonton, 124, 189; Calgary *vs.*, 116–17; effects of energy prices on, 131–32; energy production and refining in, 140

education: as factor in development of energy capitals, 50–52, 94, 186–87; in Gabon, 171, 179; in Houston, 42, 50–52, 57, 186–87; Mexico's lack of opportunities for, 157, 186–87, 249n15; Norway's strength in, 188

Eisenhower, Dwight, 89

electricity: for Alberta, 124; in Australia, 98–99, 101–2; for Calgary, 117–18, 120, 235n44; for California, 77; sources of, 98–99, 101–2, 117–18; uses of, 125, 152. *See also* utilities

Elliot, J. E., 83

enclave economy, Port-Gentil's, 162–67

energy: Canada's exports of, 236n56; demand for, 110, 194; trade in, 112–13, 122–23. *See also* specific fuels

energy capitals: balancing promotion and regulation with energy industry, 190–92; benefits for, 181–83; Calgary as, 112–17, 125–26, 137; characteristics of, 113; coal- *vs.* oil-based, 113; costs and benefits of, 143, 158, 188–89; costs of, 52, 183; development of, 2–3, 113; effects of global economy on, 140–42; effects of oil prices on, 131, 134–35, 192–93; efforts to diversify economies of, 193; factors in beneficial *vs.* detrimental effects of, 181–84; factors in development of, 4, 49–50, 91, 94, 143, 194; growing importance of natural gas in, 41; Houston's future as, 56–57; Houston as premier, 30, 43, 137, 240n32; inequities of cost and benefit of development of, 93, 145–46, 169–70, 178; latecomers as, 91, 93–94, 145; Los Angeles as, 77; other industrial cities *vs.*, 91, 112; Perth as, 93, 102, 189; Port-Gentil as, 144, 159, 174–75; recovering from recession, 137; Stavanger as, 92–93, 136, 141–42; Tampico as, 143–44, 147, 153, 155; tied to international events, 127; usefulness of concept, 181

energy companies, 102; efforts to regulate, 4, 16–17; inadequately safeguarding from dangers of natural gas, 15; judicial accommodation of, 12, 16–17; local resistance to, 4; political accommodation of, 3, 14; resisting regulations, 4, 15–16; responsibility for environmental damages, 12, 17–18. *See also* coal companies; oil companies

energy industries: impact on Western Australia, 102, 110; local involvement in, 145–46, 206n75; revenue from, 101, 236n58. *See also* coal industry; oil industry

energy prices, 25; comparison of, 24, *24*, 120; effects of, 1, 65, 134; influence on Louisiana's economy, 68–69, 75–76; OPEC's influence on, 92, 121, 130;

energy prices (*cont.*): in Western Australia, 104. *See also* coal prices; natural gas prices; oil prices

energy sources: comparison of, 198n20; fear of dependence on single, 98; search for alternative, 65, 99, 134, 193; transitioning of, 2, 5–6, 14–15, 22, 193–94

environment: Calgary's distance from energy production's, 126; consideration for, 106; oil industry transforming, 147–48, 150, 155

environmental damage, 186, 199n29; from coal, 11–14; from coal mining, 12, 17, 202nn27–28; from coke production, 12–13; as cost to energy capitals, 183; from drilling for oil and natural gas, 4, 16–17, 31, 79, 80–82; efforts to address, 109, 190; from energy industries, 69–70; from Houston's oil development, 32, 52–56; from industrial accidents, 71–72; from natural gas, 15–18, 102, 157–58; from oil industry, 59, 73, 75, 85–86, 93, 145, 177, 221n8; from petrochemical industry, 192; from tar sands processing, 125. *See also* air pollution; pollution; water pollution

environmental justice, 192; in Louisiana's petrochemical corridor, 72–75

Fadden, Arthur, 99

Fall, Albert B., 151, 154–55

Feilman, Margaret, 106, 229n38

fish: contamination of, 12, 178; killed, 70–71, 158

fishing, importance to Stavanger, 127–28

flooding: along Mississippi River, 60, *61,* 62–63; Baton Rouge considered flood proof, 62, 65; efforts to control, 61–62, 75; around Houston, 31–32, 54

Foccart, Jacques, 165–66, 168, 175

Fontenot, William, 72

forestry: decline of, 175; French control of Gabon's, 164–66; in Gabon, 161–63, 165, 170, 177, 180; Port-Gentil as enclave economy and, 162–67, 175

fossil fuels, 1; dominating energy markets, 112–13; effects on climate change, 57; uncertain future of, 194. *See also* coal; natural gas; oil

France, 167, 175, 180; colonialism of, 165, 245n9; continued influence on Gabon, 168–69, 172–73, 178; effects of colonialism on Gabon, 164–65, 170; Gabon as colony of, 144–45, 159–62, 163–64; Gabon's oil industry and, 168–69, 176–77

Frick, Henry Clay, 9–10

fuel, cities requiring smokeless, 20–22

fuel oil, 22, 248n12; availability of furnaces for, 23–24; Tampico refinery supplying, 149–50; uses of, 151–52

Fugate, G. L., 47

Gabon, 169; China and, 176–77; contamination of water supply in, 177–78; effects of French colonialism in, 144–45, 159–62, 164–65; exploitation of natural resources of, 146, 174–75, 178–79; France meddling in politics of, 165–66, 168; French colonialism in, 163–64, 245n9; indigenous people of, 162, 165, 172–73; lack of development in, 167, 178–79; oil industry in, 145, 174–75, 184, 187; opposition politics in, 172–73, 174, 175; political independence of, 165–66; Port-Gentil as energy capital of, 144, 159

Galveston Bay, 31

gasoline, 48, 56, 153

Gentil, Émile, 161–62

Glazier, Willard, 9

Glenn, John, 96–97

gold, in Western Australia, 96, 102–3

government, 40; efforts to control energy industries, 4, 16–17, 183–84, 192; stability as factor in development of energy capitals, 94, 143, 182; of "successful failed states," 183–84

government, Australian: pollution control and, 107–8; promoting industrial development, 103, 104–6, 108; purchase of natural gas by, 100–101; relations with international energy companies, 189–90

government, Canadian national, 120; promoting development of tar sands, 124–25; provincial *vs.,* 114, 119; regulating oil industry, 118, 121–22; support for oil industry, 92, 114, 121

government, Gabon's, 160; getting share of oil

industry *(cont.)*: governments promoting development of, 103–6; natural gas use by, 14, 18–19, 24–25, *25*, 100–102; petroleum-related, 33–37, 41, 86, 210n22; pollution from, 20, 69, 107–8; relation to agriculture, 64, 67, 70, 75; water usage by, 46–47, 67; WWII expansion of, 47, 53. *See also* specific industries

infrastructure, 45, 67, 105; Calgary's municipal energy, 117; for coal industry, 7–8; education system as, 50–52; for energy industries, 44–45, 93; as factor in development of energy capitals, 141, 182; for flood protection, 62–63; inadequacy of Port-Gentil's, 169, 179; for iron and steel industries, 10; supporting oil industry, 37, 39, 50, 57, 145, 150–51, 183; Tampico's, 144, 150–51, 154; transportation, 97; for transportation of natural gas, 100–101. *See also* pipelines

investment. *See* capital, financial

Iraq, invasion of Kuwait, 239n30

iron and steel industries, 202n37; fuels for, 9, 14, 19, 204n57; in Pennsylvania, 9–11; in Western Australia, 103–4, 105

iron ore, 103–5, 110

jobs, 20; in Australia's energy industry, 101; in chemical industries, 40; in coal mining, 7, 15; in energy capitals, 3, 113, 131; for foreigners *vs.* locals, 129, 133–34, 144, 145, 168, 169–70, 175, 185–86; in forestry, 175; Houston's culture and, 38–39, 50–51, 56, 213n66; in Houston's oil industry, 35–36, 141, 240n31, 240n32; increasing during recovery from recession, 137–38; influence of oil price fluctuations on, 68–69, 141; of international petroprofessionals, 113, 231n8; in Kwinana, 105, 106–7; in medicine, 42; migration for, 32–33, 35–36, 43, 53, 140, 150; in mining, 97; in offshore exploration and drilling, 41, 129–30; in oil industry, 87, 93, 133–36, 138, 144–45, 153, 168, 175, 182–83, 185–86, 239n17, 240nn31–32; in petrochemical industry, 67–68, 73; in Port-Gentil, 169–71; in refineries, 38, 56, 68; in Stavanger, 128, 130, 133–34, 136; through French colonialism, 164–65; un-

employment and, 108, 122, 135–37; workers' housing and, 105–6, 169

Johnson, Walter, 98

Johnson Space Center, 42, 137

kerosene, 149

Kimberly wilderness, 102

Knight, Ora E., 85

Knox, Frank, 83–84

Kombila, Pierre-André, 160

Kwinana, 99; pollution in, 107–9; refineries at, 98, 103, 104–5; town planning of, 105–6

labor issues: in Australian coal industry, 98; in Mexican oil industry, 148; plant safety and, 73; in Tampico, 144, 153–56

Lake Pontchartrain, 62

Lawrence, David, 22

Layman, Lenore, 104

leases: for oil drilling under residential neighborhoods, 4, 79, 223n21; for tideland oil drilling, 80

levees, along Mississippi River, 59

Los Angeles: air pollution in, 89, 227n71; deepwater harbor in, 4, 86; efforts to control oil industry in, 79–86, 191, 221n9, 224n40; as energy capital, 77, 194; environmental damage in, 53, 85–86, 221n9; growth of, 4, 86–87; oil drilling under residential neighborhoods in, 77–79, 82–83, 85, 87–88, 221n9, 223n21; oil industry around, 3–4, 77, 78, 221n1, 226n67; public opinion on oil drilling in, 82–84; race relations in, 4, 86–87; social impact of oil development in, 86–87

Louheed, Peter, 121

Louisiana, 70; agriculture in, 58–59, 63–64, 67; economy of, 58–59, 67–69, 75–76; Mississippi River's importance to, 60; petrochemical industry in, 3, 38, 65, 67, 74–75, 217n26; pro-business politics of, 191; refineries in, 67–68. *See also* petrochemical corridor, Louisiana's

Love Canal, 73

Luong, Pauline Jones, 194–95

Lynn, Robert, 98

offshore exploration and drilling, 68, 157; in Australia, 100, 102; in California, 77, 226n67; in Gabon, 167; in Gulf of Mexico, 3, 41, 134; Houston as center of, 137, 211n42; in Norway, 92–93, 128–30; state vs. federal management of, 88–89, 226n68, 227n70

Ohio River, 8, 10

oil, 43; discoveries of, 33–34, 99–100, 104, 129, 144; discovery of naphtha, 115; exports of, 125–26, 162, 174–75; importance of, 2, 127, 142, 151; public vs. private ownership of, 88, 182, 185, 195; supply and demand for, 119, 141; transition from coal to, 33, 98–99; transportation of, 3, 37, 59, 143–44, 150–51, 200n2; uses of, 39–40, 99; water pollution from, 52

oil booms: in Alberta, 115–20; around Houston, 41–42; in Los Angeles, 3–4

oil companies: in Alberta, 92; based in Calgary, 125–26, 137; based in Stavanger, 92, 128–29; governments' relations with, 81, 129, 189–90; independents vs. multinationals, 120–23, 125–26; moving into tar sands processing, 124–25; multinational, 92, 114, 120–23, 125–26, 132–33, 143; Norway's state, 92–93, 188; self-interest vs. interests of host nations, 184

oil consumption: in Asia, 137, 239n30; declining, 119, 134; domestic vs. foreign, 144, 149–51; increasing, 137, 153; Mexico's, 144, 149, 155–56, 243n25

oil curse, 145–46, 181, 183–87, 194–95

oil drilling, 31; along California beaches, 80–81, 80–82, 222n15; under residential neighborhoods, 3–4, 77–79, 82–83, 85, 221n9, 223n21, 225n54; tideland, 88–89

oil industry, 57, 78; in Altamira, 156–57; Canada's efforts to control, 121–22, 233n26; Canadian, 111–12, 114–17, 140; Canadian regulation of, 118, 121–22; competition in, 88, 115, 118, 132; concentration of, 240n32; corruption in, 154–55, 175–76; danger of economy dependent on, 183; density and diversity around Houston, 41, 57; development of, 38, 149–50, 167–70, 185–86; in development of Houston, 32–33, 33–34,

35–36; effects of declining/relocating by, 156–57, 174–75, 180; effects of prices on, 42, 134–36; environmental impact of, 52–56, 85–86; fleeing Mexico to Venezuela, 148, 153; foreign control of, 143, 168–70, 187; French influence on Gabon's, 167–69, 172–73; in Gabon, 145, 167–70, 174–77, 180, 187; governments' efforts to control, 183–84, 192; governments' relations with, 185, 195; Houston's dominance of, 2–3, 33–34; importance in global economy, 56, 127, 182; importance in Houston, 30, 37–38, 41–43, 50–51, 139–40; influence in development of energy capitals, 30, 42, 77; influence of, 68, 139–40, 185–86; infrastructure for, 39, 44–45; investments in, 35, 140; jobs in (See under jobs); LA's efforts to control, 79–86, 191, 224n40; in Mexico, 155, 185, 186; Mexico nationalizing, 144, 146, 156, 185–86; natural gas as nuisance to, 116; Norway's control of development by, 129–30, 133, 136, 187–88; oil boom along Gulf Coast, 33–37; Pennsylvania's, 5, 200n2; politicians accommodating, 37, 55; in pollution control efforts, 52–53, 80–81; pollution from, 54, 177, 221n8; production quotas for, 116, 118, 132, 134; property rights vs. regulation of, 81–83, 85–86; public mistrust of, 84–85, 89; recessions and, 43, 134–35, 155; regulation of, 65, 81, 84, 116; relation of Canadian and U.S., 118–20; relation to other industries, 42, 92; social impacts of, 86–87, 153–54; in Tampico, 149–50, 153–55; technology in, 87, 185–86; Texan vs. "foreign," 36; in Texas-Louisiana vs. Mexican Gulf Coast, 184–85; U.S., 137; waste by, 116, 119, 152–53; workers in, 49–51, 169, 229n38

Oil Is Not a Curse: Ownership and Institutions in Soviet Successor States (Luong and Weinthal), 194–95

oil prices, 41–42, 239n30; effects of fluctuations in, 68–69, 93, 99, 121–22, 127, 141, 192–93; effects of OPEC embargo on, 92, 192–93; influence on world economy, 131, 134–36, 138; OPEC's influence on, 130, 183 oil production: allowing saltwater intrusion into oil fields, 147–48, 152–53; decline in, 39, 56; environmental ill-effects of, 70; Gabon's, 167, 175; harvesting marginal reserves of, 114; Mexico's, 155, 156; new locations for: in Louisiana, 67; OPEC's influence of, 183; prices and, 130–31, 134, 138; from tar sands, 124, 241n50

oil reserves: Australia's, 102; Canada's, 237n62; Louisiana's, 75; responses to declining, 56–57; around Tampico, 144

oil revenues, 180, 184; Norway's, 132, 135–36, 142, 187–88; Port-Gentil not benefiting from, 145, 174; states' ability to control, 183–84

oil sands. *See* tar sands

Oklahoma City, oil drilling under residential neighborhoods in, 225n54

Olmsted, Frederick Law, 34

Ona, Marc, 178

Onouviet, Richard, 176

Ontario, Canada's oil industry in, 115

OPEC, 138; effects of embargo by, 92, 99, 121, 130, 192–93; effects of nationalization by, 188; influence of, 94, 130, 183; Mexico as alternative oil source to, 144; production quotas of, 134

Orange County, oil industry in, 78

Orungu, French treaty with, 162

Otis, Harrison Gray, 86

ozone, 53–54, 72

Parton, James, 14

Péan, Pierre, 175

Pearson, Weetman, 150–51, 155

Pennsylvania: coal in, 6–7, *11*, 12, 200n8; natural gas discovered in, 14, 28, 203n39, 205n63; natural gas in, 16–17, 205n67, 206n74; oil in, 16–17, 200n2

Pennsylvania Railroad, 8

Perry, Clarence, 106

Perth, 108, 188; as energy capital, 93, 99–100, 102, 189; growth of, 96–97, 110; mining around, 95–98, 103; refineries around, 98, 102–3. *See also* Australia; Western Australia

Petro-Canada, state ownership of, 121–23

petrochemical corridor, Louisiana's, 3, 38; African American neighborhoods in, 72–74; cancer rates in, 71, 73; development of, 59, 63–65, 67; environmental justice movement and, 72–75, 192; factors in site selection in, 72–73; flood protection for, 62–63, 75; government promotion of, 191–92; high incomes from, 68; industrial accidents and explosions in, 71–72; opposition to, 59; water pollution in, 70–71

petrochemical industry: Canadian, 121, 124; danger from accidents in, 73–75; effects on public health, 53–54; efforts to develop, 59, 103, 121; environmental damage from, 53–54, 70–71, 75; expansion of, 39–40, 65; around Houston, 37–40, 53, 57; in Louisiana, 59, 67, 74–75, 217n26; along Mississippi River, *66;*

Petróleo Mexicanos (PEMEX), 156

petroleum industry. *See* oil industry

philanthropy, 42, 51–52

Pierce, Henry Clay, 149

pipelines, 233n26; expanded for free trade in energy, 123; to export Canadian natural gas to U.S., 118–19, 232n22; Gabon's first, 167; around Houston, 50; around Los Angeles, 3, 78; for natural gas, 15, 22–23, 203n46; for oil, 34, 59, 65, 114, 119, 151

Pittsburgh, 200n2, 201n14; air pollution in, *13,* 13–14, 18, *19,* 205n69; coal in development of, 2, 5, 9; coal industry's influence in, 23, 206n77; coking plants around, 10–11; efforts to address air pollution, 18, 20, 26–27, 190–91, 206n77; energy-related industrial core of, 193–94; natural gas in, 23, 203n41; transitioning of energy sources in, 2, 5–6, 14–15, 26–28, 193–94

Pittsburgh Renaissance, smoke control required for, 22–23, 28

plant life, effects of air pollution on, 13

politics, 88, 100, 119, 235n44; in balancing promotion and regulation with energy industry, 190–92; Canadian regional, 111, 114–15, 123; culture as factor in energy capitals, 91–92; in Gabon, 161, 165–66, 168, 172, 175; institutions as factor in development of energy capitals, 182–84, 187, 191, 195;

politics (*cont.*): international events influencing oil prices, 130–31; over Canada's
National Energy Program, 121–22; power
of Mexican labor union in, 156; promoting growth, 35, 37, 59; regulation of oil
industry in, 52–53, 55, 121, 191; riots in
Port-Gentil and, 159–60, 172–74; of Tampico's working class, 153–54

pollution, 53; from coal, 124; from coal mining, 7, 12; from coke production, 12–13;
energy capitals balancing promotion of
energy industry with, 92, 94, 190–92;
from industries, 69, 71, 107–9; inequities
in, 199n29, 227n71; from liquefied natural
gas processing, 157–58; from oil industry,
177, 221n8; from petrochemical industry,
71–72; from refineries, 53–54, 56–57, 67, 89,
106–8, 152. *See also* air pollution; environmental damage; water pollution

pollution control, 71; delays in, 52; federal
government enforcing, 55–56; for industries in Kwinana, 107–8; laxness of, 67; opposition to, 54–55, 207n86; by petroleum
industries, 52–53. *See also* regulations

Port-Gentil, 171; development as enclave
economy, 162–67; development of oil
industry in, 167–70; economy of, 162–67,
175; effects of French colonialism in, 164–
65, 170; as energy capital, 144, 159, 174–75;
forestry and, 163–64, 175; location of, 161–
63; national government and, 174, 179;
not benefiting from oil revenues, 145, 174;
oil as curse for, 145, 184, 187; oil industry
leaving, 174–75; political unrest in, 172–74;
racial separation in, 166, 169–71; riots in,
159–60, 178; urbanization of, 169

poverty: in Gabon, 178; in Port-Gentil, 144,
159, 161, 170; in Tampico, 154, 155

Preston, E. J., 226n67

property rights: mineral rights and, 232n20;
vs. regulation of oil industry, 81–83, 85–86

public health: cancer rates and, 71, 73, 108–9;
danger from industrial accidents, 71–72,
74–75; effects of air pollution on, 27–28,
93; effects of industries in Kwinana on,
107–9; effects of petrochemical industry
on, 53–54, 74–75; effects of refineries on,

53–54, 71–72; effects of smoke on, 14, 18;
effects of transition from coal to natural
gas on, 26, 29; effects of water pollution
on, 12, 17, 70–71, 177–78

public opinion: alliances for environmental justice, 73–74; growing opposition to
pollution, 70–72, 73, 102, 107; of oil companies, 84–85, 89, 232n20; of oil drilling,
80–84, 88; of petrochemical industry, 59,
70–72

public transportation, 48–50, 123, 234n40

Quam-Wickham, Nancy, 221n9

race: inequities in Port-Gentil, 166, 169–72;
relation of pollution and neighborhoods,
72–74, 227n71

race relations: in Los Angeles, 4, 86–87; in
sugarcane plantations, 72

railroads: air pollution from, 13–14, 26; Gabon constructing, 175; around Houston,
35, 37; importance of coal revenues for,
8, 26; in Pennsylvania, 8–10, 201n14; regulations limiting smoke production by,
20, 207n86; switching to diesel engines,
2, 8, 26–27, 29, 207n88; use of fuel oil by,
149–50, 152

real estate, 77; effects of oil drilling on, 79,
82–83; effects of oil industry on, 86, 221n9;
around Houston, 42–43

refineries, 48; in Alberta, 117, 124; flooding
and, 63, 65; in Gabon, 167; along Gulf
Coast, 35–36; around Houston, 34, 36–38,
50–52; importance in Houston's economy,
2, 37–38; industrial accidents at, 71–72;
jobs in, 38, 56, 68; in Kwinana, 93, 98, 104–
8; around Los Angeles, 3, 78, 89, 221n1;
in Louisiana, 3, 65, 67–68, 70; Mexican
oil processed in U.S., 185, 248n12; around
Perth, 189; petrochemical industry and, 3,
40; in Pittsburgh, 5; pollution from, 53–54,
56–57, 67, 89, 106–8; in Port-Gentil, 169,
175; resources used by, 152; in Tampico,
144, 149, 151–52, 156–57; "topping" by,
185, 248n12; transport of oil to, 65, 200n2,
248n12

refineries, sugar, 67

regulations: on drilling for natural gas, 204n51; energy capitals balancing promotion of development with, 190–92; environmental damage and, 53, 221n9; for natural gas pipelines, 203n46; of oil industry, 83–85, 116, 191; property rights *vs.*, 81–83, 85–86; state and federal government trumping local, 81, 84–85. *See also* air pollution controls; pollution control

Rehder, John, 64

Rendjambe, Joseph, 172–73

Rice University, 51

Richards and Pratt, 119, 121, 232n23

Rihl, George L., 153

rivers, 60; in development of coal industry, 7–10; steel mills along, 9–10; Tampico's location on, 148–49, 150–51. *See also* specific rivers

river valleys, 8; iron and steel industries in, 10–11, 14

Rockefeller, John D., 36, 115

Rogers, Thomas L., 149

safety regulations, for oil drilling, 88, 130

saltwater, intrusion into oil fields, 147–48, 152–53

San Jacinto River, 46–47

Santa Barbara County, offshore drilling in, 77

Scott, Loren, 75

Scully, Cornelius, 20

segregation, in Los Angeles, 4, 86–87

Shintec, public opposition to, 73–74

Smil, Vaclav, 113

Smith, J. Spencer, 222n15

smoke, 20; from coal use, 2, 18; control required for Pittsburgh Renaissance, 22–23, 28; lower-to-ground level of household emissions of, 20–21, 27. *See also* air pollution

Smoke Control Act (Pittsburgh, 1941), 21–22, 24–26, *27*

Société Commerciale, Industrielle, et Agricole du Haut-Ogooué (SHO), 163–64

Soviet successor states, oil development in, 194–95

space industry, around Houston, 42

speculation, in oil, 3

Spindletop, oil discovered at, 33–36

St. Louis, air pollution in, 20

Standard Oil (Mobil), 35, 78, 115; refinery in Louisiana, 3, 63, 65; Texas legislature protecting local companies from, 36, 65

Statoil, 92–93, 130, 132–33, 188

Stavanger, 128, 136; economy of, 127–28, 133–36, 140; as energy capital, 92–93, 136, 141–42; oil-related jobs in, 130, 133, 239n17. *See also* Norway

steel industries. *See* iron and steel industries

Stephenson, Gordon, 105

suburbanization, 64; around Houston, 43, 48–50, 53; around Kwinana, 106–7; around Los Angeles, 4, 87; around Perth, 96–97

Suez War, 119

sugarcane, 72; considered for biofuel, 65, 75; importance in Louisiana, 58–59, 63–65, 67, 75

Tampico: decline of oil industry in, 143, 148, 156–57; development of oil industry in, 149–50; economy of, 149, 157–58; as energy capital, 143–44, 158; importance of location of, 148–49; limitations of, 157–58; oil as curse for, 184; oil industry transforming environment around, 150, 152, 155, 186; population of, *157*; port of, 158, 186; refineries in, 149, 151–52, 185; short-lived as energy capital, 147, 153, 155; social climate of, 153–54. *See also* Mexico

Tarallo, André, 176

tar sands, 131, 236n54, 237n62, 241n50; Canadian development of, 121, 124–25, 140

technology, 123; availability of fuel oil furnaces, 23–24; cities requiring smokeless equipment, 20–22, 206n82; diesel engines as, 26–27; in harvesting marginal oil reserves, 92, 114; for natural gas distribution, 203n40; for natural gas production, 126, 157; for offshore exploration and drilling, 41, 129, 132–34, 137; in oil industry, 56, 87, 137, 185–86; for pollution control, 27, 52, 86; in refineries, 39, 48

Texaco, 35, 36

Texas: government accommodating oil industry, 55, 65, 191. *See also* Houston

Texas Coastal Zone, 52; Houston's location in, 30–31, 49; industries in, 38, 40
thermal generation, 118
Thurston, George H., 203n41, 203n45
tidelands, oil drilling on California's, 80
topography: along Mississippi River, 60, *61;* alterations to Tampico's, 144; Houston's, 31–32
tourism, 59, 80, 158, 180
toxic wastes, 53, 54, 72, 152
transportation, 9–10, 97, 128; deepwater harbor in Los Angeles, 4, 86; as factor in development of energy capitals, 2–4, 33–34, 91; Houston's sprawl and, 48–50; inadequacy of Port-Gentil's, 165–66, 169, 179; on Mississippi River, 63, 67; of natural gas, 23; of oil, 3, 37, 143–44, 150–51; for oil industry, 37, 49–50, 183, 200n2; Port-Gentil as port for, 162–67; Tampico as hub for, 143–44, 150–51, 153. *See also* highway systems; railroads; rivers
Trudeau, Pierre, 121
Tucker, Raymond R., 20

United States: fossil fuels in development of, 1–2. *See also* specific cities
urban ecology, 112–13, 123
urbanization: effects of, 31, 54; Houston jobs encouraging, 38–39; of Port-Gentil, 162, 166, 169
utilities: coal consumption by, *11;* natural gas used by, 18, 23, *25,* 40; public *vs.* private ownership of, 117, 120. *See also* electricity

Vaughan, Elizabeth, 64
Venezuela, 184; oil industry fleeing Mexico to, 148, 153, 185
Villa Cecilia, split off from Tampico, 153–55

waste disposal: neighborhood opposition to, 73; from tar sands processing, 236n54
water: needed to process tar sands, 124–25; produced by drilling for natural gas, 203n49; used by refineries, 152

water pollution: from acid coal mine drainage, 12, 202n28; from energy industries, 70, 126, 152; around Houston, 32, 52, 54; from industry in Kwinana, 106–7; from liquefied natural gas processing, 157–58; of Mississippi River, 70–71; from oil and natural gas wells, 16–17; from petrochemical industry, 59; in Tampico, 152, 158. *See also* environmental damage
water pollution controls, Louisiana's, 70
watershed: around Houston, 31, 32; Mississippi River's, 60
water supply: contamination of, 70–71, 177–78; as factor in site selection for refineries, 70; Houston's, 31, 45–48, 208n6
Weinthal, Erika, 194–95
West Australia Petroleum Pty Ltd (WAPET), 99–100
Western Australia, 102; gold refining in, 102–3; infrastructure for energy industry in, 93, 100–101; minerals and mining in, 95–98, *109,* 110, 228n18; oil and natural gas in, 99–100, 190. *See also* Perth
Western Canadian Sedimentary Basin (WCSB), 111, 114, 116, 119, 122, 126
Westinghouse, George, 14
West Virginia, natural gas from, 18–19
wet gas, 116
Wilson, Charles, 119
women, in Pittsburgh's smoke control efforts, 18, *21*
World War I, Mexican oil boom in, 151–52
World War II, 26; demand for California's oil in, 83–85; expansion of petrochemical industry in, 40, 59, 65; industrial growth during, 47, 53, 103; pollution ignored in, 21, 82–83

Yergin, Daniel, 2

zoning: Houston's lack of, 48; in Los Angeles, 77, 79–80, 81–82, 85